THEORETICAL FOUNDATIONS OF VLSI DESIGN

Cambridge Tracts in Theoretical Computer Science

Managing Editor Professor C.J. van Rijsbergen, Department of Computing Science, University of Glasgow

Editorial Board

S. Abramsky, Department of Computing Science, Imperial College of Science and Technology
P.H. Aczel, Department of Computer Science, University of Manchester
J.W. de Bakker, Centrum voor Wiskunde en Informatica, Amsterdam
J.A. Goguen, Programming Research Group, University of Oxford
J.V. Tucker, Department of Mathematics and Computer Science, University College of Swansea

Titles in the series

1. G. Chaitin *Algorithmic Information Theory*
2. L.C. Paulson *Logic and Computation*
3. M. Spivey *Understanding Z*
4. G. Revesz *Lambda Calculus, Combinators and Logic Programming*
5. A. Ramsay *Topology via Logic*
6. S. Vickers *Formal Methods in Artificial Intelligence*
7. J-Y. Girard, Y. Lafont & P. Taylor *Proofs and Types*
8. J. Clifford *Formal Sematics & Pragmatics for Natural Language Processing*
9. M. Winslett *Updating Logical Databases*
10. K. McEvoy & J.V. Tucker (eds) *Theoretical Foundations of VLSI Design*
12. G. Brewka *Nonmonotonic Reasoning*
13. G. Smolka *Logic Programming over Polymorphically Order-Sorted Types*

#2137634X

THEORETICAL FOUNDATIONS OF VLSI DESIGN

Edited by
K. McEVOY
University of Leeds
and J.V. TUCKER
University College of Swansea

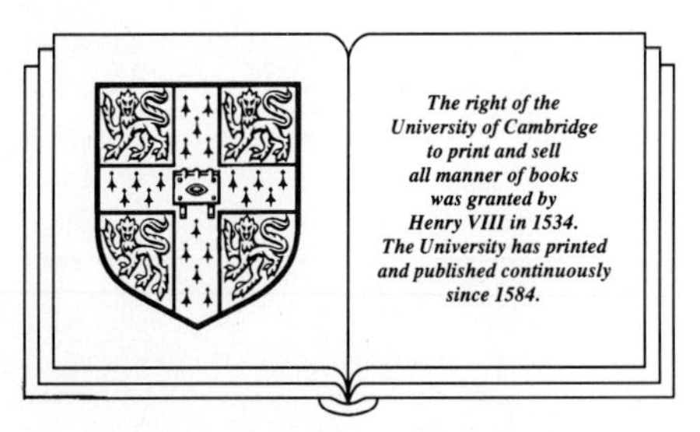

CAMBRIDGE UNIVERSITY PRESS

Cambridge

New York Port Chester Melbourne Sydney

Published by the Press Syndicate of the University of Cambridge
The Pitt Building, Trumpington Street, Cambridge CB2 1RP
40 West 20th Street, New York, NY 10011, USA
10 Stamford Road, Oakleigh, Melbourne 3166, Australia

© Cambridge University Press 1990

First published 1990

Printed in Great Britain at the University Press, Cambridge

Library of Congress cataloguing in publication data available

British Library cataloguing in publication data available

ISBN 0 521 36631 3

ANDERSONIAN LIBRARY

24. JUL 91

UNIVERSITY OF STRATHCLYDE

D
621-3819' 5835
THE

Contents

List of contributors

A. Cohn
Computer Laboratory, University of Cambridge, Pembroke Street, Cambridge, CB2 3QG.

P. M. Dew
Centre for Theoretical Computer Science and School of Computer Studies, University of Leeds, Leeds, LS2 9JT.

P. E. Dunne
Department of Computer Science, University of Liverpool, Liverpool, L69 3BX.

M. Gordon
Computer Laboratory, University of Cambridge, Pembroke Street, Cambridge, CB2 3QG.

N. A. Harman
Department of Mathematics and Computer Science, University College, Swansea, Singleton Park, Swansea, SA2 8PP.

E. S. King
School of Computer Studies, University of Leeds, Leeds, LS2 9JT.

K. McEvoy
Centre for Theoretical Computer Science and School of Computer Studies, University of Leeds, Leeds, LS2 9JT.

M. Sheeran
Department of Computing Science, Glasgow University, Glasgow, G12 8QQ.

J. V. Tucker
Department of Mathematics and Computer Science, University College, Swansea, Singleton Park, Swansea, SA2 8PP.

W. P. Weijland
Department of Computer Science, University of Amsterdam, P. O. Box 41882, 1009 DB Amsterdam, The Netherlands.

A. Williams
Centre for Theoretical Computer Science and School of Computer Studies, University of Leeds, Leeds, LS2 9JT.

H. C. Yung
Department of Electrical and Electronic Engineering, The Merz Laboratories, University of Newcastle Upon Tyne, Newcastle Upon Tyne, NE1 7RU.

PREFACE

The development of VLSI fabrication technology has resulted in a wide range of new ideas for application specific hardware and computer architectures, and in an extensive set of significant new theoretical problems for the design of hardware. The design of hardware is a process of creating a device that realises an algorithm, and many of the problems are concerned with the nature of algorithms that may be realised. Thus fundamental research on the design of algorithms, programming and programming languages is directly relevant to research on the design of hardware. And conversely, research on hardware raises many new questions for research on software. These points are discussed at some length in the introductory chapter.

The papers that make up this volume are concerned with the theoretical foundations of the design of hardware, as viewed from computer science. The topics addressed are the complexity of computation; the methodology of design; and the specification, derivation and verification of designs. Most of the papers are based on lectures delivered at our workshop on *Theoretical aspects of VLSI design* held at the Centre for Theoretical Computer Science, University of Leeds in September 1986. We wish to express our thanks to the contributors and referees for their cooperation in producing this work.

K. McEvoy
J. V. Tucker

Leeds, September 1988

1 Theoretical foundations of hardware design

K. McEVOY AND J. V. TUCKER

1 INTRODUCTION

The specification, design, construction, evaluation and maintenance of computing systems involve significant theoretical problems that are common to hardware and software. Some of these problems are long standing, although they change in their form, difficulty and importance as technologies for the manufacture of digital systems change. For example, theoretical areas addressed in this volume about hardware include

- models of computation and semantics,
- computational complexity,
- methodology of design,
- specification methods,
- design and synthesis, and
- verification methods and tools;

and the material presented is intimately related to material about software. It is interesting to attempt a comparison of theoretical problems of interest in these areas in the decades 1960–69 and 1980–89. *Plus ça change, plus c'est la même chose*?

Of course, the latest technologies permit the manufacture of larger digital systems at smaller cost. To enlarge the scope of digital computation in the world's work it is necessary to enlarge the scope of the design process. This involves the development of the areas listed above, and the related development of tools for CAD and CIM.

Most importantly, it involves the unification of the study of hardware and software. For example, a fundamental problem in hardware design is to make hardware that is independent of specific fabricating technologies. This complements a fundamental problem in software design – to make software that is independent of specific hardware (i.e., machines and peripherals). Such common problems are tackled by modelling and abstraction, using common design concepts and formal tools which are characteristic of computer science.

This volume joins a series of many-authored compendia and proceedings that record progress in the development of a comprehensive theory of VLSI computation and design. For instance, we have in mind Kung, Sproull & Steel [1981], Milne & Subrahmanyam [1986], Moore, McCabe & Urquhart [1987], Fogelman Soulie, Robert & Tchuente [1987], Makedon *et al.* [1986], Birtwhistle & Subrahmanyam [1988], Milne [1988] and Reif [1988]. In addition, there is the textbook Ullman [1984].

In this introduction to the volume we will examine some general ideas underlying a comprehensive theory of VLSI computation and design, and we will perform the useful task of attempting a literature survey.

At this stage in its development it is possible to see the subject only as a broad collection of interrelated problems and techniques. The theory of hardware is still limited by the theory of software. Perhaps the theoretical foundations are shallow, but they are also broad, and in each of the areas listed above there are significant and promising achievements. We hope that this volume encourages the reader to help contribute to improving the situation.

2 COMPUTER SCIENCE AND HARDWARE

We will make explicit certain ideas and problems that underlie the development of a comprehensive theory of computation and design for hardware within computer science.

2.1 What is computer science?

Computer science is about computations and computers. It concerns

(i) the invention and analysis of algorithms,
(ii) the design and development of programs and programming languages, and
(iii) the design and construction of computer systems, including hardware, for implementing programs and programming languages.

It is with (iii) that we are concerned in this volume, though the subjects are intimately related. A fundamental point is that computer science aims to abstract from physical devices and view its machines through the formalisms used for their operation and programming. This emphasis on formalisms – software – gives computer science some coherence and continuity in the face of changes in the physical technologies of hardware construction.

2.2 What is hardware?

We may use the term *hardware* to mean a physical system that implements a program or an *algorithm.* Other synonyms in use are *machine*, *device*, *computer*, and *digital system*. (Notice that only the last two terms are specific to computation.) Examples of hardware include the primitive calculators of Leibnitz and Pascal, the difference and analytical engines of Babbage, the devices made by electromechanical systems, and the application-specific devices and general computers manufactured by silicon technologies.

Whether old or new, specific or general, the essential feature of hardware is that, in realising or embodying the algorithm, its purpose is to process symbols and hence information. It is difficult to formulate definitions that correctly distinguish the many types of machines that process physical entities (e.g., a loom) from those that process information (e.g., a gauge). All physical systems process physical quantities: we impose on the physical system our framework of abstract information to obtain a digital or analogue computer. It is not easy to resolve but clearly the discussion underlies the distinct points of view of hardware possessed by the computer scientist and, *in our time*, the electronic engineer.

To design a computer we must design to physical and algorithmic specifications. In practice, the physical characteristics of technologies influence considerably our thinking about algorithms. For example, the von Neumann and systolic architectures are successful because of their suitability for implementation. Although in designing algorithms we are concerned with physical characteristics such as *time*, *space* or *area*, these quantities are actually basic properties of the symbolism that model in an abstract way the physical properties after which they are named. Rarely do algorithm designers model these quantities more exactly, or attempt to model other physical measures of efficiency or reliability such as *energy*, *power consumption*, *communication costs*, *operating temperatures*, *stress*, and so on. It is remarkable that simplifications used in work on algorithms are as useful to machine design as they are; or, to put it another way, result in designs that can be implemented and used at all. However, models of physical quantities such as energy, communication costs and so on are relevant to algorithm design and will therefore join the list of basic algorithmic concepts in due course.

2.3 The gap between hardware and software

A computing system is composed of hardware and software. The design of its hardware results in a physical system that realises a set of programs of programming languages. The design of its software results in symbolic systems that represent data and algorithms. There is a significant gap between our physical conception of devices and our logical conception of notations. There is a discontinuity at the bottom of well used images of the hierarchical nature of computation, as described in Bell & Newell [1971], for example.

In software the gap is seen in the comparison of the theoretical complexity of a program or computation and the empirical performance. Calculations based on models at different levels of abstraction can be refined to make estimates of the number of clock cycles required, *en route* to estimates of run-times in seconds that may be tested. This distinction is seen in the specification and verification of real-time computations (often controlling physical systems, for instance).

In hardware the gap is seen in the essential role that timing and performance play in the many notions of specification and correctness criteria for devices. In a sense, each model of computation for hardware design attempts to bridge, or more accurately, hide this gap.

The distinction between physical and logical concepts in computing is intimately related to the distinction between analogue and digital notions of computation. The notion of analogue computation is present in technologies such as neuro-computing (see Anderson & Rosenfield [1988]), and in any new chemical technology for image processing. It is present in discussions about new discrete space and discrete time models of physical and chemical systems (see Crutchfield & Kaneko [1987]). It is fundamental to long standing discussions about the nature of simulations in physics (see Feynman [1982]).

The gap in understanding is intimately related to the gap between mathematical models and their application in nature, which is one of enormous philosophical complexity. It is pleasing to think that the practical motivations of computer science lead us to technologies for hardware and software that require us to postulate borders between physics and logic, and hence raise fundamental scientific questions immediately.

3 THEORETICAL FOUNDATIONS

In the study of algorithms, and their realisation in software and hardware, there are certain fundamental concerns, including

(1) models of computation,
(2) specification,
(3) derivation and synthesis,
(4) verification,
(5) testing,
(6) maintenance.

Each specialised area of computer science – databases, theorem proving, architectures, computational geometry, VLSI, and so on – is characterised by its models of computation, and their associated methods classified under (2)–(6). In particular, each specialised area can be surveyed under these headings.

We will discuss the literature on the theoretical foundations of VLSI computations under the headings (1)–(4), combining (2) and (4); and we will neglect (5) and (6). This arrangement of three subjects – models of computation, derivation and synthesis, and specification and verification – reflects the situation as we have found it. In preparation, we will discuss the general concepts we associate with these four topics, independently of their relevance to hardware algorithms. The primary topic is models of computation.

3.1 Models of computation

A *model of computation* codifies a means of defining and composing algorithms: it defines *data*, primitive *actions* on data such as operations and tests, and methods of defining families of actions that constitute *computations*. An algorithm specifies families of computations. For example, the natural numbers together with their usual operations (e.g., successor, addition, multiplication) and the methods of composition and primitive recursion constitute a model of computation; an algorithm in this model is a *definition* of a primitive recursive function.

A model should also give *performance criteria* to evaluate the complexity of different algorithms. One fundamental idea is to count the number of primitive actions of the model involved in a computation, measuring this number as a function of input data. This method is called the *unit cost criterion* because each basic action is implicitly charged a unit. Clearly this criterion is related to time taken to compute.

To define formally a model of computation we often define a language in terms of a syntax and semantics. Conversely, a language definition incorporates a model of computation. Practical languages are often made from several disparate (even inconsistent) models of computation.

The theoretical purpose of a model of computation is

(i) to clarify the basic principles of a computing method,
(ii) to classify the algorithms based upon the method, and
(iii) to establish the scope and limits of computation using the method.

Thus a model allows us to determine whether or not a specification can be met by an algorithm; and, if performance is involved, what costs are necessarily incurred. An example of a specification that cannot be implemented is the universal function for the primitive recursive functions, which is recursive but not primitive recursive. The performance property is recorded by *lower bound theorems* for performance criteria. For example, when C.D. Thompson first devised a model for circuits (in Thompson [1980]), he was able to prove that $AT^2 = \Omega(n^2 \log^2 n)$ for sorting n elements; in Thompson [1983] some of the conditions on the models are relaxed, and consequently the lower bound on sorting must be weakened to $AT^2 = \Omega(n^2 \log n)$.

In its origins, however, a model may have been devised for one or more of the following purposes:

(i) to systematise algorithm development for an applications area;
(ii) to systematise algorithm development for an implementing technology;
(iii) to allow the analysis of computational properties.

For example, the systolic algorithm model satisfies (i) and (ii), supporting signal processing and VLSI technology; and the arbitrary interleaving model of concurrency satisfies all of (i), (ii) and (iii), supporting multiprocessing on a von Neumann computer, and the analysis of non-determinism and the independence of parallel actions.

3.2 Specification

A *specification* of an algorithm is an independent and abstract description of properties of the algorithm, or of the problem it is intended to solve. The statements making up a specification concern its inputs, outputs, and efficiency, for example. A specification is intended as a record of

(i) information relevant for users of an algorithm, and
(ii) requirements relevant for designers of an algorithm.

These two uses ensure that a specification is as fundamental as the algorithm itself. In connection with (i), specifications are used in the modelling of the task to be accomplished by the algorithm. In connection with (ii) specifications are used in confirming the correctness of the algorithm by both empirical testing and mathematical verification.

The precision in the description of a specification should be comparable with the precision in the description of an algorithm. Precision and rigour are indispensable in the process of algorithm design, which involves classifying properties, and their ramifications concerning the user's task and the designer's resources. Formally defined specifications complement formally defined algorithms. If specifications are machine processable then they can be animated, transformed, tested and verified with the assistance of computers.

A theoretical distinction between the general concept of a specification and the general concept of an algorithm is hard to draw. A specification can be very detailed, and indistinguishable from a coding of an algorithm. This attention to detail is common in practical work with specifications, and is an insidious problem. The point is that a specification is of use when it is an abstract description of some properties of an algorithm, or set of algorithms.

Given a specification method for a model of computation, important theoretical questions arise about its *expressiveness*:

Soundness or consistency problem Can every specification be realised by a set of algorithms based on the model?

Adequacy problem Can every set of algorithms based on the model be defined by a specification?

To formulate precisely and answer these questions for any specific model of computation and specification method involves considerable theoretical research. For example, the scope and limits of the algebraic specification methods for computable data types are surveyed in Meseguer & Goguen [1985].

An important theoretical use of specifications is in defining notions of equivalence for algorithms. Notice that if algorithms A and A' satisfy specification S then they are equivalent as far as S is concerned. More generally, given a specification method M, it is important to study the following equivalence relation on algorithms: given algorithms A and A' define that A is *equivalent* to A' *under* M if, and only if, for every specification S based on M, A satisfies S if and only if A' satisfies S.

Nevertheless, it is useful to disconnect the study of specifications and their use from that of algorithms. This attention to the theory of specifications is an original and important contribution to computer science from the field of programming methodology. An extensive study in the context of hardware is the chapter by Harman and Tucker contained in this volume.

3.3 Derivation

A *derivation* of an algorithm A from a specification S is a process of defining a sequence

$$A_0, A_1, \ldots, A_n$$

of algorithms in which $A = A_n$, A_0 satisfies the specification, and the *transformation* or *refinement* of A_i to A_{i+1} for $i = 0, \ldots, n-1$ preserves the specification S. The sequence is called a *derivation*; the transformations or refinements are said to be *specification preserving* or *correctness preserving*, and the process is also called the *stepwise refinement* or *synthesis* of the algorithm.

The notion of derivation is *very* general. Typically, a derivation arises in the solution of a problem, represented by the specification S. Here A_0 is some simple first algorithm that meets S but is not satisfactory; perhaps it is inefficient, sequential and unsuited to implementation in hardware. The transformations result in a complicated last algorithm A_n that is satisfactory; perhaps it is efficient, concurrent, and readily implemented. Many notions are involved in derivations, such as top-down design, automatic synthesis and compilation.

Among the basic concerns are

(i) transformation methods for developing algorithms for a given model of computation,

(ii) logical systems for formulating and proving that transformations preserve correctness,

(iii) automatic tools for processing derivations.

These concerns guide much theoretical research on models of computation, and specification and programming languages, throughout computer science. For example, in programming methodology, the concern (ii) for correctness is analysed by refinement calculi, such as the weakest precondition calculus described in Dijkstra [1976]; this has led to significant theoretical understanding of the process of derivation (see Back [1980] and Back [1981]) and its practical extension (see Back & Sere [1989], Back & Kurki-Suonio [1988] and Chandy & Misra [1988]).

Of course, the original example of a theoretically well-founded and practically well-developed formal derivation process is the theory of boolean algebra and its applications to circuit design.

3.4 Verification

A *verification* that an algorithm A meets a specification S is a process of defining a sequence

$$P_0, P_1, \ldots, P_n$$

of statements in which P_n asserts that A satisfies S, and each statement P_i is either an assumption about the model of computation and specification, or the result of deduction from statements preceding P_i in the sequence. The sequence is called a *proof of correctness.* A verification can be performed independently of a derivation, although a derivation ought to determine a verification.

Verification techniques must be founded upon mathematical theories which include mathematical models of computation and specification, but they can be divided according to the nature of the method of proof:

- informal methods, which are based on standard mathematical concepts, techniques and reasoning; or
- formal methods, which are based on formally defined languages with associated proof rules, and are often standard logical systems of mathematical logic.

The advantages of informal mathematical methods are that they are focussed on the human understanding of the essential technical points in the proof, they are understandable by a wide audience of people with mathematical training, and they are independent of specific logical and computer systems. The advantage of formal methods is that formal specifications and proofs are machine processable, and so formal proofs can be contructed, or at least checked, by computers. This is significant for raising the standard of rigour in a verification, and for solving the large problems that arise in specifications and verifications of algorithms of practical interest. The informal and formal methods are quite distinct but complement one another, of course; this distinction is true of both the nature of proofs and the talents necessary to construct them.

Currently there is renewed interest in the use of automatic theorem provers and proof checkers in the verification of software and hardware. It is essential that these computer systems should be based upon formally defined frameworks for doing proofs. The formal proof framework is usually some established formal logic. Examples of logics used in theorem proving software are

- higher order logic,
- first order logic,
- equational logic,
- temporal logic,
- Church's type theory,
- Martin-Löf's type theory.

Thus the algorithms and their specifications must be described directly, or compiled into such a logical language. Automatic theorem proving originates in attempts to prove theorems of mathematics and logic, and each of the above logics was first implemented for this purpose. The first implementation of a program verifier is reported in King [1969]. For basic historical and contemporary surveys, and source material, see Siekmann & Wrightson [1983] and Bledsoe & Loveland [1984].

Let us consider the terms *theorem prover* and *proof checker.* Strictly speaking we imagine a *theorem prover* to be a system that inputs a statement and, if the statement is true, returns a proof of the statement; if the statement is false, or cannot be proved, the theorem prover could react in several disciplined ways – it may give a proof of its negation, or simply reply that it cannot find a proof (it should not, of course, search *ad infinitum* for a proof that does not exist). A *proof checker*, however, inputs a statement and a proof of that statement, and returns information concerning the validity of the proof.

This distinction needs further analysis. First there is the distinction between the truth and falsity of a statement, and its provability or non-provability in a formalised logical theory. This distinction is fundamental in mathematical logic, and is analysed in terms of various notions of *completeness* and *incompleteness* of formal theories.

A formal theory T arises from the codification of certain properties of a model of computation M_0 about which one wants to reason. The statements and proof rules of T are true of M_0. However, the theory T is more abstract and possesses a semantics M defined by the following so-called completeness condition:

$$P \text{ is provable in } T \text{ if, and only if, } P \text{ is true of } M.$$

Since T is true of M_0 by design, we expect that

$$P \text{ is provable in } T \text{ implies } P \text{ is true of } M_0,$$

but that the converse can fail, namely

$$P \text{ is true of } M_0 \text{ does not imply } P \text{ is provable in } T.$$

Thus it is essential to distinguish carefully between the notions of true statements and provable statements. For example, it can be proved using the theorems of K. Gödel, that given any formal theory designed to reason about specifications concerning algorithms on the natural numbers $\{0, 1, 2, \ldots\}$ there are specifications and algorithms that are true but which cannot be proved in T.

The algorithmic notions of theorem prover and proof checker described above are better described in terms of *decision procedures.* A decision procedure for *provability* of statements in a theory is an algorithm that given any statement decides whether or not the statement has a proof in the theory. For some simple theories, including propositional calculi, there are decision procedures; however, for most basic theories there do not exist decision procedures. For example, according to Church's Theorem (Church [1936]) there does not exist a decision procedure for the predicate calculus. For any reasonable logical theory there is a semidecision procedure that given a statement returns a proof, if one exists, but can fail to return a message if such a proof does not exist.

A decision procedure for *proofs* in a theory is an algorithm that given a statement *and* a proof, decides whether or not the proof is a proof of the statement in the theory. For any reasonable logical theory, there is a decision procedure for proofs. Interestingly, in common formulations of Floyd–Hoare logic, designed to be complete for computations on the natural numbers, there is no decision procedure for proofs; this is because of the use of the set of all first order statements that are true of the natural numbers as an oracle: see Apt [1981]. The logical foundations of Floyd–Hoare logic as seen from theorem proving are studied in Bergstra & Tucker [1984].

In using a system to prove a theorem it is to be expected that a combination of these concepts will be needed in a process of interactive proof development. There must be an informal proof involving a tree of lemmas, a formal theory concerning the model of computation, and a selection of theorem provers and proof checkers that can be used for proving lemmas and checking proofs involving lemmas.

4 LITERATURE SURVEY

In this section we offer a survey of the literature on theoretical foundations of VLSI and hardware design. The survey is divided into three parts to reflect the nature of the research which has been carried out in this subject. These parts concern

- the development of models of hardware devices in Section 4.1,
- the development of derivations in Section 4.2, and
- the development of techniques for verification of hardware devices in Section 4.3.

We hope we have been sufficiently comprehensive to give an accurate impression of current research directly relevant to the foundations of hardware. We know there are omissions, and expect some to be unfortunate.

4.1 Models

A large number of models of computer hardware have been presented in the literature. Some of these have been formal models for scientific analysis, and others have been more informal models for engineering applications. It is natural that both the form and the properties of a model, as first presented in the literature, reflect the reason for the formulation of the model, and historically a theoretical model of hardware has been developed for one of three reasons:

- to clarify the basic principles of a design method;
- to allow for formal specification or verification; or
- to analyse the complexity of certain problems or algorithms.

We consider, in turn, the models which fall into these three categories. Systolic computation has, since the first use of the term in Kung & Leiserson [1979], received a great deal of attention, and so this subject is treated in a section of its own.

4.1.1 Models for design The earliest models of interest here are probably those developed in the late thirties, in which basic techniques of boolean algebra are applied to the analysis of switching circuits; see Shannon [1938], Nakasima [1936] and Shestakov [1938].Other important models used in the design of hardware and digital systems can be surveyed by consulting some of the major texts in this field: Keister, Ritchie & Washburn [1951] is a fascinating text, largely concerned with designing networks of electro-mechanical switches; Flores [1963] contains a thorough treatment of computer arithmetic; Harrison [1965] contains an excellent account of the mathematical theory of switching circuits; Clare [1973] was the first text to use T. E. Osbourne's algorithmic state machine (ASM) notation; Hill & Peterson [1973] contains an early application of a hardware description language (AHPL) for simulation and description of digital systems; Mead & Conway [1980] was the first text on structured digital systems design in VLSI; Ercegovac & Lang [1985] is a good modern treatment of digital system design which stresses the current (mainly) informal, but highly structured, design methodologies; other interesting works include Lewin [1968], Lewin [1977], Mano [1979] and Winkel & Prosser [1980].

A design language μFP for regular arrays is presented in Sheeran [1983] and Sheeran [1985]. μFP is a functional programming language (it is an adaptation of FP; Backus [1978]) in which each construct in the language (both the primitives and the combining forms) has an associated geometric 'layout', so that the construction of a functional program in μFP carries with it, as an immediate by-product, the construction of a corresponding layout. In order to successfully model hardware, an operator μ to model feedback or state is added to FP. A μFP program represents a function on streams of data. Basic functions might be full adders and nand gates, and μFP programs are designed by applying higher order functions such as *map* and *reduce* to these primitives. The 'clean' mathematical semantics of functional programs facilitate the design of an initial program which is correct (involving little or no use of the μ-operator); this program can then be transformed to one which is more suitable for implementation in VLSI by the application of algebraic laws (which are proven to be correct). The advantage of this method of design is that the set of higher order functions which are available as combining forms to the designer is restricted to a small set of simple functions which have geometric representations which are guaranteed to produce the regular layouts which are the aim of VLSI design; arbitrary user-defined higher-order functions are not permitted, so that, for example, spaghetti-type wiring cannot be introduced. Also, whereas local communication is implicit, distant communication must be explicitly described, thereby discouraging long communication wires. In the chapter by Sheeran in this volume the functional nature of the language is replaced by a relational approach. In Sheeran [1988] retiming transformations are examined in this model. This approach to design has also been developed in Jones & Luk [1987] and Luk & Jones [1988].

Patel, Schlag & Ercegovac [1985] have also adapted Backus' FP to the design and evaluation of hardware algorithms. A program in νFP is a function which maps objects to objects, where objects are either atomic or sequences of objects (undefined values are treated). The approach to design is that the algorithm is first specified within νFP at a level of abstraction that is high enough to aid validation, and then it is refined to a level at which it is easily interpreted. Next this νFP function is mapped (by an interpreter) into an intermediate form (IF) which reflects the planar topology of the function, and then this IF is mapped to a fixed geometry by selecting and resolving relative position constraints (compaction). The result can be displayed on a graphics terminal. The νFP system can estimate performance parameters for FP programs.

Boute [1986] has developed a theory of digital systems which is based on transformational reasoning, and a functional programming style to support this reasoning. The functional programming language SASL (Turner [1979]) is augmented with a simple type description language, and this language is used to describe digital systems at all levels. It is argued that for many results in the theory of digital systems, transformational reasoning is more appropriate than the standard deductive style – important considerations here are compositionality and the flow of information in a system.

Milne [1985] presents the calculus CIRCAL as a model for circuit description and design. A CIRCAL term describes the behaviour of an agent, and terms are constructed by applications to primitive terms of the four primitive operators of guarding, choice, non-determinism and termination, and the two derived operators of concurrent composition and abstraction. CIRCAL is equipped with an acceptance semantics (which was developed from the observational semantics of Hennessy & Milner [1980]), and a set of laws; the laws have been proved to be sound with respect to the observational semantics. The principal use of CIRCAL is to prove that two systems of computing agents have equivalent behaviours (one system may be a specification); as term equality implies semantic equivalence, proofs involve manipulating terms using the laws until the terms are equal (use of the laws avoids constant recourse to the underlying acceptance semantics). It is emphasised in Milne [1986] that CIRCAL can be used to represent all the levels in the VLSI design hierarchy, including the geometric features of a mask-level design. It is clear that if adjacent levels in the design hierarchy can both be satisfactorily represented in the same framework, then formal verification of a design refinement will be a more tractable task. CIRCAL has the same roots as CCS (i.e., concurrent processes, as defined in Milne & Milner [1979]), and also has many similarities with CSP (Hoare [1985]). However (among other differences) CIRCAL permits the representation of simultaneous occurrences of distinct actions – this allows concurrency to be modelled directly without recourse to the arbitrary interleaving semantics necessary in CCS and CSP. Typed CIRCAL is introduced in Milne & Pezzè [1988] to improve techniques for data abstraction. A framework not dissimilar to that of Milne is described in Subrahmanyam [1983]; the calculus used here is based on CCS.

Delgado Kloos [1987] has presented a language STREAM for the description of digital circuits. The syntax is defined using BNF notation, and a denotational semantics is given. The language is parameterised over the primitive agents, and a number of compositional operators are defined; also, some behaviour preserving transformation rules are given. STREAM can be directly interpreted as a data flow language, but a simple procedural language is also introduced as an interface between hardware and software; the procedural parts can be modelled within STREAM itself. In Delgado Kloos [1987] the language is used to describe circuits at three standard levels of abstraction: the register transfer level, the gate level and the switch level.

Brookes [1984] derives a semantics for VLSI circuits by applying fixed point reasoning to a directed graph model. A function is associated with each arc (the function defines the output on that arc of the node which the arc is incident from, as a function of its inputs) and the fixed point solution of this set of functions is the function which describes the semantics of the system. Weights are added to the arcs so that semantic proofs of the retiming results of Leiserson & Saxe [1981] can be given. The method is illustrated by proving correct a finite impulse response filter and a palindrome recogniser.

Cardelli & Plotkin [1981] describe an algebraic approach to hierarchical VLSI design in which networks are described by expressions in a many-sorted algebra. The algebra is then embedded as an abstract data type in the functional programming language ML.

In the study of communicating processes the directed graph is often seen as an intuitive and helpful model in which to construct designs; in formal treatments of top-down design the algebraic definition of data type, and homomorphisms of algebras, provide a natural framework. In his chapter in this volume, McEvoy presents a hierarchical model for the design of synchronous algorithms which incorporates both of these ideas, together with a formal state transition semantics. Refinements of data types are discussed, together with retiming results.

4.1.2 Models for verification The formal verification of a physical device can only be carried out within a model of its behaviour. However, the length of this section reflects the situation that in the study of verification of hardware, considerations on the nature of the model itself are often seen as secondary to the development of the verification technique; the use of more general purpose models is discussed in Section 4.3, and in this section we restrict attention to those models designed specifically for modelling hardware devices. These models are important, as it must not be forgotten that if the semantics of the model do not correctly describe the behaviour of the device then a proof of correctness in that model no longer guarantees that the device behaves correctly. Many of the models of circuits which are used do not fully reflect the complicated electrical and timing properties of VLSI chips, and consequently within such a model it is possible to formally verify a chip which operates incorrectly (such an example is described in Camilleri, Gordon & Melham [1986]). For example, many models describe devices as functional units computing outputs in response to their inputs; however the direction of current flow through a cMOS transmission gate is determined solely by the driving capability of the circuitry at either end. Nevertheless, if formal reasoning techniques are to be tractable then such models must abstract from physical detail; the physical effects which should be modelled are those which affect the computational behaviour of the device.

In Bergstra & Klop [1983] an axiomatic semantics is given for restoring logic circuits, both statically and dynamically. As an example, the Muller C-element is discussed in detail. It is shown that a consistent circuit reacts in an unambiguous way on new inputs.

Bryant [1981] and Bryant [1984] define a switch-level model which describes the logical behaviour of digital systems implemented in metal oxide semi-conductor (MOS) technology. The model is based on a ternary logic, where the third value represents an uninitialised, uncertain or invalid logic level; the following characteristics of MOS circuits have been successfully described in the model: ratioed, complementary, and precharged logic; dynamic and static storage; (bidirectional) pass transistors; busses; charge-sharing; and sneak paths. Boolean gate models cannot satisfactorily model the field-effect transistors and dynamic memory used in MOS technology. An algorithm for a logic simulator based on this model is also described in Bryant [1984] and this has been implemented in the simulator MOSSIM2. Bryant [1986] further stresses the importance of accurate models; a simple CMOS *nor* gate is used to show how state in circuits can cause boolean models to give incorrect results.

Winskel [1988] also defines a model of MOS circuits which describes circuit behaviour more closely than most other computational models (e.g., capacitances and resistances are contained within the model); a circuit is described as a term in a language derived from CCS. The model is different from Bryant's in that it is compositional – if a term t is formed by applying an operator to terms t_1 and t_2 then the behaviour of the circuit represented by T can be derived solely from the behaviours of the circuits represented by t_1 and t_2.

Musser *et al.* [1988] also criticise the over-simplicity of many hardware models when compared to the physical operation of the hardware devices themselves; they present a model which is adequate to describe techniques such as bidirectionality, gate capacitance and charge-sharing.

4.1.3 VLSI complexity models The late seventies and early eighties witnessed intense research activity in models for the complexity analysis of VLSI computation. This activity was initiated by the realisation that the traditional complexity measure for the amount of hardware required – a component count, or transistor count – was not necessarily a realistic measure for VLSI, as the physical quantity of interest is the area of the chip, and there need not be a simple linear relationship between the number of transistors on a chip and the area of the chip. Indeed for many algorithms the area used for communication, determined by wires between transistors, is the dominant factor in the area of any chip implementing the algorithm. So area replaced component count as a complexity measure. Thompson [1980] presented the *grid model* for a measure of area.

However, area is just one aspect of a model of a VLSI chip. Fundamental precepts for a complexity model of a VLSI chip were laid down in Mead & Rem [1979], Thompson [1980] and Brent & Kung [1981] although the universal validity of some of these suppositions was disputed by other authors. The models were not dissimilar to finite automata, with the basic assumptions that VLSI circuits are digital devices with each node and each wire having a facility for storing a fixed finite amount of data.

By assigning absolute minima for both the dimension of a node and a wire, and for propagation times for signals through a node and a wire, asymptotic lower bounds on certain metrics for the solution of problems such as sorting, multiplication and fast Fourier transforms were derived. These lower bound results are described in Hromkovic [1987]. Apart from computation time T and area A, common complexity measures on these models are AT (used by Mead & Rem [1979], and 'related' to the quantity of energy required), and AT^2 (introduced by Thompson [1980], and 'related' to the quantity of power dissipated), and period of computation P (introduced by Vuillemin [1983]). The most interesting arguments based on these models are the information flow, or bandwidth, arguments introduced by Thompson [1980] to derive lower bounds on the complexity measure AT^2. Vuillemin [1983] noted that these information flow arguments depended upon the computation of certain permutations, and he used some simple group theory to generalise the results to the class of transitive functions (defined in terms of transitive permutation groups).

These measures A, T, AT, AT^2 and P are simple, theoretically interesting, useful, and very abstract. The models and performance criteria need further refinement and theoretical analysis, motivated by an interest in more detailed information. For example, in Kissin [1982] there is an account of power consumption. In Vitanyi [1988] there is an appraisal of the cost of communication and wiring.

There are other assumptions on the models of interest. For example, hypotheses concerning the *i/o* schedule. Can each input be read once only? Are the ports at which each output should appear predetermined, and should each output appear at a different port? Is the schedule of inputs and outputs fixed and independent of the inputs? For discussion of these and related matters see Savage [1981a], Kedem & Zorat [1981], Lipton & Sedgewick [1981] and Chazelle & Monier [1981a]; Baudet [1983] presents a review of much of this work on VLSI complexity. For a theoretical analysis of more general synchronous timing schedules see Meinke [1988]. A question on which there appears to be much dispute is: what is the most realistic model for the minimum time for transmitting a signal along a wire? See Chazelle & Monier [1981b], Mead & Conway [1980] and Chazelle & Monier [1985]. This particular question is examined in the chapter by Dew, King, Tucker and Williams in this volume.

The development of lower bounds for the VLSI solution of certain problems, with respect to performance criteria, provides an optimal standard measure against which algorithms to solve the problem can be evaluated. This encourages research into the invention of such asymptotically efficient algorithms – for example, Thompson [1980] (sorting and fast Fourier transform), Baudet, Preparata & Vuillemin [1983] (convolution), Brent & Kung [1981] (multiplication), Abelson & Andreae [1980] (multiplication), Yung & Allen [1984] (multiplication). A study of lower bounds for A and AT^2 based on reduction to computational prototype problems is Hornick & Sarrafzadeh [1987].

Much theoretical research on circuits has concentrated on models of circuits which implement boolean functions, which are functions of the form $f : \{0,1\}^m \to \{0,1\}^n$ for natural numbers m and n. The study of the complexity theory of boolean networks for boolean functions dates back to Shannon [1938], and (excepting the disparity between gate count and chip area) much of this tradition is still relevant to VLSI complexity. An excellent elementary introduction to the basic mathematics is Harrison [1965], and comprehensive treatises on the subject are Wegener [1987] and Dunne [1988]. Recent results on boolean network complexity are presented by P. E. Dunne in this volume.

Kramer & Leeuwen [1983] give some results relating VLSI complexity upper bounds to the work of Lupanov [1958] on the boolean network complexity of functions. The recent work of Furst, Saxe & Sipser [1984] and Hastad [1986] on the complexity of bounded-depth networks has an interesting consequence for the design of PLAs computing arithmetic and other functions. For integer multiplication a PLA which iterates for only a constant number of times (when regarded as a sequential circuit) must have an exponential size, but integer additions can be realised by a PLA without feedback having polynomial size. The addition result is proved in Chandra, Stockmeyer & Vishkin [1984]. In particular, planar boolean networks are used by Savage [1981b] to relate VLSI complexity and network complexity.

4.1.4 The systolic model The term systolic was first applied to algorithms in the paper Kung & Leiserson [1979]. In the introductory paper Kung [1982] it is explained how the systolic approach to the design of algorithms is proposed as a method of design which would avoid the *i/o bottleneck*, which is the most significant barrier in attempts to speed up many algorithms when executed on standard *von Neumann* architectures. Many designs for parallel algorithms accentuate the problem of the von Neumann bottleneck, as every action on a processor must be preceded by a read from memory and followed by a write to memory. The systolic approach is to ensure that every time a read from memory occurs, the data is passed to every processor which can make use of the data before any write to memory is carried out. This requires extensive pipelining of the computation. In this way the ratio of computation time to time for *i/o* traffic is increased, and so the *i/o* bottleneck is alleviated.

In Kung [1982] a family of systolic arrays for solving the convolution problem is described, and these examples remain the archetypal examples of systolic arrays.

Such examples are important as it is now over ten years since the adjective systolic was first applied to algorithms, and yet a precise, universally accepted definition of the term systolic array has still not emerged (see the prologue of Moore, McCabe & Urquhart [1987]). Despite this situation, systolic arrays have been recognised as a major development in design methodology for VLSI, and they have been the focus of a large amount of research activity. These activities could be categorised into four sections.

- The invention and development of systolic algorithms – both algorithms to solve problems arising from new technologies and new fields of interest, and new algorithms for the solution of established and well known problems. (No attempt is made to list examples here, but this includes such developments as bit-level algorithms, e.g., McCanny & McWhirter [1982], and two-level pipelining, e.g., Kung & Lam [1984]).
- The principal features of systolic designs are regularity and simplicity, and so they characterise a class of algorithms for which it is possible to attempt the goal of algorithm synthesis from some high level specification, or (more typically) from some algorithm described at a high level of abstraction. These techniques are discussed in Section 4.2, and indeed most of the techniques discussed in that section were developed for systolic design.
- The simplicity of systolic arrays has also encouraged their use as a testbed for formal verification techniques; these are discussed in Section 4.3.
- A research project at CMU has been the development of a computer with a systolic architecture – the Warp (e.g., Annaratone *et al.* [1987]).

It is the purpose of this section to discuss the work (particularly that of a formal nature) which has concentrated on the definition of the systolic model of computation. The following 'criteria' to characterise a class of synchronous parallel algorithms, which were to be termed systolic, were put forward in Kung [1982] and are quoted here:

(i) 'the design makes multiple use of each data item';
(ii) 'the design uses extensive concurrency' (i.e., pipelining or multiprocessing);
(iii) 'there are only a few types of simple cells';
(iv) 'data and control flows are simple and regular'.

These criteria would achieve the goal of an increase in data throughput, without requiring a corresponding increase in memory bandwidth. The properties of regularity and simplicity of design had already been recognised as being essential characteristics for the good design of VLSI components (e.g., Mead & Conway [1980]), and so it was the first property which would truly characterise systolic systems – this multiple use of each data item is achieved by *pumping* data around every processor, and the name systolic is drawn from an analogy with the heart pumping blood around the organs of the body. To quote again from Kung [1982], 'For implementation and proof purposes, rigorous notation other than informal pictures (as used in this article) for specifying systolic designs is desirable.'

The thesis Leiserson [1981] presented a more formal definition of systolic. He recognised that a systolic solution to a problem consists not of a single array, but rather of a whole class of arrays – one array for each problem size n. An array is represented by a *machine graph*, which is a graph in which the nodes represent finite state machines (typically Moore machines, but in the semi-systolic case, Mealy machines *). These definitions provide some tools to help define the terms regular and simple in this setting. We require that within a particular class of arrays, as the array size n grows the following numbers are bounded above by some constant for the class: the number of states in any state machine (represented on a node), the number of input symbols in any state machine, and the degree of any node. These restrictions do prevent the complexity of an algorithm for a particular problem growing with the problem size, but they do not capture all the properties of a systolic array (e.g., regularity). Furthermore in his important work on *retiming* (see also Leiserson & Saxe [1981]) these conditions are dropped, and a systolic system is represented as a graph where the semantics of the nodes are purely functional (i.e., there is no internal state), and delays (providing storage between clock cycles and therefore state) are represented on edges; the condition that there is at least one delay on each edge is now termed *the systolic condition.* This property ensures that in a hardware implementation of a systolic algorithm, where the synchronous communication regime is controlled by a global clock, the length of the clock cycle is independent of the size of the array.

That a systolic array is a class of arrays rather than a single array is further reflected in Kung [1984]. He requires that such a computing network 'may be extended indefinitely', and uses a *pipelinability* measure on this 'pumping' of data in the network. As the array size grows the *speed-up factor* must be linear in the size of the array, where the speed-up factor is defined to be the processing time in a single processor divided by the processing time in the whole array processor. The terms 'temporal locality' (more than one delay on each edge in the signal flow graph) and 'spatial locality' (local interconnection only) are also introduced.

---

* In a Moore machine the output is a function of the state only, whereas in a Mealy machine the output may depend on the current input as well as the state.

This concept that a larger problem size should be solved by simply adding one or more cells to a smaller array is termed *modularity* in the more recent introductory paper Quinton [1987b]. If the memory bandwidth is not to increase with the size of the array then clearly the number of cells which can perform *i/o* operations ('boundary cells') must be limited; Quinton calls this property *i/o* closeness, giving as an example value $O(\sqrt{n})$ operations in a two-dimensional array.

There has been much work on the synthesis of systolic algorithms from recurrence equations or nested for loops (see Section 4.2); Miranker & Winkler [1984] have used this approach to characterise systolic algorithms through their regular data flow patterns. If a computation can be represented as a number of nested **for** loops in a standard assignment based language, then a point in that computation can be represented as a vector giving the values of the control variables being incremented in the loops. With each such point is associated a 'dependence vector' defining the other points in the computation on which its value depends (these are determined by the terms on the right-hand side of the assignment statements). The essence of a systolic computation is that the collection of dependence vectors emanating from each point are simply translations of each other.

Related to this work is that of Kung & Lin [1984], who consider a model of synchronous parallel computation where, essentially, a network is represented by an $n \times n$ matrix with non-zero a_{ij} entry if the processor v_i is to receive an output from the processor v_j; the non-zero entries are in the form of the z-notation common in engineering, so the value of v_j which is used by v_i may be that of several clock cycles previous to the current one. This approach allows algorithms to be manipulated algebraically, but it cannot be said to make any more precise the definition of the systolic model of computation. The authors state two conditions – temporal locality, and that 'the value of a node at any time never has to be sent to more than one node simultaneously, thus no broadcasting or fan-out is needed'.

In Melhem & Rheinboldt [1984] an abstract model of synchronous computation is defined on arbitrary directed multigraphs. A network is called 'general systolic' (in the sense of temporally local) if the semantics of the nodes are *causal operators* in that the output of a node at any clock cycle depends only on inputs to that node in those clock cycles (strictly) previous to the current one. The regularity condition is introduced as a homogeneity condition, and essentially defines that each node in a *homogeneous set* of nodes has precisely the same *i/o* behaviour (edges are coloured to ensure strong typing). This model is more formal than those mentioned previously, and it supports verification techniques.

One of the most thorough studies to date of the nature of the systolic model of computation is to be found in Li & Jayakumar [1986]. Jagadish *et al.* [1986] have given more precise definitions to many of the terms which have been used very loosely in standard papers in this subject, and Li & Jayakumar [1986] build on this work, together with that of Miranker & Winkler [1984] and Moldovan [1983], to analyse systolic arrays in terms of their extendability into a 2-D grid and the interconnections between the wavefronts in the pipelined computation (wavefronts were introduced by Weiser & Davis [1981] and Kung *et al.* [1982]). In this way the approach captures most, if not all, of the systolic properties originally outlined by Kung and Leiserson. An embeddable structure consists of an initial structure together with an extension rule, and an embeddable structure is defined to be systolic if for some constant c, all extensions can be embedded in the two-dimensional grid (as used by Thompson [1980]) in such a way that no edge is longer than c. It is then shown that if an embeddable structure satisfies the following four conditions then it is systolic if and only if $b \geq a$: (i) homogeneity – all vertices have in-degree b and out-degree a; (ii) adjacency – the inter-wavefront interconnections are between successive wavefronts only; (iii) bounded growth – if n_i is the number of vertices in the wavefront W_i then $n_{i+1} = \lceil n_i \times (a/b) \rceil + c$ for a constant c; (iv) uniformity – the successor function $f_i(w, i)$ is invariant of w and monotonic in i.

Bartha [1987] has extended the work of Elgot [1975] and Elgot & Shepherdson [1982] on flowchart schemes to give an equational axiomatization of systolic systems. Systolic schemes differ from ordinary schemes in having a non-negative integer, a (possibly zero) delay, associated with each edge.

The study of the particular type of pipelined computation now characteristic of systolic did not originate with the work of Kung and Leiserson. In the epilogue of Moore, McCabe & Urquhart [1987], Witte [1972] is credited as one of the first to draw attention to the potential for arrays of 'LSI $Ax + b$' modules for the efficient computation of signal processing tasks. One of the most important early theoretical papers in this field was Cole [1969] on real-time computation by iterative arrays. Cole states that 'a real-time computer must generate the nth output at the nth step of the computation'. This type of condition is an important characteristic of systolic algorithms, and is tied in with the work on wavefronts mentioned above; more general synchronous algorithms will not satisfy this condition.

4.2 Algorithm synthesis and program transformation

The automatic synthesis of efficient algorithms from high-level specifications of problems using formally verified transformations is the ambitious aim of much current research in computer science. The terms synthesis and derivation are not usually precisely defined, so that what some may term algorithm synthesis, others may dismiss as mere program transformation, or compilation, or algorithm refinement. True synthesis should produce an algorithm from a pure specification, but then this specification can often be viewed as an algorithmic solution to a higher level specification. These problems have been approached in VLSI design, and form a crucial part of the even wider aim of silicon compilation – the automatic production of VLSI chips from high-level specifications of problems.

Johnsson *et al.* [1981] describe a method for the design of networks to evaluate mathematical expressions such as finite impulse response, matrix operations and the discrete Fourier transform. The basic technique is the manipulation of mathematical equations at the mathematical level – mainly through the use of the delay operator z on data streams, using the fact that this operator satisfies certain simple mathematical identities. In this way several mathematical expressions for the same function can be derived, and it is shown how each expression 'represents' a different network. Although the manner of this representation is clear to the reader, it is not defined in any formal sense, and no attempt is made to describe how the networks might be synthesised from the equations; neither are the limits of application of the method investigated.

A similar technique, also based on z-operators, is presented in Kung & Lin [1984]. However, this work is expressed in a more general mathematical framework, with the manipulation of the equations being based on algebraic manipulation of matrices. Equivalence proofs are given. Kung and Lin also present the retiming transformations (see below) in this framework. The correspondence between the algebraic presentation and the z-graph presentation (graphs or networks) is also more closely explained, but the method still does not represent the automatic generation of a network from an equation.

Retiming transformations have proved simple but useful tools in the design of synchronous and systolic algorithms. One of the most significant problems in the design of pipelined synchronous algorithms is ensuring that at each clock cycle each processor receives precisely those inputs which it needs. Retiming transformations can remove such problems for the designer. The technique is that the algorithm is designed using only combinational logic, so avoiding difficult timing problems; then retiming transformations can be applied to introduce pipelining, which will allow a reduction in the clock cycle and so increase throughput. Retiming transformations were introduced in Leiserson & Saxe [1981] with a result called the Retiming Lemma (see also Leiserson & Saxe [1983]). The work is extended in Leiserson, Rose & Saxe [1983] and Leiserson & Saxe [1986] where algorithms are presented for retiming a circuit so that the length of the clock cycle is minimal, and for pipelining combinational circuitry using the minimal number of registers. The Cut Theorem was introduced in Kung & Lam [1984] and Kung [1984]. The Retiming Lemma and the Cut Theorem have since been presented by many authors in many different frameworks, and in McEvoy & Dew [1988] these two retiming transformations are proved to be equivalent using mathematical techniques. Retiming transformations are discussed in the chapter by McEvoy in this volume.

These two transformational techniques are in fact closely related. Both involve the manipulation of delay to transform synchronous algorithms, but whereas the z-transform techniques manipulate delay within the mathematical equation (the specification?), the retiming transformations manipulate delay within the network itself. Neither approach can be classified as synthesis as the step of producing the network is left to the designer – although in each case the whole approach is designed to make this step as simple and obvious as possible.

Weiser & Davis [1981] and Johnsson & Cohen [1981] contain extensions of their previously mentioned work. In particular, in Weiser & Davis [1981] the z-operator is used to define several operators on wavefronts, which can then be used to design algorithms or networks. Such transformations on wavefronts are built into a prototype program *Sys* described in Lam & Mostow [1985]. This program will produce systolic designs from pieces of Algol 60-like code which are nested **for** loops, annotated with phrases such as 'in parallel' and 'in place'.

In VLSI the most successful solutions to the problem of algorithm synthesis have proposed synthesis of systolic arrays from either recurrence equations or nested **for** loop program fragments (the two are essentially different presentations of the same specifications). The basic ideas for this type of work were developed in Karp, Miller & Winograd [1967]. There are many working in this field, but Quinton [1987a] gives a good mathematical explanation of the approach, which he calls dependence mapping. A set of recurrence equations is *uniform* if there exists a set of constant vectors (called dependence vectors) θ_i such that for each element x of the domain, $f_1(x) = g(f_1(x-\theta_1), \ldots, f_p(x-\theta_p))$, $f_2(x) = f_2(x-\theta_2), \ldots, f_p(x) = f_p(x-\theta_p)$. So a function value must be computed for every point in the domain, and the problem lies in fixing a particular processor array and mapping each point in the problem domain into a particular processor at a particular time in such a way that the points on which it depends have already been computed and are available to it. This requires the definition of (i) a timing function $t : D \to \mathbf{N}$ such that $t(z) > t(z - \theta_i)$, for $i = 1, \ldots, p$, and only finitely many points are mapped onto each time instant; and (ii) an allocation function which satisfies the condition that no processor is asked to carry out two computations at the same time (so the function g determines the level of abstraction of the processor). Quinton gives necessary and sufficient conditions under which affine time functions and allocation functions are guaranteed to exist. He also presents a procedure for determining the timing function when the domain is a convex hull. DIASOL (see Gachet, Joinnault & Quinton [1987]) is a software system which uses these techniques to construct systolic arrays from uniform recurrence equations.

Delosme & Ipsen [1987] consider the use of these techniques for a more general class of recurrence equations – coupled recurrence equations and recurrence equations in which the dependence mapping is affine rather than just translational. Also, rather than just enumerating possible arrays, the designer may specify a number of constraints (e.g., number of processors, or *i/o* bandwidth) and these constraints are used to formulate a linear integer programming problem which can lead to an optimal solution.

Li & Wah [1985] describe a method for the design of systolic arrays from linear recurrence equations in which three parameters – velocity of data flow, spatial distribution of data and period of computation – can be entered as parameters in constraint equations in order to produce an optimal solution.

The techniques of Quinton are also extended into a larger class of recurrence equations in Rajopadhye, Purushothaman & Fujimoto [1986]. In a recurrence equation with linear dependencies the translational dependency $x - \theta_i$ of the uniform recurrence relation can be replaced by a linear dependency $A_i z + b_i$ where A and b are square and one-dimensional matrices respectively. In this case the architectures produced may no longer be automatically systolic, but a specific pipelining stage can be introduced to produce a systolic architecture.

Moldovan (Moldovan [1982], Moldovan [1987]) has developed a software system ADVIS for the design of systolic arrays. The initial specification is a program fragment of nested **for** loops, but the basic techniques are similar to those described for recurrence equations. Mirankler & Winkler [1984] also offer an approach of synthesis of systolic arrays from nested **for** loops – indeed here the translational nature of the dependence vectors is used as the very definition of systolic.

Mongenet & Perrin [1987] have also extended Quinton's work. Here several systolic arrays can be obtained from the same dependency vectors by applying affine transformations to the time orderings. Inductive problems can also be solved. Cappello & Steiglitz [1984] derive from a recurrence equation a particular canonical geometric representation of an algorithm (which includes a representation of time). Various geometric transformations are then applied to produce further designs.

Chen has also presented methods for the synthesis of parallel algorithms from recurrence equations (see Chen [1983], Chen [1986] and Chen [1988]). An n-dimensional directed acyclic graph (or dependency graph) is obtained by data parallel interpretation of the equations, and some optimal dependency matrices B are found. Then all possible communication matrices C of an $(n-1)$-dimensional communication matrix are found. In this way minimising the number of communications in the design (finding an optimal B) is separated from the choice of the network (choosing C).

Huang & Lengauer [1987] present a method for the derivation of systolic algorithms which is significantly different from the above techniques, although sharing some of the basic ideas. The method is based on transformations of traces (or executions) of programs. The initial input to the process is a simple (sequential) 'refinement' program for the solution of a problem. A trace is derived for this program, and each of the basic statements of the program is analysed with respect to certain semantic properties (e.g., idempotence, commutativity, independence and neutrality). Using this information, an algorithm *Transform* derives a parallel trace which has the same semantic behaviour as the sequential trace (this equivalence has been mechanically checked using the Boyer–Moore theorem prover). (The detection of parallelism plays a central role in this scheme.) It remains to express this parallel trace on an architecture; if the user provides a layout of a network then the flow of the data through the network will be synthesised by the system. The final network is specified by the functions *step*, *place*, *flow* and *pattern*. *Step* determines when basic statements are to be executed (by a mapping into $\mathbf{N}$); *place* determines where basic statements are to be executed (by mapping them into $\mathbf{N}^d$, for some dimension d); *flow* specifies the direction of data movement (by mapping the program variables into the range of *place*, and so determining where the outputs of a processor go); *pattern* lays out the initial pattern of the data (by mapping the program variables into the range of place). In Huang & Lengauer [1987] the method is limited to systolic arrays in which processors are connected only by unidirectional channels to processors that occupy neighbouring points; under these conditions the functions *flow* and *pattern* are calculated if *step* and *place* are given. Further conditions necessary for *place*, *flow* and *pattern* to be well-defined are stated and proved in three theorems. The system is automated in a graphics system which can also display two-dimensional processor layouts and simulate sequences of execution steps on them.

Rao [1985] has studied *regular iterative algorithms* (RIAs) in depth, examining both the synthesis of such algorithms and their implementation on processor arrays. Some of this work is built upon that of Karp, Miller & Winograd [1967], and it generalises the work of Quinton and others into a much larger class of algorithms (covering essentially all iterative algorithms). The most important condition in the definition of an RIA is that there exists a finite set of constants $\{D_j : j \in J\}$ such that for every I in the index space (and each $j \in J$), $x(I)$ is directly dependent on $x(I - D_j)$. The systolic algorithms are characterised as (equivalent to) a subclass of the RIAs, but they exhibit certain pipelinability features that are not characteristic of every RIA. Results such as the systolic conversion lemma of Leiserson & Saxe [1981] and the general linear scheduling problem are examined in this setting. Techniques for optimising systolic arrays by reducing extrinsic and intrinsic iteration levels are considered. The computability and causality of RIAs are examined, together with algorithms for deriving scheduling functions (which must be linear with respect to the index vectors), and efficient algorithms for implementing RIAs in a similar, possibly lower dimensional, lattice of processors.

Rao [1985] dates work on iterative algorithms back to Atrubin [1958], McCluskey [1958], Hennie [1958], Unger [1963], Waite [1967] and Karp, Miller & Winograd [1967]. Hennie's thesis (Hennie [1961]) and the book Hennie [1968] contain some remarks on the space-time representation of sequential circuits. Waite [1967] studied the problem of determining the existence of a path between two nodes in an iterative network – corresponding to examining the dependence of variables in an RIA. An iterative array is similar to a finite state machine and Waite [1967] uses techniques similar to those of McNaughton & Yamada [1960] to construct a regular expression from which a series of linear programming problems can be formulated. The most important contribution is that of Karp, Miller & Winograd [1967] which contains many significant results. Many well known and important algorithms can be formulated in the RIA idiom – e.g., Euclid's greatest common divisor algorithm, many of Newton's finite difference approximation algorithms, and Gauss' algorithms for the inversion of regular matrices. More recently, the use of RIAs for the numerical solution of partial and ordinary differential equations has been noted by Thomas [1940] and Karp, Miller & Winograd [1967]. Moreover, the discretised versions of the Darlington synthesis procedure Darlington [1939], the Brune synthesis procedure Brune [1931] and many other classical network synthesis procedures lead to RIAs for solving the digital filtering problem. A unifying framework that contains all these RIAs as special cases has been presented in Rao & Kailath [1984] and Rao & Kailath [1985].

Chandy & Misra [1986] apply general program development techniques to the design of systolic algorithms – systolic algorithms are represented as programs, and these programs are derived from invariants. This approach is part of a general purpose theory of parallel program design *and* verification developed in Chandy & Misra [1988]. The theory is based upon so-called *unbounded nondeterministic iterative transformations* which comprise a model of specification and verification, and for which the model is called *UNITY*. A unity description combines multiple assignments and a fair nondeterministic selection mechanism. These simple components are well suited to formal reasoning about correctness in program development, and are sufficiently abstract to be applicable to programs destined for sequential machines, synchronous and asynchronous shared-memory multiprocessor systems, and message-based distributed systems.

An independent and earlier general theory of parallel program development has been established by R. J. R. Back, R. Kurki-Suonio and their associates. Their theory is based on *action systems.* These are slightly more general in form than unity systems, but are otherwise the same concepts in all essential respects. The methods and case studies of the theory focus on the refinement of action systems rather than their specifications, however. Examples of the application of the theory of action systems to hardware design are not available at the present. See papers Back & Kurki-Suonio [1983], Back & Kurki-Suonio [1988], Back & Sere [1988] and Back & Sere [1989].

The multiple assignment makes the notation well-suited to the development of systolic algorithms. In this connection we note that concurrent assignments are also suited to tools for the animation and testing of general synchronous algorithms: see Martin & Tucker [1987], Martin & Tucker [1988], Thompson & Tucker [1988], Martin [1989], and Thompson [1987].

Burns & Martin [1988] also apply techniques of program transformation to the design of circuits. However in this case self-timed circuits are produced from concurrent programs. First program transformations are used to decompose the concurrent program into one consisting entirely of instances of well-defined basic processes (the source and target languages are well-defined and based on CSP as in Hoare [1985]); this program can then be compiled directly into a self-timed circuit.

There has been considerable work at Eindhoven on trace theory and its application to VLSI. A trace structure may be viewed as a specification of a system. The alphabet of the trace structure is a set of symbols where each symbol represents a type of communication possible between the system and its environment; a trace is a word on this alphabet, and represents a possible sequence of communications between the system and its environment. The trace set is a language on this alphabet, and describes all possible patterns of communication between the system and its environment. The mathematical theory of traces has been investigated in (for example) Rem [1981], Rem, Snepscheut & Udding [1983], Snepscheut [1983], Snepscheut [1985] and Rem [1987]. Trace theory is related to CSP, and concurrency is modelled using arbitrary interleaving of processes. Properties of traces which are of interest include abstraction properties such as projection, and composition properties such as weaving. Applications of trace theory include the definition and derivation of delay-insensitive circuits from programs (Udding [1984], Snepscheut [1985] and Ebergen [1987]). Snepscheut [1985] describes how finite state machines are used in the derivation of delay-insensitive circuits (or, more correctly, schematics) from trace structures. Rem [1987] isolates the trace properties of data independence, transparency and conservatism as central notions in the study of systolic arrays.

This section on synthesis has concentrated on the synthesis of systolic arrays, as this subject would appear to be the most mathematically developed. Johnson [1984] describes a method for the synthesis of circuit designs from recursion equations. The method is very much based on the functional programming idiom and the applicative style – the circuit designs themselves are expressions in a functional programming language, and the synthesis is program transformation through a collection of correctness preserving rewriting rules. Investigation into this method continues in Johnson, Bose & Boyer [1988].

4.3 Verification and specification

In recent years the need for verification and certification of hardware devices has provided a major motivation for the study of the theory of VLSI. The complexity of VLSI devices is continually increasing, and there is growing concern as to the confidence one can place in the correctness of such complex devices, especially since increasingly these chips are being used as control devices in safety critical applications. Another important factor may be the economic cost of discovering design errors in the testing phase after fabrication. The only realistic methods with which to *guarantee* correctness of a device are founded on theoretical methods for verification. The only alternative is testing, and as the testing of every possible input is not feasible, testing can never guarantee correctness, but only guarantee incorrectness. It is reasonable to claim that the only way to fully guarantee that a device will output y when x is input is to actually input x and see what happens; however, how can you be sure that this result is not dependent upon the particular state which the device was in at the particular time when x was input? Repeating the test may yield a different answer. However, the verification of a device can only be carried out within a theoretical model of the device, and just as testing does not guarantee correct behaviour under inputs which have not been tested, so theoretical methods will not guarantee correctness under logical and physical conditions which are not correctly described in the model. The gap between logic and physics is the significant problem. The key to all these problems is specification. If the use to be made of the device is not accurately specified, then the results of any verification, validation or testing are worthless.

So verification consists of three requirements: specification of the task which the device must fulfil, description of the device in some model, and proof that the device 'meets' its specification. The nature of these requirements depends very much on the techniques being used, and some of these are described below. Many of these techniques are adapted from those of software verification. It is arguable that the techniques appear to have been more immediately successful in the hardware medium; this may be because the virtues of regularity and simplicity in design are more generally adhered to in hardware design, or perhaps it is just that hardware designs are, in general, simpler than software designs. Indeed, given the nature of the models which computer scientists use, it is not surprising that the techniques of software verification are immediately applicable to hardware verification; the computer scientist naturally models the hardware device at an algorithmic level (as a digital device), so in such a model verification of the device becomes verification of an algorithm (typically this algorithm is described in a programming language which consists of primitive processes communicating over wires).

4.3.1 Formal logic and machine assisted verification Currently the most widely used of such computer systems is HOL, a system due to M. Gordon, and based on Church's simple theory of types, augmented by an ε-operator (Church [1940]). HOL is a system for generating proofs in higher order logic (Gordon [1985] and Gordon [1988]). The HOL system is implemented on top of Cambridge LCF (Paulson [1987]), which is itself a development of LCF (Gordon, Milner & Wadsworth [1979]). Both of these systems use the functional programming language ML (Cousineau, Huet & Paulson [1986]) as a metalanguage; e.g., in HOL theorem is an ML abstract data type. The successful development of HOL is measured by the fact that it could be used in the verification of VIPER (Cullyer [1988]), the first production microprocessor for which significant parts of the design of the chip were formally verified (Cohn [1988]). This interesting achievement included the verification of state transitions of the machine. Its aim was to study the structuring and management of large (rather than intricate) proofs involving similar components. A discussion of the meaning, in theory and practice, of a completely verified computer system is a complex task which is made pressing by experimental research of this kind.

HOL describes the behaviour of a device as an expression in higher order logic. A primitive device is modelled by a relation which has one argument for each 'external wire' or 'pin' of the device. As the logic is higher-order, the variables may vary over streams or signals of data, and so the primitive devices may have state. The behaviour of the conjunction of two devices is the conjunction of their two predicates, given that any wire joining the two devices must be represented by the same variable in each predicate. The existential introduction rule will allow the hiding of such wires in the expressions. The chapter by Cohn and Gordon in this volume demonstrates the use of HOL in verifying a small example – a counter.

HOL has grown out of earlier work of Gordon. In Gordon [1981a] a model based on λ-calculus, but with a naming mechanism close to CCS (Milner [1980]), is used to describe a microcoded computer and digital circuits at the switch level (see also Gordon [1981b]). Work on LCF (Gordon, Milner & Wadsworth [1979]) led into the system LCF-LSM (Gordon [1983a]) which is used in Gordon [1983b] to verify a computer. (The verification of this computer has been repeated in HOL by Joyce; see Joyce [1988] and Joyce, Birtwhistle & Gordon [1986].) LCF-LSM was the system used for the initial verification of VIPER (see Cullyer [1988]).

Many of the ideas used in HOL are present in other models which use logic for verification. In particular, Hanna is credited as the first to propose the use of higher order logic to verify hardware devices (Hanna [1983]). In Hanna & Daeche [1986] a theory based on analogue wave forms and gate behaviours is used to verify a D-flipflop. Such proofs can be mechanised within their Veritas software system – this system also takes the LCF approach, but here the metalanguage is a purely functional language, MV.

Work using automatic theorem provers based on first order logic includes Hunt's verification of the FM8501 microprocessor (Hunt [1986]) using the Boyer–Moore theorem prover (Boyer & Moore [1979] and Boyer & Moore [1988]). Recently, in Purushothaman & Subrahmanyam [1988] and Purushothaman & Subrahmanyam [1989] verification techniques for certain systolic algorithms have been presented using uniform recurrence relations and the Boyer–Moore theorem prover.

The fragments of first order logic underlying logic programming can also be employed in hardware verification. A notable early example of this is contained in Barrow [1984]. Some recent examples of the enhancement of the 'logic' of Prolog for verification are Tiden [1988] and Simonis, Nguyen & Dincbas [1988].

The verification of hardware using systems based on equations and rewriting techniques has been considered; the systems include OBJ3 (Goguen & Winkler [1988]), REVE (Lescanne [1983]), RRL (Kapur, Sivakumar & Zhang [1986]), and AFFIRM-5 (Musser & Cyrluk [1985]), for example. An early and independent observation of the possible role of rewriting, in the case of simple boolean expressions and combinational expressions, was made in Chandrasekhar, Privitera & Conradt [1987]. Other works are Narendran & Stillwell [1988] and Musser *et al.* [1988]. Stavridou [1988] has investigated the use of the specification language OBJ in conjunction with the theorem prover REVE for verification of an adder as a case study. A substantial study of the basic mathematical and practical techniques involved in using OBJ3 for the verification of hardware is Goguen [1988].

Stavridou, Barringer & Edwards [1988] is a useful comparative study of several of the above specification and verification tools for hardware.

In all of the theorem provers, proof assistants and proof checkers mentioned, the user plays an essential interactive role in verification. An attempt at a fully automatic verification process, based on a decision algorithm for a class of systolic algorithms and specifications, is made in Abdulla [1989]. The method uses integer linear programming.

Other authors to have used the techniques of formal logic in hardware verification include Shostak and Eveking. These methods are based in first order logic. Shostak (Shostak [1983a] and Shostak 1983b]) has adapted to circuit graphs (a model which can represent electrical properties such as voltage and current) Floyd's method of program verification by loop assertions in flowgraphs. One of the main contributions of Eveking [1985] is a formalisation of design abstraction (e.g., data abstraction and time abstraction) in terms of the interpretation of one formal theory in another. Methods for formalizing such abstractions within HOL are presented in Melham [1988] and Herbert [1988].

Another approach to the use of logical systems in hardware specification and verification is through temporal logic. Early papers to explore these ideas were Bochmann [1982] and Malachi & Owicki [1981]. Bochmann uses the standard temporal operators 'henceforth' and 'eventually' for the specification of liveness and termination properties of processes, and introduces a **while** operator for the specification of memory and safeness properties – an arbiter example is given together with a discussion of a verification of an implementation. Malachi & Owicki [1981] use similar techniques to specify self-timed systems such as those defined by Seitz [1980]. Fujita, Tanaka & Moto-oka [1983] also make use of temporal logic in the verification of hardware.

Moszkowski [1983] has developed a sophisticated temporal predicate logic for reasoning about hardware (see also Moszkowski [1985]). The language is used to describe circuits at various levels of abstraction, from detailed quantitative timings to functional behaviour; case studies have been examined ranging from simple delay elements up to the Am2901 ALU bit slice. The language has been developed into Tempura (Moszkowski [1986]), a programming language based on interval temporal logic.

Starting with Clarke & Mishra [1984] and Mishra & Clarke [1985], a substantial study of hardware verification using temporal logic has been undertaken by E. Clarke and associates. Browne *et al.* [1985] describe an automatic verification system for sequential circuits in which specifications are expressed in a propositional temporal logic of branching time – CTL, computation tree logic. The system first builds a finite state machine (which can be used as a finite Kripke structure) for the circuit, and then it uses a linear time algorithm to determine the truth of a temporal formula with respect to the state machine. This verification tool is now one of a number of programs that stimulate systems and interface to VLSI fabrication tools. There is a common interface for these tools in a language SML (Browne, Clarke & Dill [1986]). From an SML description a finite state machine is produced, and the VLSI tools can be applied to this finite state machine.

4.3.2 Mathematical methods There are some general models of concurrent systems that apply directly, or can be customized, to model and verify hardware. For example, there are various axiomatic theories of process algebra, such as CCS, CSP and ACP, which have been employed in design work (recall Section 4.1.1) and verification. The origins of these models of concurrent computation is the arbitrary interleaving semantics of operating systems constructs.

Hennessy [1986] describes a method for the verification of systolic algorithms which uses an operator calculus based on CCS (Milner [1980]). In addition to the standard CCS operators, the calculus contains a synchronous parallel operator $\times$, so that in the process $P_1 \times P_2$, P_1 may only perform an action if P_2 performs an action simultaneously. A systolic array is created through use of the operator $\times$ to connect cells together. Both specifications and implementations can be described in this language, and for two examples – a palindrome recogniser and a sorter – careful proofs are given that systolic algorithms correctly implement their specifications. The proofs involve transforming the specification into the implementation using a suite of syntactically presented transformations together with a form of induction called Fixpoint Induction. Typically the transformations are axioms such as the commutativity law for a given operator.

Weijland [1987] and Mulder & Weijland [1987] have presented a method of proof of correctness of systolic algorithms which is similar to that of Hennessy, but which is based on the calculus ACP_τ, the Algebra of Communicating Processes with silent steps (Bergstra & Klop [1985]) rather than CCS. The main difference with the work of Hennessy is that the concurrent algorithms are described using an *asynchronous* method of communication, so that the corresponding circuits are delay-insensitive. Concise correctness theorems for algorithms computing matrix-vector multiplication

(Weijland [1987]) and sorting (Mulder & Weijland [1987]) are formulated and are proved, relative to a bisimulation semantics, using the axioms and laws for ACP_τ together with induction. In the chapter by Weijland in this book similar techniques are used to verify a palindrome recogniser.

We have mentioned the verification by derivation methods of Chandy & Misra [1988] for synchronous algorithms in Section 4.2. Gribomont [1988] uses classical loop invariant methods to verify systolic arrays. This is achieved by first deriving an equivalent sequential program from the systolic array. The complexity of these two procedures is greatly reduced by the choice of the original language in which the systolic array is defined – SCSP is a small language which is derived from CSP (Hoare [1985]), but does not require handshaking communication.

A family of general models of concurrency based on algorithms for hardware has been developed by B. C. Thompson and J. V. Tucker. A characteristic of devices is their dependency on one or more clocks. A clock and its clock cycles introduce a simple and explicit notion of operations taking place in parallel, as well as a simple notion of determinism, into modelling. A *synchronous concurrent algorithm* is defined to be a network of processors and channels, computing and communicating in parallel, and synchronised by a global clock. Such an algorithm processes infinite streams of data. Examples of synchronous concurrent algorithms include clocked hardware, systolic arrays, neural nets, cellular automata, and coupled map lattice dynamical systems. In the theory, the data, the functions implemented by the basic modules, the clock, and the streams are collected to form a many sorted algebra, representing the components by an abstract data type. The architecture and algorithm is represented in various *equivalent* models of synchronous computing over this algebra, including (i) recursive functions, (ii) concurrent assignments and loops, (iii) graphs. The different computational models are complementary in that (i) suits verification, (ii) suits simulation, and (iii) suits architecture. The objective is to establish a mathematical theory of synchronous hardware that unifies the study of many disparate architectural models. As a pure mathematical theory it contains concepts and methods for design that are independent of particular specification and verification formalisms, and their computer implementations.

The general mathematical theory of synchronous concurrent algorithms based on many-sorted algebras and recursive functions is intimately connected with many-sorted first order logic, and forms of second order logic to allow for streams. In particular, it is intimately connected with equational logics and term rewriting techniques. In detailed case studies of verifications of convolution, correlation, matrix, sorting algorithms (in Hobley, Thompson & Tucker [1988], Harman, Hobley & Tucker [to appear], Thompson [1987]), and of graphics devices (in Eker & Tucker [1987] and Eker & Tucker [1989]), the logical principles used are equational, but are complicated by limited forms of induction, conditionals and streams. In Thompson & Tucker [1989], a sequence of theorems shows that if a set of basic modules or cells can be equationally axiomatised then *any* synchronous concurrent algorithm made from the cells can be equationally defined, and if its specification is equationally definable then the correctness problem can be formulated in terms of equations, and hence tackled by many logical and verification systems.

The mathematical theory has been used in the verification of synchronous algorithms by machine, using the proof development system *Nuprl*, due to R. Constable. Nuprl is based on a logical system called *constructive type theory*, due to P. Martin-Löf, and is intended to prove theorems of mathematics, and to develop programs from specifications: see Constable *et al.* [1986]. Its application to hardware is reported in Derrick, Lajos & Tucker [1989].

A theory of specification of synchronous devices, that can be considered independently of the theory of implementation and verification, has been developed using many-sorted algebras and recursive functions. The basic paper on this topic is the chapter by Harman and Tucker in this volume. Further work on devices, such as a UART and a RISC, is reported in Harman & Tucker [1988b], Harman & Tucker [1988a], and Harman [1989].

A fundamental problem is to classify disparate models of synchronous computation over algebras, and to establish the scope and limits of computation by means of an appropriate Church–Turing Thesis for deterministic concurrent computation over streams. Papers on these subjects are Thompson & Tucker [1985], Thompson & Tucker [1988], Meinke & Tucker [1988], Thompson & Tucker [1989] and Tucker [1989]; and theses are Thompson [1987] and Meinke [1988].

Acknowledgements
We thank P. Abdulla, K. Meinke, V. Stavridou, and B. C. Thompson for information on the literature.

REFERENCES

Abdulla [1989]
P. Abdulla (1989) 'Automatic verification of a class of systolic hardware circuits', Technical Report, Department of Computer Systems, Uppsala University, Sweden.

Abelson & Andreae [1980]
H. Abelson and P. Andreae (1980) 'Information transfer and area–time tradeoffs for VLSI multiplication', *Comm. ACM,* **23**, 20–3.

Anderson & Rosenfield [1988]
J. A. Anderson and E. Rosenfield (1988) *Neurocomputing*, MIT Press.

Annaratone *et al.* [1987]
M. Annaratone, E. Arnould, T. Gross, H. T. Kung, M. Lam, O. Menzilcioglu and J. A. Webb (1987) 'The Warp Computer: architecture, implementation and performance', *IEEE Trans. Computers,* **C-36**, 1523–38.

Apt [1981]
K. R. Apt (1981) 'Ten years of Hoare's logic: a survey – Part I', *ACM Trans. Programming Languages and Systems,* **3**, 431–83.

Atrubin [1958]
A. J. Atrubin (1958) 'A study of several planar iterative switching circuits', S. M. Thesis, Department of Electrical Engineering, MIT.

Back [1980]
R. J. R. Back (1980) 'Correctness preserving program refinements: proof theory and applications', Mathematical Centre Tracts, CWI, Amsterdam.

Back [1981]
R. J. R. Back (1981) 'On correct refinement of programs', *J. Computer and System Sciences,* **23**, 49–68.

Back & Kurki-Suonio [1983]
R. J. R. Back and R. Kurki-Suonio (1983) 'Decentralisation of process nets with centralised control', 131–42 in *Principles of Distributed Computing, 2nd Symposium, Montreal,* ACM SIGACT-SIGOPS.

Back & Kurki-Suonio [1988]
R. J. R. Back and R. Kurki-Suonio (1988) 'Distributed co-operation with action systems', *Trans. Programming Languages and Systems,* **10**, 513–54.

Back & Sere [1988]
R. J. R. Back and K. Sere (1988) 'An exercise in deriving parallel algorithms: Gaussian elimination', Report on Computer Science and Mathematics 65, Ser. A, Abo Akademi, SF-20520 Abo, Finland.

Back & Sere [1989]
R. J. R. Back and K. Sere (1989) 'Stepwise refinement of action programs', Report on Computer Science and Mathematics 78, Ser. A, Abo Akademi, SF-20520 Abo, Finland.

Backus [1978]
J. Backus (1978) 'Can programming be liberated from the von Neumann style? A functional style and its algebra of programs', *Comm. ACM,* **21**, 613–41.

Barrow [1984]
H. G. Barrow (1984) 'Proving the correctness of digital hardware designs', *VLSI Design,* **5**, 64–77.

Bartha [1987]
M. Bartha (1987) 'An equational axiomatisation of systolic systems', *Theoretical Computer Science,* **55**, 265–89.

Baudet [1983]
G. Baudet (1983) 'Design and complexity of VLSI algorithms', in *Foundations of Computer Science IV: Distributed Systems Part 1, Algorithms and Complexity,* ed. J. W. de Bakker & J. van Leeuwen, Mathematical Centre Tracts 158, Mathematisch Centrum, Amsterdam.

Baudet, Preparata & Vuillemin [1983]
G. M. Baudet, F. P. Preparata and J. E. Vuillemin (1983) 'Area time optimal circuits for convolution', *IEEE Trans. Computers,* **C-32**, 684–8.

Bell & Newell [1971]
C. G. Bell and A. Newell (1971) *Computer Structures: Readings and Examples,* McGraw–Hill, New York.

Bergstra & Klop [1983]
J. A. Bergstra and J. W. Klop (1983) 'A proof rule for restoring logic circuits', *Integration,* **1**, 161–78.

Bergstra & Klop [1985]
J. A. Bergstra and J. W. Klop (1985) 'Algebra of communicating processes with abstraction', *Theoretical Computer Science,* **37**, 77–121.

Bergstra & Tucker [1984]
J. A. Bergstra and J. V. Tucker (1984) 'The axiomatic semantics of programs based on Hoare's logic', *Acta Informatica,* **21**, 293–320.

Birtwhistle & Subrahmanyam [1988]
G. Birtwhistle and P. A. Subrahmanyam, ed. (1988) *VLSI specification, verification and synthesis*, Kluwer, Dordrecht.

Bledsoe & Loveland [1984]
W. W. Bledsoe and D. W. Loveland, ed. (1984) *Automated Theorem Proving: After 25 years*, Contemporary Mathematics 29, American Mathematical Society, Providence, RI.

Bochmann [1982]
G. V. Bochmann (1982) 'Hardware specification with temporal logic: an example', *IEEE Trans. Computers,* **C-31**, 223–331.

Boute [1986]
R. T. Boute (1986) 'Current work on the semantics of digital systems', 99–112 in *Formal Aspects of VLSI Design*, ed. G. Milne & P. A. Subrahmanyam, North-Holland, Amsterdam.

Boyer & Moore [1979]
R. S. Boyer and J. Strother Moore (1979) *A Computational Logic*, Academic Press.

Boyer & Moore [1988]
R. S. Boyer and J. Strother Moore (1988) *A Computational Logic Handbook*, Academic Press.

Brent & Kung [1981]
R. P. Brent and H. T. Kung (1981) 'Area–time complexity of binary multiplication', *J. ACM,* **28**, 521–34.

Brookes [1984]
S. D. Brookes (1984) 'Reasoning about synchronous systems', Technical Report CMU-CS-84-145, Department of Computer Science, Carnegie-Mellon University, Pittsburgh, Pennsylvania 15213.

Browne, Clarke & Dill [1986]
M. C. Browne, E. M. Clarke and D. L. Dill (1986) 'Automatic circuit verification using temporal logic: two new examples', 113–24 in *Formal Aspects of VLSI Design*, ed. G. J. Milne & P. A. Subrahmanyam, North-Holland, Amsterdam.

Browne *et al.* [1985]
M. C. Browne, E. M. Clarke, D. L. Dill and B. Mishra (1985) 'Automatic verification of sequential circuits using temporal logic', 98–113 in *Computer Hardware Description Languages and their Applications, Seventh International Conference, Tokyo*, ed. C. J. Koomen & T. Moto-oka, North Holland, Amsterdam.

Brune [1931]
O. Brune (1931) 'Synthesis of a finite two-terminal network whose driving point impedance is a prescribed function of frequency', *J. Mathematical Physics,* **10**, 191–236.

Bryant [1981]
R. E. Bryant (1981) 'A switch-level model of MOS logic circuits', 329–40 in *VLSI 81, Edinburgh*, ed. J. P. Gray, Academic Press, London.

Bryant [1984]
R. E. Bryant (1984) 'A switch-level model and simulator for MOS digital systems', *IEEE Trans. Computers,* **C-33**, 160–77.

Bryant [1986]
R. E. Bryant (1986) 'Can a simulator verify a circuit?', 125–36 in *Formal Aspects of VLSI*, ed. G. J. Milne & P. A. Subrahmanyam, North-Holland, Amsterdam.

Burns & Martin [1988]
S. M. Burns and A. J. Martin (1988) 'Synthesis of self-timed circuits by program transformation', 99–116 in *The Fusion of Hardware Design and Verification, Proceedings IFIP WG 10.2 Working Conference, University of Strathclyde*, ed. G. J. Milne, North Holland, Amsterdam.

Camilleri, Gordon & Melham [1986]
A. Camilleri, M. Gordon and T. Melham (1986) 'Hardware verification using higher-order logic', in *From HDL Descriptions to Guaranteed Correct Circuit Designs, IFIP International Working Conference*, Grenoble, France.

Cappello & Steiglitz [1984]
P. R. Cappello and K. Steiglitz (1984) 'Unifying VLSI array design with linear transformations of space-time', in *Advances in Computing Research, 2: VLSI Theory*, ed. F. Preperata, Jai Press Inc., Greenwich, Connecticut.

Cardelli & Plotkin [1981]
L. Cardelli and G. Plotkin (1981) 'An algebraic approach to VLSI design', 173–82 in *VLSI 81, Edinburgh*, ed. J. P. Gray, Academic Press, London.

Chandra, Stockmeyer & Vishkin [1984]
A. K. Chandra, L. J. Stockmeyer and U. Vishkin (1984) 'Constant depth reducibility', *SIAM J. Computing,* **13**, 423–39.

Chandrasekhar, Privitera & Conradt [1987]
M. S. Chandrasekhar, J. P. Privitera and K. W. Conradt (1987) 'Application of term rewriting techniques to hardware design verification', 277–82 in *Proceedings of 24th Design Automation Conference*, ACM & IEEE.

Chandy & Misra [1986]
K. M. Chandy and J. Misra (1986) 'Systolic algorithms as programs', *Distributed Computing,* **1**, 177–83.

Chandy & Misra [1988]
K. M. Chandy and J. Misra (1988) *Parallel Program Design*, Addison-Wesley.

Chazelle & Monier [1981a]
B. Chazelle and L. Monier (1981) 'A model of computation for VLSI with related complexity results', 318–25 in *Proceedings 13th Symposium on Theory of Computing*, ACM.

Chazelle & Monier [1981b]
B. Chazelle and L. Monier (1981) 'Towards more realistic models of VLSI computation', in *Proceedings 2nd Caltech Conference on VLSI.*

Chazelle & Monier [1985]
B. Chazelle and L. Monier (1985) 'A model of computation for VLSI with related complexity results', *J. ACM,* **32**, 573–88.

Chen [1983]
M. C. Chen (1983) 'Space-time algorithms: semantics and methodology', PhD Thesis, California Institute of Technology.

Chen [1986]
M. C. Chen (1986) 'A design methodology for synthesizing parallel algorithms and architectures', *J. Parallel and Distributed Computing,* **3**, 461–91.

Chen [1988]
M. C. Chen (1988) 'The generation of a class of multipliers: synthesizing highly parallel algorithms in VLSI', *IEEE Trans. Computers,* **37**, 329–38.

Church [1936]
A. Church (1936) 'A note on the Entscheidungsproblem', *J. Symbolic Logic,* **1**, 40–1 and 101–2.

Church [1940]
A. Church (1940) 'A formulation of the simple theory of types', *J. Symbolic Logic,* **5**, 56–8.

Clare [1973]
C. R. Clare (1973) *Designing Logic Systems Using State Machines*, McGraw-Hill.

Clarke & Mishra [1984]
E. Clarke and B. Mishra (1984) 'Automatic verification of asynchronous circuits', in *Proceedings of CMU Workshop on Logics of Programs*, ed. E. Clarke & D. Kozen, Lecture Notes in Computer Science 164, Springer Verlag.

Cohn [1988]
A. Cohn (1988) 'A proof of correctness of the VIPER microprocessor: the first levels', 22–72 in *VLSI Specification, Verification and Synthesis*, ed. G. Birtwhistle & P. A. Subrahmanyam, Kluwer Academic Publishers, Boston.

Cole [1969]
S. N. Cole (1969) 'Real-time computation by n-dimensional iterative arrays of finite-state machines', *IEEE Trans. Computers,* **C-18**, 349–65.

Constable *et al.* [1986]
R. L. Constable, S. F. Allen, H. M. Bromley, W. R. Cleaveland, J. F. Cremer, R. W. Harper, D. J. Howe, T. B. Knoblock, N. P. Mendler, P. Panangaden, J. T. Sasaki and S. F. Smith (1986) *Implementing Mathematics with the Nuprl Proof Development System*, Prentice-Hall, Englewood Cliffs, NJ.

Cousineau, Huet & Paulson [1986]
G. Cousineau, G. Huet and L. Paulson (1986) 'The ML handbook', INRIA, France.

Crutchfield & Kaneko [1987]
J. P. Crutchfield and K. Kaneko (1987) 'Phenanenology of spatio-temporal chaos', in *Directions in Chaos*, ed. H. Bai-lin, World Scientific Publishing, Singapore.

Cullyer [1988]
W. J. Cullyer (1988) 'Implementing safety-critical systems: the VIPER microprocessor', 1–26 in *VLSI Specification, Verification and Synthesis*, ed. G. Birtwhistle & P. A. Subrahmanyam, Kluwer Academic Publishers, Boston.

Darlington [1939]
S. Darlington (1939) 'Synthesis of reactance 4-ports which produce prescribed insertion loss characteristics', *J. Mathematical Physics,* **18**, 257–355.

Delgado Kloos [1987]
C. Delgado Kloos (1987) *Semantics of Digital Circuits*, Lecture Notes in Computer Science 285, Springer Verlag.

Delosme & Ipsen [1987]
J. Delosme and I. Ipsen (1987) 'Efficient systolic arrays for the solution of Toeplitz systems: an illustration of a methodology for the construction of systolic architectures in VLSI', 37–46 in *Systolic Arrays, Proceedings First International Conference, Oxford*, ed. W. Moore, A. McCabe & R. Urquhart, Adam Hilger.

Derrick, Lajos & Tucker [1990]
J. Derrick, G. Lajos and J. V. Tucker (1990) 'Specification and verification of synchronous concurrent algorithms using the Nuprl proof development system', Technical Report, Centre for Theoretical Computer Science, University of Leeds, to appear.

Dijkstra [1976]
E. W. Dijkstra (1976) *A Discipline of Programming*, Prentice-Hall International, London.

Dunne [1988]
P. E. Dunne (1988) *The Complexity of Boolean Networks*, Academic Press.

Ebergen [1987]
J. C. Ebergen (1987) 'Translating programs into delay-insensitive circuits', PhD Thesis, Department of Mathematics and Computer Science, Eindhoven University of Technology.

Eker & Tucker [1987]
S. M. Eker and J. V. Tucker (1987) 'Specification, derivation and verification of concurrent line drawing algorithms and architectures', 449–516 in *Theoretical Foundations of Computer Graphics and CAD*, ed. R. A. Earnshaw, Springer Verlag, Berlin.

Eker & Tucker [1989]
S. M. Eker and J. V. Tucker (1989) 'Specification and verification of synchronous concurrent algorithms: a case study of the pixel planes architecture', in *Parallel Processing for Vision and Display*, ed. P. M. Dew, R. A. Earnshaw & T. R. Heywood, Addison Wesley.

Elgot [1975]
C. C. Elgot (1975) 'Monadic computation and iterative algebraic theories', 175–230 in *Logic Colloquium '73*, ed. H. E. Rose, North Holland, Amsterdam.

Elgot & Shepherdson [1982]
C. G. Elgot and J. C. Shepherdson (1982) 'An equational axiomatisation of the algebra of reducible flowchart schemes', in *Calvin C. Elgot, Selected Papers*, ed. S. L. Bloom, Springer Verlag, Berlin.

Ercegovac & Lang [1985]
M. D. Ercegovac and T. Lang (1985) *Digital Systems and Hardware: Firmware Algorithms*, Wiley.

Eveking [1985]
H. Eveking (1985) 'The application of CHDLs to the abstract specification of hardware', 167–78 in *Computer Hardware Description Languages and their Applications, Seventh International Conference, Tokyo*, ed. C. J. Koomen & T. Moto-oka, North-Holland, Amsterdam.

Feynman [1982]
R. P. Feynman (1982) 'Simulating physics with computers', *Int. J. Theoretical Physics,* **21**, 467–88.

Flores [1963]
I. Flores (1963) *The Logic of Computer Arithmetic*, Prentice-Hall.

Fogelman Soulie, Robert & Tchuente [1987]
F. Fogelman Soulie, Y. Robert and M. Tchuente (1987) *Automata Networks in Computer Science: Theory and Applications*, Manchester University Press.

Fujita, Tanaka & Moto-oka [1983]
M. Fujita, M. Tanaka and T. Moto-oka (1983) 'Computer Hardware Description Languages and their Applications', 103–14 in *Sixth International Conference, CMU*, ed. T. Uehara & M. Barbacci, North Holland, Amsterdam.

Furst, Saxe & Sipser [1984]
M. Furst, J. B. Saxe and M. Sipser (1984) 'Parity, circuits and the polynomial time hierarchy', *Math. Systems Theory,* **17**, 13–27.

Gachet, Joinnault & Quinton [1987]
P. Gachet, B. Joinnault and P. Quinton (1987) 'Synthesising systolic arrays using DIASTOL', 25–36 in *Systolic Arrays, Proceedings First International Conference, Oxford*, ed. W. Moore, A. McCabe & R. Urquhart, Adam Hilger.

Goguen [1988]
J. A. Goguen (1988) 'OBJ as a theorem prover with applications to hardware verification', Technical Report SRI-CSL-88-4R2, Computer Science Laboratory, SRI International, Menlo Park, California 94025.

Goguen & Winkler [1988]
J. A. Goguen and T. Winkler (1988) 'Introducing OBJ3', Technical Report SRI-CSL-88-9, Computer Science Laboratory, SRI International, Menlo Park, California 94025.

Gordon [1981a]
M. Gordon (1981) 'A model of register transfer systems with applications to microcode and VLSI correctness', Technical Report CSR-82-81, Department of Computer Science, University of Edinburgh.

Gordon [1981b]
M. Gordon (1981) 'A very simple model of sequential behaviour of nMOS', 85–94 in *VLSI 81, Edinburgh*, ed. J. P. Gray, Academic Press, London.

Gordon [1983a]
M. Gordon (1983) 'LCF-LSM, a system for specifying and verifying hardware', Technical Report No 41, Computer Laboratory, University of Cambridge.

Gordon [1983b]
M. Gordon (1983) 'Proving a computer correct with the LCF-LSM hardware verification system', Technical Report No 42, Computer Laboratory, University of Cambridge.

Gordon [1985]
M. Gordon (1985) 'HOL: a machine orientated formulation of higher order logic', Technical Report 68, Computer Laboratory, University of Cambridge.

Gordon [1988]
M. Gordon (1988) 'HOL: a proof generating system for higher-order logic', 73–128 in *VLSI Specification, Verification and Synthesis*, ed. G. Birtwhistle & P. A. Subrahmanyam, Kluwer, Dordrecht.

Gordon, Milner & Wadsworth [1979]
M. Gordon, R. Milner and C. Wadsworth (1979) *Edinburgh LCF*, Lecture Notes in Computer Science 78, Springer Verlag.

Gribomont [1988]
E. P. Gribomont (1988) 'Proving systolic arrays', 185–99 in *CAAP '88*, ed. M. Dauchet & M. Nivat, Lecture Notes in Computer Science 299, Springer Verlag.

Hanna [1983]
F. K. Hanna (1983) 'Overview of the VERITAS Project', Electronics Laboratories Technical Report, University of Kent.

Hanna & Daeche [1986]
F. K. Hanna and N. Daeche (1986) 'Specification and verification using higher-order logic: a case study', 179–213 in *Formal Aspects of VLSI*, ed. G. J. Milne & P. A. Subrahmanyam, North-Holland, Amsterdam.

Harman [1989]
N. A. Harman (1989) 'Formal specification of digital systems', PhD Thesis, School of Computer Studies, University of Leeds.

Harman, Hobley & Tucker [to appear]
N. A. Harman, K. M. Hobley and J. V. Tucker (to appear) 'The formal specification of a digital correlator II: algorithm and architecture verification', Centre for Theoretical Computer Science Report, University of Leeds.

Harman & Tucker [1988a]
N. A. Harman and J. V. Tucker (1988) 'Formal specifications and the design of verifiable computers', 500–3 in *Proceedings of UK IT Conference, held under the auspices of the Information Engineering Directorate of the Department of Trade and Industry*, IEE.

Harman & Tucker [1988b]
N. A. Harman and J. V. Tucker (1988) 'Clocks, retimings and the formal specification of a UART', 375–96 in *The Fusion of Hardware Design and Verification, Proceedings IFIP WG 10.2 Working Conference, University of Strathclyde*, ed. G. J. Milne, North-Holland, Amsterdam.

Harrison [1965]
M. A. Harrison (1965) *Introduction to Switching and Automata Theory*, McGraw-Hill.

Hastad [1986]
J. Hastad (1986) 'Almost optimal lower bounds for small depth circuits', 6–20 in *Proceedings 18th ACM Symposium on Theory of Computing.*

Hennessy [1986]
M. Hennessy (1986) 'Proving systolic algorithms correct', *ACM Trans. Programming Languages and Systems,* **8**, 344–87.

Hennessy & Milner [1980]
M. Hennessy and R. Milner (1980) *On Observing Non-determinism and Concurrency*, Lecture Notes in Computer Science 85, Springer Verlag, Berlin.

Hennie [1958]
F. C. Hennie (1958) 'Analysis of one-dimensional iterative logical circuits', S. M. Thesis, Department of Electrical Engineering, MIT.

Hennie [1961]
F. C. Hennie (1961) *Iterative Arrays of Logical Circuits*, MIT Press and John Wiley.

Hennie [1968]
F. C. Hennie (1968) *Finite State Models for Logical Machines*, John Wiley, New York.

Herbert [1988]
J. Herbert (1988) 'Temporal abstraction of digital designs', 1–26 in *The Fusion of Hardware Design and Verification, Proceedings IFIP WG 10.2 Working Conference, University of Strathclyde*, ed. G. J. Milne, North Holland, Amsterdam.

Hill & Peterson [1973]
F. J. Hill and G. R. Peterson (1973) *Digital Systems: Hardware Organisation and Design*, Wiley.

Hoare [1985]
C. A. R. Hoare (1985) *Communicating Sequential Processes*, Prentice-Hall, London.

Hobley, Thompson & Tucker [1988]
K. M. Hobley, B. C. Thompson and J. V. Tucker (1988) 'Specification and verification of synchronous concurrent algorithms: a case study of a convolution algorithm', 347–74 in *The Fusion of Hardware Design and Verification, Proceedings IFIP WG 10.2 Working Conference, University of Strathclyde*, ed. G. J. Milne, North Holland, Amsterdam.

Hornick & Sarrafzadeh [1987]
S. Hornick and M. Sarrafzadeh (1987) 'On problem transformability in VLSI', *Algorithmica,* **2**, 97–111.

Hromkovic [1987]
J. Hromkovic (1987) 'Lower bound techniques for VLSI algorithms', 2–25 in *Trends, Techniques and Problems in Computer Science*, ed. A. Kelemenova *et al.*, Lecture Notes in Computer Science 281, Springer Verlag.

Huang & Lengauer [1987]
C.-H. Huang and C. Lengauer (1987) 'The derivation of systolic implementations of programs', *Acta Informatica,* **24**, 595–632.

Hunt [1986]
W. A. Hunt (1986) 'The FM5801: a verified microprocessor', Technical Report 47, University of Texas at Austin.

Jagadish *et al.* [1986]
H. V. Jagadish, R. G. Mathews, T. Kailath and J. A. Newkirk (1986) 'A study of pipelining in computer arrays', *IEEE Trans. Computers,* **C-35**, 431–40.

Johnson [1984]
S. D. Johnson (1984) *Synthesis of Digital Designs from Recursion Equations*, MIT Press.

Johnson, Bose & Boyer [1988]
S. Johnson, B. Bose and C. Boyer (1988) 'A tactical framework for hardware design', 349–84 in *VLSI, Specification, Verification and Synthesis*, ed. G. Birtwhistle & P. A. Subrahmanyam, Kluwer Academic Publishers, Boston.

Johnsson & Cohen [1981]
L. Johnsson and D. Cohen (1981) 'A mathematical approach to modelling the flow of data and control in computational networks', 213–24 in *VLSI Systems and Computations*, ed. H. T. Kung, R. Sproull & G. Steel, Computer Science Press, Rockville, Md, USA.

Johnsson *et al.* [1981]
L. Johnsson, U. Weiser, D. Cohen and A. L. Davis (1981) 'Towards a formal treatment of VLSI arrays', 375–98 in *Proceedings 2nd Caltech Conference on VLSI.*

Jones & Luk [1987]
G. Jones and W. Luk (1987) 'Exploring designs by circuit transformation', 91–8 in *Systolic Arrays, Proceedings First International Conference, Oxford*, ed. W. Moore, A. McCabe & R. Urquhart, Adam Hilger.

Joyce [1988]
J. Joyce (1988) 'Formal verification and implementation of a microprocessor', 129–58 in *VLSI Specification, Verification and Synthesis*, ed. G. Birtwistle & P. A. Subrahmanyam, Kluwer Academic Publishers, Boston.

Joyce, Birtwhistle & Gordon [1986]
J. Joyce, G. Birtwhistle and M. Gordon (1986) 'Proving a computer correct in higher order logic', Technical Report 100, Computer Laboratory, University of Cambridge.

Kapur, Sivakumar & Zhang [1986]
D. Kapur, G. Sivakumar and H. Zhang (1986) 'RRL: a rewrite rule laboratory', in *Proceedings 8th International Conference on Automated Deduction, Oxford.*

Karp, Miller & Winograd [1967]
R. M. Karp, R. E. Miller and S. Winograd (1967) 'The organisation of computations for uniform recurrence equations', *J. ACM,* **14**, 563–90.

Kedem & Zorat [1981]
Z. M. Kedem and A. Zorat (1981) 'Replication of inputs may save computational resources in VLSI', 52–60 in *VLSI Systems and Computations*, ed. H. T. Kung, B. Sproull & G. Steel, Computer Science Press, Rockville, Md, USA.

Keister, Ritchie & Washburn [1951]
W. Keister, A. E. Ritchie and S. H. Washburn (1951) *The Design of Switching Circuits*, van Nostrand.

King [1969]
J. C. King (1969) 'A program verifier', PhD Thesis, Carnegie-Mellon University.

Kissin [1982]
G. Kissin (1982) 'Measuring energy consumption in VLSI circuits: a foundation', 99–104 in *Proceedings of 14th Symposium on the Theory of Computing*, ACM.

Kramer & Leeuwen [1983]
M. Kramer and J. van Leeuwen (1983) *The VLSI Complexity of Boolean Functions*, Lecture Notes in Computer Science 171, Springer Verlag, Berlin.

Kung [1982]
H. T. Kung (1982) 'Why systolic architectures?', *IEEE Computer,* **15**, 37–46.

Kung & Lam [1984]
H. T. Kung and M. S. Lam (1984) 'Wafer-scale integration and two-level pipelined implementations of systolic arrays', *J. Parallel and Distributed Computing,* **1**, 32–63.

Kung & Leiserson [1979]
H. T. Kung and C. E. Leiserson (1979) 'Systolic arrays (for VLSI)', 256–82 in *Sparse Matrix Proceedings 1978*, ed. I. S. Duff & G. W. Stewart, SIAM.

Kung & Lin [1984]
H. T. Kung and W. T. Lin (1984) 'An algebra for systolic computation', 141–60 in *Elliptic Problem Solvers 2, Proceedings Conference, Monterey, Jan. 1983*, ed. G. Birkhoff & A. L. Schoenstadt, Academic Press.

Kung, Sproull & Steel [1981]
H. T. Kung, R. Sproull and G. Steel, ed. (1981) *VLSI Systems and Computations*, Computer Science Press, Rockville,Md, USA.

Kung [1984]
S. Y. Kung (1984) 'On supercomputing with systolic/wavefront array processors', *Proc. IEEE,* **72**, 867–84.

Kung *et al.* [1982]
S. Y. Kung, K. S. Arun, R. J. Gal-Ezer and D. V. Bhaskar Rao (1982) 'Wavefront array processor: language, architecture, and applications', *IEEE Trans. Computers,* **C-31**, (Special Issue on Parallel and Distributed Computers), 1054–66.

Lam & Mostow [1985]
M. Lam and J. Mostow (1985) 'A Transformational model of VLSI systolic design', *IEEE Computer,* **18**, 42–52.

Leiserson [1981]
C. E. Leiserson (1981) 'Area-efficient VLSI computation', PhD Dissertation, Department of Computer Science, Carnegie-Mellon University.

Leiserson, Rose & Saxe [1983]
C. E. Leiserson, F. M. Rose and J. B. Saxe (1983) 'Optimising synchronous circuitry by retiming', 87–116 in *Third Caltech Conference on VLSI*, ed. R. Bryant, Springer Verlag.

Leiserson & Saxe [1981]
C. E. Leiserson and J. B. Saxe (1981) 'Optimising synchronous systems', 23–36 in *22nd Annual Symposium on Foundations of Computer Science, Nashville.*

Leiserson & Saxe [1983]
C. E. Leiserson and J. B. Saxe (1983) 'Optimising synchronous systems', *J. VLSI and Computer Systems,* **1**, 41–68.

Leiserson & Saxe [1986]
C. E. Leiserson and J. B. Saxe (1986) 'Retiming synchronous circuitry', Technical Report, Systems Research Center, Digital Equipment Corporation, 130 Lytton Avenue, Palo Alto, California 94301.

Lescanne [1983]
P. Lescanne (1983) 'Computer experiments with the REVE term rewriting system generator', in *Proceedings 10th Symposium on Principles of Programming Languages,* ACM, Austin, Texas.

Lewin [1968]
D. Lewin (1968) *Logical Design of Switching Circuits,* Nelson.

Lewin [1977]
D. Lewin (1977) *Computer Aided Design of Digital Systems,* Edward Arnold.

Li & Wah [1985]
G. Li and B. W. Wah (1985) 'The design of optimal systolic arrays', *IEEE Trans. Computers,* **C-34**, 66–77.

Li & Jayakumar [1986]
H. F. Li and R. Jayakumar (1986) 'Systolic structures: a notion and characterization', *J. Parallel and Distributed Computing,* **3**, 373–97.

Lipton & Sedgewick [1981]
R. J. Lipton and R. Sedgewick (1981) 'Lower bounds for VLSI', 300–7 in *Proceedings 13th Symposium on Theory of Computing,* ACM.

Luk & Jones [1988]
W. Luk and G. Jones (1988) 'From specification to parameterised architectures', 267–88 in *The Fusion of Hardware Design and Verification, Proceedings IFIP WG 10.2 Working Conference, University of Strathclyde,* ed. G. J. Milne, North Holland, Amsterdam.

Lupanov [1958]
O. B. Lupanov (1958) 'On a method of circuit synthesis' [in Russian], *Isvestia VUZ (Radiofizika)*, **1**, 120–40.

Makedon *et al.* [1986]
F. Makedon, K. Melhorn, T. Papatheodorou and P. Spirakis, ed. (1986) *VLSI Algorithms and Architectures*, Lecture Notes in Computer Science 227, Springer Verlag, Berlin.

Malachi & Owicki [1981]
Y. Malachi and S. Owicki (1981) 'Temporal specification of self-timed systems', 203–12 in *VLSI Systems and Computations*, ed. H. T. Kung, R. Sproull & G. Steel, Computer Science Press, Rockville, Md, USA.

Mano [1979]
M. M. Mano (1979) *Digital Logic and Computer Design*, Prentice-Hall.

Martin [1989]
A. R. Martin (1989) 'The specification and simulation of synchronous concurrent algorithms', PhD Thesis, School of Computer Studies, University of Leeds.

Martin & Tucker [1987]
A. R. Martin and J. V. Tucker (1987) 'The concurrent assignment representation of synchronous systems', 369–86 in *Parallel Architectures and Languages Europe Vol II: Parallel Languages*, ed. A. J. Nijman & P. C. Treleaven, Lecture Notes in Computer Science 259, Springer Verlag.

Martin & Tucker [1988]
A. R. Martin and J. V. Tucker (1988) 'The concurrent assignment representation of synchronous systems', *Parallel Computing,* **9**, 227–56.

McCanny & McWhirter [1982]
J. V. McCanny and J. G. McWhirter (1982) 'Completely iterative, pipelined multiplier array suitable for VLSI', *IEE Proc.,* **129**, Pt G, 40–6.

McCluskey [1958]
E. J. McCluskey (1958) 'Iterative combinational switching networks – general design considerations', *IRE Trans. Electronic Computers,* **EC-7**, 285–91.

McEvoy & Dew [1988]
K. McEvoy and P. M. Dew (1988) 'The cut theorem: a tool for design of systolic algorithms', *Int. J. Computer Mathematics,* **25**, 203–33.

McNaughton & Yamada [1960]
R. McNaughton and H. Yamada (1960) 'Regular expressions and state graphs for automata', *IRE Trans. Electronic Computers,* **EC-9**, 39–47.

Mead & Conway [1980]
C. Mead and L. Conway (1980) *Introduction to VLSI Systems*, Addison-Wesley.

Mead & Rem [1979]
C. A. Mead and M. Rem (1979) 'Cost and performance of VLSI computing structures', *IEEE J. Solid-State Circuits,* **SC-14**, 455–62.

Meinke [1988]
K. Meinke (1988) 'A graph-theoretic model of synchronous concurrent algorithms', PhD Thesis, School of Computer Studies, University of Leeds.

Meinke & Tucker [1988]
K. Meinke and J. V. Tucker (1988) 'Specification and representation of synchronous concurrent algorithms', 163–80 in *Concurrency '88*, ed. F. H. Vogt, Lecture Notes in Computer Science 335, Springer Verlag.

Melham [1988]
T. Melham (1988) 'Abstraction mechanisms for hardware verification', 267–93 in *VLSI Specification, Verification and Synthesis*, ed. G. Birtwistle & P. A. Subrahmanyam, Kluwer Academic Publishers, Boston.

Melhem & Rheinboldt [1984]
R. G. Melhem and W. C. Rheinboldt (1984) 'A mathematical model for the verification of systolic networks', *SIAM J. Computing,* **13**, 541–65.

Meseguer & Goguen [1985]
J. Meseguer and J. A. Goguen (1985) 'Initiality, induction and computability', 459–541 in *Algebraic Methods in Semantics*, ed. M. Nivat & J. C. Reynolds, Cambridge University Press.

Milne [1985]
G. J. Milne (1985) 'CIRCAL and the representation of communication, concurrency and time', *ACM Trans. Programming Languages and Systems,* **7**, 270–98.

Milne [1986]
G. J. Milne (1986) 'Towards verifiably correct VLSI design', 1–23 in *Formal Aspects of VLSI Design*, ed. G. Milne & P. A. Subrahmanyam, North-Holland, Amsterdam.

Milne [1988]
G. J. Milne, ed. (1988) *The Fusion of Hardware Design and Verification, Proceedings IFIP WG 10.2 Working Conference, University of Strathclyde*, North-Holland, Amsterdam.

Milne & Milner [1979]
G. Milne and R. Milner (1979) 'Concurrent processes and their syntax', *J. ACM,* **26**, 302–21.

Milne & Pezzè [1988]
G. J. Milne and M. Pezzè (1988) 'Typed CIRCAL: A high-level framework for hardware verification', 117–38 in *The Fusion of Hardware Design and Verification, Proceedings IFIP WG10.2 Working Conference, University of Strathclyde*, ed. G. J. Milne, North-Holland, Amsterdam.

Milne & Subrahmanyam [1986]
G. J. Milne and P. A. Subrahmanyam (1986) *Formal Aspects of VLSI Design*, North-Holland, Amsterdam.

Milner [1980]
R. Milner (1980) *A Calculus of Communicating Systems*, Lecture Notes in Computer Science 92, Springer Verlag, Berlin.

Miranker & Winkler [1984]
W. L. Miranker and A. Winkler (1984) 'Spacetime representations of computational structures', *Computing,* **32**, 93–114.

Mishra & Clarke [1985]
B. Mishra and E. Clarke (1985) 'Hierarchical verification of asynchronous circuits using temporal logic', *Theoretical Computer Science,* **38**, 269–91.

Moldovan [1982]
D. I. Moldovan (1982) 'On the analysis and synthesis of VLSI algorithms', *IEEE Trans. Computers,* **C-31**, 1121–6.

Moldovan [1983]
D. I. Moldovan (1983) 'On the design of algorithms for VLSI systolic arrays', *Proc. IEEE,* **71**, 113–20.

Moldovan [1987]
D. I. Moldovan (1987) 'ADVIS: a software package for the design of systolic arrays', *IEEE Trans. Computer-Aided Design,* **CAD-6**, 33–40.

Mongenet & Perrin [1987]
C. Mongenet and G. R. Perrin (1987) 'Synthesis of systolic arrays for inductive problems', 260–77 in *Proceedings Parallel Architectures and Languages Europe Vol. 1*, ed. J. W. de Bakker, A. J. Nijman & P. C. Treleaven, Lecture Notes in Computer Science 258, Springer Verlag, Berlin.

Moore, McCabe & Urquhart [1987]
W. Moore, A. McCabe and R. Urquhart, ed. (1987) *Systolic Arrays, Proceedings of the First International Conference, Oxford, July 1986*, Adam Hilger.

Moszkowski [1983]
B. C. Moszkowski (1983) 'A temporal logic for multi-level reasoning about hardware', 79–90 in *Proceedings of the 6th International Symposium on Computer Hardware Description Languages and their Applications*, ed. T. Uehara & M. Barbacci, North-Holland, Amsterdam.

Moszkowski [1985]
B. C. Moszkowski (1985) 'A temporal logic for multi-level reasoning about hardware', *IEEE Computer,* **18**, 10–19.

Moszkowski [1986]
B. C. Moszkowski (1986) *Executing Temporal Logic Programs*, Cambridge University Press.

Mulder & Weijland [1987]
J. C. Mulder and W. P. Weijland (1987) 'Verification of an algorithm for log-time sorting by square comparison', Centre for Mathematics and Computer Science Report CS-R8729, Amsterdam.

Musser & Cyrluk [1985]
D. R. Musser and D. A. Cyrluk (1985) 'AFFIRM-85 reference manual', General Electric Corporate Research and Development Center, Schenectady, NY 12301.

Musser *et al.* [1988]
D. Musser, P. Narendran and W. Premerlani (1988) 'BIDS: A method for specifying bidirectional hardware devices', 217–34 in *VLSI Specification, Verification and Synthesis*, ed. G. Birtwhistle & P. A. Subrahmanyam, Kluwer Academic Publishers, Boston.

Nakasima [1936]
A. Nakasima (1936) 'The theory of relay circuits', *Nippon Elec. Communication Engineering,* May, 197–226.

Narendran & Stillwell [1988]
P. Narendran and J. Stillwell (1988) 'Formal verification of the Sobel image processing chip', in *Proceedings 25th ACM & IEEE Design Automation Conference*, Computer Society Press, Anaheim, California.

Patel, Schlag & Ercegovac [1985]
D. Patel, M. Schlag and M. Ercegovac (1985) 'νFP: An environment for the multilevel specification, analysis and synthesis of hardware algorithms', 238–55 in *Functional Programming Languages and Computer Architectures*, ed. J.-P. Jouannaud, Lecture Notes in Computer Science 201, Springer Verlag, Berlin.

Paulson [1987]
L. Paulson (1987) *Logic and Computation: Interactive Proof with Cambridge LCF*, Cambridge University Press.

Purushothaman & Subrahmanyam [1988]
S. Purushothaman and P. A. Subrahmanyam (1988) 'Reasoning about systolic algorithms', *J. Parallel and Distributed Computing,* **5**, 669–99.

Purushothaman & Subrahmanyam [1989]
S. Purushothaman and P. A. Subrahmanyam (1989) 'Mechanical certification of systolic algorithms', *J. Automated Reasoning,* **5**, 67–91.

Quinton [1987a]
P. Quinton (1987) 'The systematic design of systolic arrays', 261–302 in *Automata Networks in Computer Science*, ed. F. Fogelman Soulie, Y. Robert & M. Tchuente, Manchester University Press.

Quinton [1987b]
P. Quinton (1987) 'An introduction to systolic architectures', 387–400 in *Future Parallel Computers*, ed. P. Treleaven & M. Vanneschi, Lecture Notes in Computer Science 272, Springer Verlag.

Rajopadhye, Purushothaman & Fujimoto [1986]
S. V. Rajopadhye, S. Purushothaman and R. M. Fujimoto (1986) 'On synthesising systolic arrays from recurrence equations with linear dependencies', 488–503 in *Proceedings Sixth Annual Conference Foundations of Software Technology and Theoretical Computer Science, New Delhi*, ed. K. V. Nori, Lecture Notes in Computer Science 241, Springer Verlag, Berlin.

Rao [1985]
S. K. Rao (1985) 'Regular iterative algorithms and their implementations on processor arrays', PhD Thesis, Department of Electrical Engineering, Stanford University.

Rao & Kailath [1984]
S. K. Rao and T. Kailath (1984) 'Digital filtering in VLSI', Technical Report, Information Systems Laboratory, Department of Electrical Engineering, Stanford University.

Rao & Kailath [1985]
S. K. Rao and T. Kailath (1985) 'VLSI arrays for digital signal processing: part I – a model identification approach to digital filter realisations', *IEEE Trans. Circuits and Systems,* **32**, 1105–18.

Reif [1988]
J. H. Reif, ed. (1988) *VLSI Algorithms and Architectures*, Lecture Notes in Computer Science 319, Springer Verlag, Berlin.

Rem [1981]
M. Rem (1981) 'The VLSI challenge: complexity bridling', 65–73 in *VLSI 81, Edinburgh*, ed. J. P. Gray, Academic Press, London.

Rem [1987]
M. Rem (1987) 'Trace theory and systolic computations', 14–33 in *Proceedings Parallel Architectures and Languages Europe Vol. 1*, ed. J. W. de Bakker, A. J. Nijman & P. C. Treleaven, Lecture Notes in Computer Science 258, Springer Verlag, Berlin.

Rem, Snepscheut & Udding [1983]
M. Rem, J. L. A. Van de Snepscheut and J. T. Udding (1983) 'Trace theory and the definition of hierarchical components', 225–40 in *Proceedings Third Caltech Conference on Very Large Scale Integration*, ed. R. Bryant, Springer Verlag, Berlin.

Savage [1981a]
J. E. Savage (1981) 'Area time tradeoffs for matrix multiplication and related problems in VLSI models', *J. Computer and System Sciences,* **22**, 230–42.

Savage [1981b]
J. E. Savage (1981) 'Planar circuit complexity and the performance of VLSI algorithms', 61–8 in *Proceedings of VLSI Systems and Computations*, ed. H. T. Kung, B. Sproull & G. Steele, Computer Science Press, Rockville,Md, USA.

Seitz [1980]
C. L. Seitz (1980) 'System timing', Chapter 7 of Mead and Conway [1980].

Shannon [1938]
C. E. Shannon (1938) 'A symbolic analysis of relay and switching circuits', *AIEE Trans. on Communications and Electronics,* **57**, 713–23.

Sheeran [1983]
M. Sheeran (1983) 'μFP: an algebraic design language', PhD Thesis, Oxford University (also appears as Technical Report PRG-39, Programming Research Group, Oxford University).

Sheeran [1985]
M. Sheeran (1985) *The Design and Verification of Regular Synchronous Circuits*, Preprint, Programming Research Group, University of Oxford.

Sheeran [1988]
M. Sheeran (1988) 'Retiming and slowdown in Ruby', 289–308 in *The Fusion of Hardware Design and Verification, Proceedings IFIP WG 10.2 Working Conference, University of Strathclyde*, ed. G. J. Milne, North-Holland, Amsterdam.

Shestakov [1938]
V. I. Shestakov (1938) 'Some mathematical methods for the construction and simplification of two-terminal electrical networks of class A' [in Russian], Dissertation, Lomonosov State University.

Shostak [1983a]
R. E. Shostak (1983) 'Verification of VLSI designs', 185–206 in *Proceedings Third Caltech Conference on Very Large Scale Integration*, ed. R. Bryant, Springer Verlag, Berlin.

Shostak [1983b]
R. E. Shostak (1983) 'Formal verification of circuit designs', 13–30 in *Computer Hardware Description Languages and their Applications, Sixth International Conference, CMU*, ed. T. Uehara & M. Barbacci, North-Holland, Amsterdam.

Siekmann & Wrightson [1983]
J. Siekmann and G. Wrightson, ed. (1983) *Automation of Reasoning, Volumes I and II*, Springer Verlag.

Simonis, Nguyen & Dincbas [1988]
H. Simonis, N. Nguyen and M. Dincbas (1988) 'Verification of digital circuits using CHIP', 421–42 in *The Fusion of Hardware Design and Verification, Proceedings IFIP WG 10.2 Working Conference, University of Strathclyde*, ed. G. J. Milne, North-Holland, Amsterdam.

Snepscheut [1983]
J. L. A. van de Snepscheut (1983) 'Deriving circuits from programs', 241–56 in *Proceedings Third Caltech Conference VLSI*, ed. R. Bryant, Springer Verlag, Berlin.

Snepscheut [1985]
J. L. A. van de Snepscheut (1985) *Trace Theory and VLSI Design*, Lecture Notes in Computer Science 200, Springer Verlag, Berlin.

Stavridou [1988]
V. Stavridou (1988) 'Specifying in OBJ, verifying in REVE and some ideas about time', Technical Report CSD-TR-605, Department of Computer Science, Royal Holloway and Bedford New College, London.

Stavridou, Barringer & Edwards [1988]
V. Stavridou, H. Barringer and D. Edwards (1988) 'Formal specification and verification of hardware: a comparative case study', in *Proceedings 25th ACM & IEEE Design Automation Conference*, Computer Society Press, Anaheim, California.

Subrahmanyam [1983]
P. A. Subrahmanyam (1983) 'Synthesizing VLSI circuits from behavioural specifications, a very high-level silicon compiler and its theoretical basis', 195–213 in *VLSI 83, Trondheim, Norway*, ed. F. Anceau & E. J. Aas, North-Holland, Amsterdam.

Thomas [1940]
J. M. Thomas (1940) 'Orderly differential systems', *Duke Mathematical J.*, **7**, 249–90.

Thompson [1987]
B. C. Thompson (1987) 'A mathematical theory of synchronous concurrent algorithms', PhD Thesis, Department of Computer Studies, University of Leeds.

Thompson & Tucker [1985]
B. C. Thompson and J. V. Tucker (1985) 'Theoretical considerations in algorithm design', 855–78 in *Fundamental Algorithms for Computer Graphics*, ed. R. A. Earnshaw, NATO ASI Series, Vol. F17, Springer Verlag, Berlin.

Thompson & Tucker [1988]
B. C. Thompson and J. V. Tucker (1988) 'A parallel deterministic language and its application to synchronous concurrent algorithms', 228–31 in *Proceedings of UK IT Conference, held under the auspices of the Information Engineering Directorate of the Department of Trade and Industry*, IEE.

Thompson & Tucker [1989]
B. C. Thompson and J. V. Tucker (1989) 'Synchronous concurrent algorithms', Centre for Theoretical Computer Science, University of Leeds.

Thompson [1980]
C. D. Thompson (1980) 'A complexity theory for VLSI', PhD Thesis, Department of Computer Science, Carnegie-Mellon University.

Thompson [1983]
C. D. Thompson (1983) 'The VLSI complexity of sorting', *IEEE Trans. Computers,* **C-32**, 1171–84.

Tiden [1988]
E. Tiden (1988) 'Symbolic verification of switch-level circuits using a prolog enhanced with unification in finite algebras', 465–86 in *The Fusion of Hardware Design and Verification, Proceedings IFIP WG 10.2 Working Conference, University of Strathclyde*, ed. G. J. Milne, North-Holland, Amsterdam.

Tucker [1990]
J. V. Tucker (1990) 'Theory of computation and specification over abstract data types, and its applications', in *Lectures at NATO Summer School on Logic, Algebra and Computation, Marktoberdorf*, Springer Verlag, to appear.

Turner [1979]
D. A. Turner (1979) 'SASL reference manual (revised version)', Technical Report, University of Kent.

Udding [1984]
J. T. Udding (1984) 'Classification and composition of delay-insensitive circuits', PhD Thesis, Department of Mathematics and Computer Science, Eindhoven University of Technology.

Ullman [1984]
J. Ullman (1984) *Computational Aspects of VLSI*, Computer Science Press, Rockville, Md, USA.

Unger [1963]
S. H. Unger (1963) 'Pattern recognition using two-dimensional bilateral iterative combinational switching circuits', in *Proceedings Polytechnic Institute of Brooklyn Symposium on Mathematical Theory of Automata*, Polytechnic Press, Brooklyn, New York.

Vitanyi [1988]
P. M. B. Vitanyi (1988) 'Locality, communication and interconnect length in multicomputers', *SIAM J. Computing,* **17**, 659–72.

Vuillemin [1983]
J. E. Vuillemin (1983) 'A combinatorial limit to the computing power of VLSI circuits', *IEEE Trans. Computers,* **C-32**, 294–300.

Waite [1967]
W. M. Waite (1967) 'Path detection in multi-dimensional iterative arrays', *J. ACM*, **14**, 300–10.

Wegener [1987]
I. Wegener (1987) *The Complexity of Boolean Functions*, John Wiley.

Weijland [1987]
W. P. Weijland (1987) 'A systolic algorithm for matrix–vector multiplication', 143–60 in *Computing Science in the Netherlands*, CWI, Amsterdam.

Weiser & Davis [1981]
U. Weiser and A. Davis (1981) 'A wavefront notation tool for VLSI array design', 226–34 in *Proceedings of VLSI Systems and Computations*, ed. H. T. Kung, B. Sproull & G. Steele, Computer Science Press, Rockville,Md, USA.

Winkel & Prosser [1980]
D. Winkel and F. Prosser (1980) *The Art of Digital Design*, Prentice-Hall.

Winskel [1988]
G. Winskel (1988) 'A compositional model of MOS circuits', 323–48 in *VLSI Specification, Verification and Synthesis*, ed. G. Birtwistle & P. A. Subrahmanyam, Kluwer Academic Publishers, Boston.

Witte [1972]
B. F. Witte (1972) 'Utility concepts of an $Ax + b$ microelectronic module', 47–55 in *Region Six Conference Record, San Diego CA*, IEEE.

Yung & Allen [1984]
H. C. Yung and C. R. Allen (1984) 'VLSI implementation of an optimised hierarchical multiplier', *IEE Proc.*, **131**, 56–66.

Part 1

Formal methods and verification

This part describes three different approaches to the use of formal methods in the verification and design of systems and circuits.

Chapter 2 describes the stages involved in the verification of a counter using a mechanized theorem prover.

The next chapter describes a mathematical model of synchronous computation within which formal transformations which are useful in the design process can be defined.

Chapter 4 describes verification in a different framework – that of the algebra of communicating processes.

2 A mechanized proof of correctness of a simple counter

AVRA COHN AND MIKE GORDON

1 INTRODUCTION

The VIPER microprocessor designed at the Royal Signals and Radar Estasblishment (RSRE) is probably the first commercially produced computer to have been developed using modern formal methods. Details of VIPER can be found in Cullyer [1985, 1986, 1987] and Pygott [1986]. The approach used by W. J. Cullyer and C. Pygott for its verification is explained in Cullyer & Pygott [1985], in which a simple counter is chosen to illustrate the verification techniques developed at RSRE. Using the same counter, we illustrate the approach to hardware verification developed at Cambridge, which formalizes Cullyer and Pygott's method. The approach is based on the HOL system, a version of LCF adapted to higher-order logic (Camilleri *et al.* [1987], Gordon [1983, 1985]). This research has formed the basis for the subsequent project to verify the whole of VIPER to register transfer level (Cohn [1987, 1989]).

In Cullyer and Pygott's paper, the implementation of the counter is specified at three levels of decreasing abstractness:

- As a state-transition system called the *host machine*;
- As an interconnection of functional blocks called the *high level design*;
- As an interconnection of gates and registers called the *circuit.*

Ultimately, it is the circuit that will be built and *its* correctness is the most important. However, the host machine and high level design represent successive stages in the development of the implementation and so one would like to know if they too are correct. It is possible, at least in principle, for the circuit to be correct even if the host machine (and/or the high level design) were wrong; this could happen if the circuit failed to implement the host (and/or the high level design) but did happen to implement the counter. If this were the case, then the hierarchical proof method described in this chapter would fail.

Cullyer and Pygott formalize the state transitions of the counter and the host machine in the language LSM, the object language of the now obsolete LCF_LSM system. LSM is an *ad hoc* notation developed by M. Gordon for specifying sequential machines.* Cullyer and Pygott formalize the high level design in both LSM and the hardware description language ELLA;† but they only formalize the circuit in ELLA. They prove informally that the state transitions of the counter are correctly implemented by certain sequences of transitions of the host machine, and that the host machine is correctly implemented by the high level design. The verification that the high level design is correctly implemented by the circuit is done using a novel ELLA-based simulation technique called 'intelligent exhaustion' (Pygott [1986]).

In this chapter we formulate and prove a single correctness statement that must be true if the circuit is to implement the counter. The proof consists of three main lemmas which (respectively) establish the correctness of the host machine Section 4), the equivalence of the high level design to the host machine (Section 6), and the equivalence of the circuit to the high level design (Section 8). These lemmas are combined to yield the correctness of the counter in Section 9.

All proofs were generated, and mechanically checked, using the HOL system (Gordon [1987]). HOL is a version of LCF (Gordon *et al.* [1979], Paulson [1987]) for higher order logic. Therefore, to perform the proofs, the LSM and ELLA specifications had to be translated into higher order logic. In the case of LSM this was trivial, as the two notations are very similar (both are derived from LCF's object language PPLAMBDA). In the case of ELLA, more thought was needed, but the translation was still straightforward.

In what follows, proofs are presented informally; the *formal* proofs (sequences of basic inference steps) have, as mentioned, all been generated mechanically using the HOL system (Gordon [1987]). This system is based on LCF which was developed by Robin Milner (Gordon *et al.* [1979]). LCF consists of a logic called PPLAMBDA interfaced to a functional programming language called ML. The user invokes ML functions to set up goals, reduce them to subgoals, and so on. Milner's LCF approach to proof (which HOL inherits) guarantees the soundness of the theorem proving tools. For more details see Gordon *et al.* [1979] and Paulson [1987]. HOL differs from LCF in that the logic to which ML is interfaced is higher order logic rather than PPLAMBDA.

* Work by Hanna & Daeche [1985], Halpern *et al.* [1983] and Gordon [1986] has shown that pure logic is both a more general and a more elegant specification notation than LSM; and furthermore, it does not require the *ad hoc* and insecure inference rules of LSM.

† ELLA is a hardware description language supported by a number of tools including a high performance simulator. Currently, the ELLA system does not contain any tools for formal proof.

2 SPECIFICATION OF THE COUNTER

Cullyer and Pygott's counter is a device with two inputs and one output:

The input lines `loadin` and `func` carry 6-bit and 2-bit words respectively; the output line `count` carries 6-bit words. The counter stores a 6-bit word that can be read via the output line `count`. The counter behaves like a finite state machine; its next state is determined by the value input at `func`.

- If `func = 0`: The value stored in the counter is unchanged.
- If `func = 1`: The value stored in the counter is the value input at `loadin`.
- If `func = 2`: The value stored in the counter is incremented once.
- If `func = 3`: The value stored in the counter is incremented twice.

This behaviour can be formalized by defining a function `COUNTER` which takes as argument a triple consisting of the stored value and the two input values, and returns as result the new stored value. Before giving the formal definition of this function we briefly summarize HOL notation.

2.1 A brief summary of HOL notation

This section gives a very brief summary of HOL notation – sufficient, we hope, to enable the reader to understand the rest of the chapter.

A *term* of higher-order logic can be one of four kinds:

(1) A *variable* (e.g., `x`, `x1`, `count`);
(2) A *constant* (e.g., `1`, `T`, `#011001`);
(3) A *function application* of the form $t_1 t_2$, where t_1 is called the *operator* and t_2 the *operand*;
(4) An *abstraction* of the form `\x.t` where `x` is called the *bound variable* and `t` the *body*. (The character '\' is the ASCII representation of λ.)

To make terms more readable, HOL allows a number of alternative syntactic forms. The term $\mathtt{t}_1\mathtt{t}_2\cdots\mathtt{t}_n$ abbreviates $(\cdots(\mathtt{t}_1\mathtt{t}_2)\cdots\mathtt{t}_n)$ – that is, function application associates to the left. Some two-argument function constants can be infixed; for example, $\mathtt{t}_1/\backslash\mathtt{t}_2$ can be written instead of $/\backslash\mathtt{t}_1\mathtt{t}_2$. Some terms of the form `f(\x.t)` can be written as `f x.t`; when this is the case, `f` is called a *binder*. For example, `!` (the ASCII representation of ∀) is a binder, so `!x.t` can be written instead of `!(\x.t)`.

The HOL system uses the characters `~`, `\/`, `/\`, `==>`, `!` and `?` to represent the logical symbols ¬, ∨, ∧, ⊃, ∀ and ∃, respectively, where `\/`, `/\` and `==>` are infixes and `!` and `?` are binders. A term of the form $\mathtt{t=>t}_1\mathtt{|t}_2$ is a *conditional* with *if*-part `t`, *then*-part $\mathtt{t}_1$ and *else*-part $\mathtt{t}_2$; it abbreviates $\mathtt{COND\ t\ t}_1\ \mathtt{t}_2$. A term of the form $\mathtt{let\ x=t}_1\ \mathtt{in\ t}_2$, where `x` is a variable and $\mathtt{t}_1$ and $\mathtt{t}_2$ are terms abbreviates $(\backslash\mathtt{x.t}_2)\mathtt{t}_1$. The two truth values *true* and *false* are represented by the boolean constant symbols `T` and `F` respectively. Constant terms of the form $\#\mathtt{b}_{n-1}\cdots\mathtt{b}_0$ (where $\mathtt{b}_i$ is either `0` or `1`) denote n-bit words in which $\mathtt{b}_0$ is the least significant bit.

All HOL terms have a *type*. We write `t:ty` if term `t` has type `ty`. For example, we can write `T:bool` and `F:bool` to indicate that the truth values `T` and `F` have type `bool` (for boolean). Numbers have type `num` and n-bit words have type $\mathtt{word}_n$ (see below regarding these families of types). For example, `3:num` and `#101101:word6`.

If `ty` is a type then `(ty)list` (also written `ty list`) is the type of lists whose components have type `ty`. If $\mathtt{ty}_1$ and $\mathtt{ty}_2$ are types then $\mathtt{ty}_1\mathtt{->ty}_2$ is the type of functions whose arguments have type $\mathtt{ty}_1$ and results $\mathtt{ty}_2$. The cartesian product operator is denoted by `#`, so that $\mathtt{ty}_1\#\mathtt{ty}_2$ is the type of pairs whose first components have type $\mathtt{ty}_1$ and second, $\mathtt{ty}_2$. If $\mathtt{t}_1$ and $\mathtt{t}_2$ have types $\mathtt{ty}_1$ and $\mathtt{ty}_2$, then $(\mathtt{t}_1,\mathtt{t}_2)$ denotes the pair whose first component is $\mathtt{t}_1$ and whose second component is $\mathtt{t}_2$; the brackets can be omitted. The infix ‘,’ associates to the right so that (omitting the brackets) $\mathtt{t}_1,\mathtt{t}_2,\ldots,\mathtt{t}_n$ denotes $(\mathtt{t}_1,(\mathtt{t}_2,\ldots,\mathtt{t}_n))$. The product operator `#` associates to the right and binds more tightly than the operator `->`. For example, the function we shall use to model the counter has the type `word6#word6#word2->word6` which abbreviates `(word6#(word6#word2))->word6`. Types may contain type variables; these are `*`, `**`, `***` and so on, and can be instantiated to any types.

The HOL system has a number of predefined constants. The ones used in this chapter are listed below. Constants that operate on words come in families parameterized by the word length. For example, HOL has separate constants `VAL1`, `VAL2`, `VAL3` etc.; a typical member of this family is denoted by $\mathtt{VAL}_n$:

- `V:(bool)list->num` converts a list of truth-values to a number;
- `EL:num->(*)list->*` selects a member of a list: $\texttt{EL i [t}_{n-1}\texttt{;} \cdots \texttt{;t}_0\texttt{] = t}_i\texttt{;}$
- $\texttt{VAL}_n\texttt{:word}_n\texttt{->num}$ converts an n-bit word to a number;
- $\texttt{BIT}_n\texttt{:word}_n\texttt{->num->bool}$ selects a bit of a word: $\texttt{BIT}_n \texttt{ \#b}_{n-1}\cdots\texttt{b}_0\texttt{i = b}_i\texttt{;}$
- $\texttt{BITS}_n\texttt{:word}_n\texttt{->(bool)list}$ converts an n-bit word to a list of booleans;
- $\texttt{WORD}_n\texttt{:num->word}_n$ converts a number to an n-bit word;
- $\texttt{TUPLE}_n\texttt{:(*)list->*\#}\cdots\texttt{\#*}$ converts a list of length n to an n-tuple;
- $\texttt{V}_n\texttt{:bool\#}\cdots\texttt{\#bool->word}_n$ converts an n-tuple of truth-values to an n-bit word;
- $\texttt{VS}_n\texttt{:(num->bool)\#}\cdots\texttt{\#(num->bool)->(num->word}_n\texttt{)}$ converts an n-tuple of functions to an n-bit valued function: $\texttt{VS}_n\texttt{(s}_1\texttt{,}\cdots\texttt{,s}_n\texttt{)x = V}_n\texttt{ (s}_1\texttt{x,}\cdots\texttt{,s}_n\texttt{ x).}$

For a discussion of why *higher order* logic in particular is appropriate for hardware specification see Hanna & Daeche [1985] or Gordon [1986]. An example of a higher order predicate used in the proof described here is `NEXTTIME`, defined by

```
NEXTTIME (x1,x2) f =
(x1 < x2) /\ (f x2) /\ (!x. (x1 < x) /\ (x < x2) ==> ~f x)
```

This is explained in Section 4.

Because we do not give any detailed proofs in this chapter, the inference rules of higher order logic are not described here. A good introduction to the subject is Andrews [1986]; see also Gordon [1987].

2.2 HOL specification of the counter

The definition of `COUNTER`, below, is the same (except for minor notational changes) as the one given in Cullyer and Pygott's paper.

```
COUNTER(count,loadin,func) =
 let funcnum = VAL2 func  in
 let value   = VAL6 count in
 ((funcnum=0) => count  |
  (funcnum=1) => loadin |
  (funcnum=2) => ((value=63) => WORD6 0 | WORD6(value+1)) |
  (funcnum=3) => ((value=63) => WORD6 1 |
                  (value=62) => WORD6 0 | WORD6(value+2)) | ARB)
```

The differences between this definition and the LSM definition in the RSRE paper are:

(1) `let`, `in` and `=` are used in place of `LET`, `IN` and `==`, respectively, and
(2) The conditional has an extra arm containing an arbitrary value `ARB`. (In both HOL and LSM, conditionals must always have an *else*-clause.)

3 SPECIFICATION OF THE HOST MACHINE

Cullyer and Pygott explain the behaviour of the host machine with the following state transition diagram.

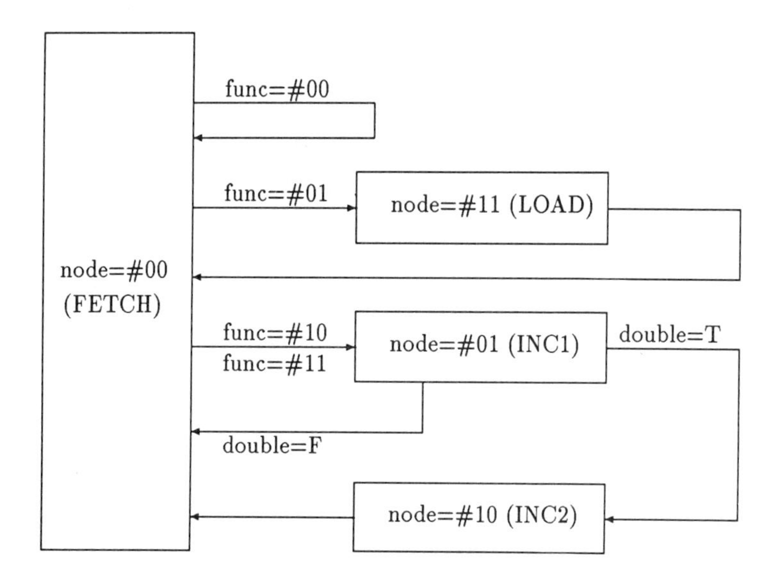

The state of the host machine is fully specified by a triple `(count,double,node)` where

- `count` is the 6-bit word being stored by the counter,
- `double` is a 1-bit boolean flag which determines whether the counter is to add one or two,
- `node` is a 2-bit word determining in which of the four nodes of the above diagram the host is at any particular time.

A state `(count,double,node)` is called *primary* if `node=#00`. The actions taken when control reaches a node are represented in HOL by the functions `FETCH`, `LOAD`, `INC1` and `INC2`. These actions are as follows.

(1) `FETCH`: `func` determines the next state.
(2) `INC1`: `count` is incremented; `double` determines the next state.
(3) `INC2`: `count` is incremented; the next state is primary. (A state with `node=#10` is reached only if the counter is to be incremented twice, the first increment being done at the preceding node `#01`).
(4) `LOAD`: `count` becomes `loadin`; the next state is primary. (Note that the value stored is the value input when node `#11` is reached; it is *not* the value input during the preceding primary state.)

The HOL functions representing the four nodes are defined below. Some of these definitions make use of an auxiliary function ADD1 for incrementing a 6-bit word. ADD1 is defined by

```
ADD1(x) =
let xval = VAL6 x in
((xval=63) => WORD6 0 | WORD6(xval+1))
```

The HOL definitions that follow are based on those given in Cullyer & Pygott [1985].

```
FETCH(count,double,loadin,func) =
 let twice     = EL 0 (BITS2 func) in
 let funcnum   = VAL2 func         in
 let fetchnode = WORD2 0           in
 let inclnode  = WORD2 1           in
 let loadnode  = WORD2 3           in
 ((funcnum=0) => (count,twice,fetchnode) |
  (funcnum=1) => (count,twice,loadnode)  |
                 (count,twice,inclnode))

LOAD(count,double,loadin,func) =
 let twice     = EL 0 (BITS2 func) in
 let fetchnode = WORD2 0           in
 (loadin,twice,fetchnode)

INC1(count,double,loadin,func) =
 let twice     = EL 0 (BITS2 func) in
 let fetchnode = WORD2 0           in
 let inc2node  = WORD2 2           in
 (double => (ADD1 count,twice,inc2node) |
            (ADD1 count,twice,fetchnode))

INC2(count,double,loadin,func) =
 let twice     = EL 0 (BITS2 func) in
 let fetchnode = WORD2 0           in
 (ADD1 count,twice,fetchnode)
```

A state transition function NEXT for the host machine can now be defined:

```
NEXT((count,double,node),loadin,func) =
 let nodenum = VAL2 node in
 ((nodenum=0) => FETCH(count,double,loadin,func) |
  (nodenum=1) => INC1 (count,double,loadin,func) |
  (nodenum=2) => INC2 (count,double,loadin,func) |
  (nodenum=3) => LOAD (count,double,loadin,func) | ARB)
```

If the host machine is in state (count,double,node) and the values being input are loadin and func, then the next state of the machine is the value of

```
NEXT((count,double,node),loadin,func).
```

In the next section, we outline what it means for the function NEXT to correctly implement the specification COUNTER.

4 CORRECTNESS OF THE HOST MACHINE

We use the variable state to range over host machine states (i.e., triples of the form (count,double,node)). The functions COUNT, DOUBLE and NODE select the first, second and third components of a state, respectively. These functions satisfy the equation

```
state = (COUNT(state),DOUBLE(state),NODE(state))
```

The host machine may require several transitions to implement a single transition of COUNTER. Suppose $\mathtt{state}_0,\ldots,\mathtt{state}_n$ is a sequence of states such that $\mathtt{NODE(state}_0\mathtt{)=\#00}$, $\mathtt{NODE(state}_n\mathtt{)=\#00}$ and $\mathtt{NODE(state}_i\mathtt{)}$ is not equal to #00 for $0<i<n$. Suppose also that $\mathtt{loadin}_0,\ldots,\ \mathtt{loadin}_n$ and $\mathtt{func}_0,\ldots,\mathtt{func}_n$ are sequences of inputs. Then the function NEXT is a correct implementation of COUNTER if whenever

$$\mathtt{NEXT(state_0,loadin_0,func_0)\ =\ state_1}$$
$$\mathtt{NEXT(state_1,loadin_1,func_1)\ =\ state_2}$$
$$\vdots$$
$$\mathtt{NEXT(state_{n-1},loadin_{n-1},func_{n-1})\ =\ state_n}$$

then

$$\mathtt{COUNTER(COUNT(state_0),loadin_1,func_0)\ =\ COUNT(state_n)}$$

Note that the value read in is the value input on the second cycle of the computation (i.e., $\mathtt{loadin}_1$, not $\mathtt{loadin}_0$). Primary states are states whose NODE-component is equal to #00. The host machine is therefore correct if the state transitions specified by COUNTER correspond to those sequences of state transitions specified by NEXT that start and end at primary states and have no primary states in between.

In HOL we can represent the sequence of states $\mathtt{state}_0$, $\mathtt{state}_1$, ... , $\mathtt{state}_n$ by a function state_sig such that $\mathtt{state\_sig(0)=state}_0$, $\mathtt{state\_sig(1)=state}_1$, etc. The function state_sig is a state-valued *signal*; it has type num->(word6#bool#word2). In general, a ty-valued signal is a function of type num->ty where numbers represent time.

A state-valued signal state_sig will represent a sequence of states of the implementation if

```
!t. state_sig(t+1) = NEXT(state_sig(t),loadin_sig(t),func_sig(t))
```

The implementation is correct if whenever a primary state occurs at time t (i.e., NODE(state_sig(t))=#00) and the *next* time a primary state occurs is at time t', then

```
COUNT(state_sig(t')) =
COUNTER(COUNT(state_sig(t)),loadin(t+1),func(t))
```

To formalize this, we define the function NEXTTIME such that NEXTTIME(t,t')f is true if and only if t' is the next time after t that f is true. For example, the term NEXTTIME(t,t')(\x.NODE(state_sig x)=#00) is true if and only if t' is the next time after t that the value of node is #00. NEXTTIME is defined by

```
NEXTTIME (x1,x2) f =
(x1 < x2) /\ (f x2) /\ (!x. (x1 < x) /\ (x < x2) ==> ~f x)
```

Using the function NEXTTIME, we can formulate the *partial correctness* of the host as:

```
(!t. state_sig(t+1) = NEXT(state_sig(t),loadin(t),func(t))) /\
(NODE(state_sig(t))=#00) /\
NEXTTIME(t,t')(\x.NODE(state_sig(x))=#00)
==>
(COUNT(state_sig(t')) =
 COUNTER(COUNT(state_sig(t)),loadin(t+1),func(t)))
```

In addition to this partial correctness property, we must also prove *termination*, namely that the implementation always returns to a primary state when started in a primary state. This can be expressed as

```
(!t. state_sig(t+1) = NEXT(state_sig(t),loadin(t),func(t))) /\
(NODE(state_sig(t))=#00)
==>
?t'.NEXTTIME(t,t')(\x.NODE(state_sig(x))=#00)
```

Partial correctness and termination can be combined into a single statement of total correctness:

```
(!t. state_sig(t+1) = NEXT(state_sig(t),loadin(t),func(t))) /\
(NODE(state_sig(t))=#00)
==>
?t'.NEXTTIME(t,t')(\x.NODE(state_sig(x))=#00) /\
    (COUNT(state_sig(t')) =
     COUNTER(COUNT(state_sig(t)),loadin(t+1),func(t)))
```

In the next section we outline a formal proof of the total correctness of the host machine. Detailed HOL commands for generating this proof are given in Cohn & Gordon [1986].

5 VERIFICATION OF THE HOST MACHINE

The goal is to prove that

```
(!t. state_sig(t+1) = NEXT(state_sig(t),loadin(t),func(t))) /\
(NODE(state_sig(t))=#00)
==>
?t'.NEXTTIME(t,t')(\x.NODE(state_sig(x))=#00) /\
    (COUNT(state_sig(t')) =
     COUNTER(COUNT(state_sig(t)),loadin(t+1),func(t)))
```

The first step is to expand out the definitions (and simplify the results) to derive the four lemmas shown below. Each of these lemmas describes the possible state transitions from a node of the host machine; taken together they completely formalize the complete state transition diagram.

For node `#00`:

```
|- NEXT((count,double,#00),loadin,func) =
    count,EL 0(BITS2 func),
    ((VAL2 func = 0) => #00 | ((VAL2 func = 1) => #11 | #01))
```

For node `#01`:

```
|- NEXT((count,double,#01),loadin,func) =
    ((VAL6 count = 63) => #000000 | WORD6((VAL6 count)+1)),
    EL 0(BITS2 func),(double => #10 | #00)
```

For node `#10`:

```
|- NEXT((count,double,#10),loadin,func) =
   ((VAL6 count = 63) => #000000 | WORD6((VAL6 count)+1)),
   EL 0(BITS2 func),#00
```

For node #11:

```
|- NEXT((count,double,#11),loadin,func) = loadin,EL 0(BITS2 func),#00
```

Using these lemmas, we next derive four theorems of the form

```
|- (!t. state_sig(t+1) = NEXT(state_sig t,loadin_sig t,func_sig t)) ==>
   (NODE(state_sig t) = w) ==>
   (COUNT(state_sig(t+1))  = ...  )  /\
   (DOUBLE(state_sig(t+1)) = ...  )  /\
   (NODE(state_sig(t+1))   = ...  )
```

where '...' stands for terms giving the values of the three components of the state at time `t+1` in terms of their values at time `t`, and the values of the inputs `loadin` and `func` at time `t`.

The four theorems below correspond to the four nodes of the state transition diagram: For node #00:

```
|- (!t. state_sig(t+1) = NEXT(state_sig t,loadin_sig t,func_sig t)) ==>
   (NODE(state_sig t) = #00) ==>
   (COUNT(state_sig(t+1)) = COUNT(state_sig t)) /\
   (DOUBLE(state_sig(t+1)) = EL 0(BITS2(func_sig t))) /\
   (NODE(state_sig(t+1)) =
    ((VAL2(func_sig t) = 0) => #00 |
     ((VAL2(func_sig t) = 1) => #11 | #01)))
```

For node #01:

```
|- (!t. state_sig(t+1) = NEXT(state_sig t,loadin_sig t,func_sig t)) ==>
   (NODE(state_sig t) = #01) ==>
   (COUNT(state_sig(t+1)) =
    ((VAL6(COUNT(state_sig t)) = 63) => #000000 |
     WORD6((VAL6(COUNT(state_sig t)))+1))) /\
   (DOUBLE(state_sig(t+1)) = EL 0(BITS2(func_sig t))) /\
   (NODE(state_sig(t+1)) = (DOUBLE(state_sig t) => #10 | #00))
```

For node #10:

```
|- (!t. state_sig(t+1) = NEXT(state_sig t,loadin_sig t,func_sig t)) ==>
   (NODE(state_sig t) = #10) ==>
   (COUNT(state_sig(t+1)) =
    ((VAL6(COUNT(state_sig t)) = 63) => #000000 |
     WORD6((VAL6(COUNT(state_sig t)))+1))) /\
   (DOUBLE(state_sig(t+1)) = EL 0(BITS2(func_sig t))) /\
   (NODE(state_sig(t+1)) = #00)
```

For node #11:

```
|- (!t. state_sig(t+1) = NEXT(state_sig t,loadin_sig t,func_sig t)) ==>
   (NODE(state_sig t) = #11) ==>
   (COUNT(state_sig(t+1)) = loadin_sig t) /\
   (DOUBLE(state_sig(t+1)) = EL 0(BITS2(func_sig t))) /\
   (NODE(state_sig(t+1)) = #00)
```

These theorems describe single state transitions of the host. We use them to derive descriptions of the effect of each complete path through the spanning tree of the state transition diagram shown below. The four paths through this tree are called A, B, C and D (following Cullyer & Pygott [1985]). The four theorems following the diagram correspond to the respective paths.

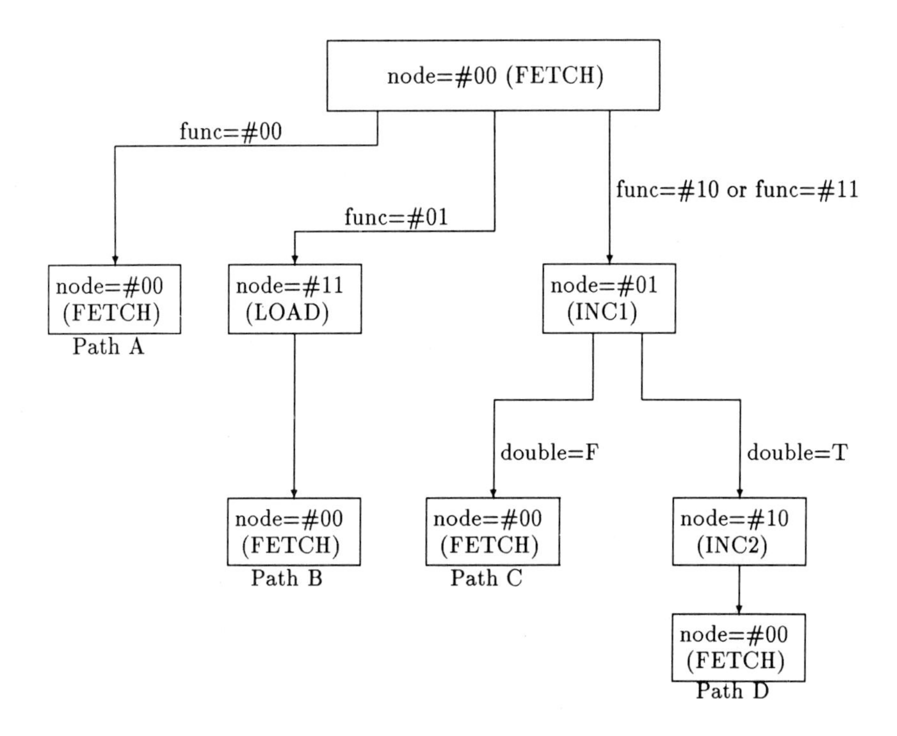

For Path A:

```
|- (NODE(state_sig t) = #00) ==>
   (!t. state_sig(t+1) = NEXT(state_sig t,loadin_sig t,func_sig t)) ==>
   (func_sig t = #00) ==>
   (COUNT(state_sig(t+1)) = COUNT(state_sig t)) /\
   (DOUBLE(state_sig(t+1)) = F) /\
   (NODE(state_sig(t+1)) = #00)
```

For Path B:

```
|- (NODE(state_sig t) = #00) ==>
   (!t. state_sig(t+1) = NEXT(state_sig t,loadin_sig t,func_sig t)) ==>
   (func_sig t = #01) ==>
   ((COUNT(state_sig(t+1)) = COUNT(state_sig t)) /\
    (DOUBLE(state_sig(t+1)) = T) /\
    (NODE(state_sig(t+1)) = #11)) /\
   (COUNT(state_sig((t+1)+1)) = loadin_sig(t+1)) /\
   (DOUBLE(state_sig((t+1)+1)) = EL 0(BITS2(func_sig(t+1)))) /\
   (NODE(state_sig((t+1)+1)) = #00)
```

For path C:

```
|- (NODE(state_sig t) = #00) ==>
   (!t. state_sig(t+1) = NEXT(state_sig t,loadin_sig t,func_sig t)) ==>
   (func_sig t = #10) ==>
   ((COUNT(state_sig(t+1)) = COUNT(state_sig t)) /\
    (DOUBLE(state_sig(t+1)) = F) /\
    (NODE(state_sig(t+1)) = #01)) /\
   (COUNT(state_sig((t+1)+1)) =
    ((VAL6(COUNT(state_sig t)) = 63) => #000000 |
     WORD6((VAL6(COUNT(state_sig t)))+1))) /\
   (DOUBLE(state_sig((t+1)+1)) = EL 0(BITS2(func_sig(t+1)))) /\
   (NODE(state_sig((t+1)+1)) = #00)
```

and for path D:

```
|- (NODE(state_sig t) = #00) ==>
   (!t. state_sig(t+1) = NEXT(state_sig t,loadin_sig t,func_sig t)) ==>
   (func_sig t = #11) ==>
   ((COUNT(state_sig(t+1)) = COUNT(state_sig t)) /\
    (DOUBLE(state_sig(t+1)) = T) /\
    (NODE(state_sig(t+1)) = #01)) /\
   ((COUNT(state_sig((t+1)+1)) =
     ((VAL6(COUNT(state_sig t)) = 63) => #000000 |
      WORD6((VAL6(COUNT(state_sig t)))+1))) /\
    (DOUBLE(state_sig((t+1)+1)) = EL 0(BITS2(func_sig(t+1)))) /\
    (NODE(state_sig((t+1)+1)) = #10)) /\
   (COUNT(state_sig(((t+1)+1)+1)) =
    ((VAL6
      ((VAL6(COUNT(state_sig t)) = 63) => #000000 |
       WORD6((VAL6(COUNT(state_sig t)))+1)) = 63) => #000000 |
     WORD6
     ((VAL6
       ((VAL6(COUNT(state_sig t)) = 63) => #000000 |
        WORD6((VAL6(COUNT(state_sig t)))+1))) +
      1))) /\
   (DOUBLE(state_sig(((t+1)+1)+1)) = EL 0(BITS2(func_sig((t+1)+1)))) /\
   (NODE(state_sig(((t+1)+1)+1)) = #00)
```

Using these theorems, along with the definition of NEXTTIME and some elementary theorems of arithmetic, we can derive four theorems that give the time taken to traverse the four spanning tree paths:

For Path A:

```
|- (!t. state_sig(t+1) = NEXT(state_sig t,loadin_sig t,func_sig t)) /\
   (NODE(state_sig t) = #00) /\
   (func_sig t = #00) ==>
   NEXTTIME(t,t+1)(\x. NODE(state_sig x) =#00)
```

For Path B:

```
|- (!t. state_sig(t+1) = NEXT(state_sig t,loadin_sig t,func_sig t)) /\
   (NODE(state_sig t) = #00) /\
   (func_sig t = #01) ==>
   NEXTTIME(t,(t+1)+1)(\x. NODE(state_sig x) =#00)
```

For Path C:

```
|- (!t. state_sig(t+1) = NEXT(state_sig t,loadin_sig t,func_sig t)) /\
   (NODE(state_sig t) = #00) /\
   (func_sig t = #10) ==>
   NEXTTIME(t,(t+1)+1)(\x. NODE(state_sig x) =#00)
```

For Path D:

```
|- (!t. state_sig(t+1) = NEXT(state_sig t,loadin_sig t,func_sig t)) /\
   (NODE(state_sig t) = #00) /\
   (func_sig t = #11) ==>
   NEXTTIME(t,((t+1)+1)+1)(\x. NODE(state_sig x) =#00)
```

The host machine terminates if

```
(!t. state_sig(t+1) = NEXT(state_sig t,loadin_sig t,func_sig t)) /\
(NODE(state_sig t) = #00)
==>
(?t'. NEXTTIME(t,t')(\x. NODE(state_sig x) =#00))
```

This follows easily by case analysis on the term func_sig t, followed by rewriting using the four path theorems.

To prove correctness we need the (intuitively obvious) theorem that for any f and x1 there is a unique x2 such that NEXTTIME(x1,x2)f. Hence,

```
|- !f x1 x2. NEXTTIME(x1,x2)f /\ NEXTTIME(x1,x3)f ==> (x2=x3)
```

From this fact and the path lemmas, it easily follows that

```
|- (!t. state_sig(t+1) = NEXT(state_sig t,loadin_sig t,func_sig t)) /\
   (NODE(state_sig t)=#00) /\
   NEXTTIME (t,t') (\x. NODE(state_sigx)=#00)
   ==>
   (t' = ((func_sig t = #00) => t+1           |
          (func_sig t = #01) => (t+1)+1       |
          (func_sig t = #10) => (t+1)+1       |
          (func_sig t = #11) => ((t+1)+1)+1 | ARB))
```

Using this theorem, the definition of COUNTER, the path theorems, some case analysis and some arithmetic, we can deduce the partial correctness of the host machine, namely

```
(!t. state_sig(t+1) = NEXT(state_sig t,loadin_sig t,func_sig t)) /\
(NODE(state_sig t)=#00) /\
NEXTTIME (t,t') (\x. NODE(state_sig x)=#00)
==>
(COUNT(state_sig t') =
 COUNTER(COUNT(state_sig t),loadin_sig(t+1),func_sig t))
```

The total correctness of the host machine follows easily from its partial correctness and termination.

The large number of omitted details of this proof can be found in Cohn & Gordon [1986].

6 SPECIFICATION AND VERIFICATION OF THE HIGH LEVEL DESIGN

The high level design consists of the three registers, COUNT_LATCH, DOUBLE_LATCH and NODE_LATCH, together with some combinational control logic called COUNTLOGIC:

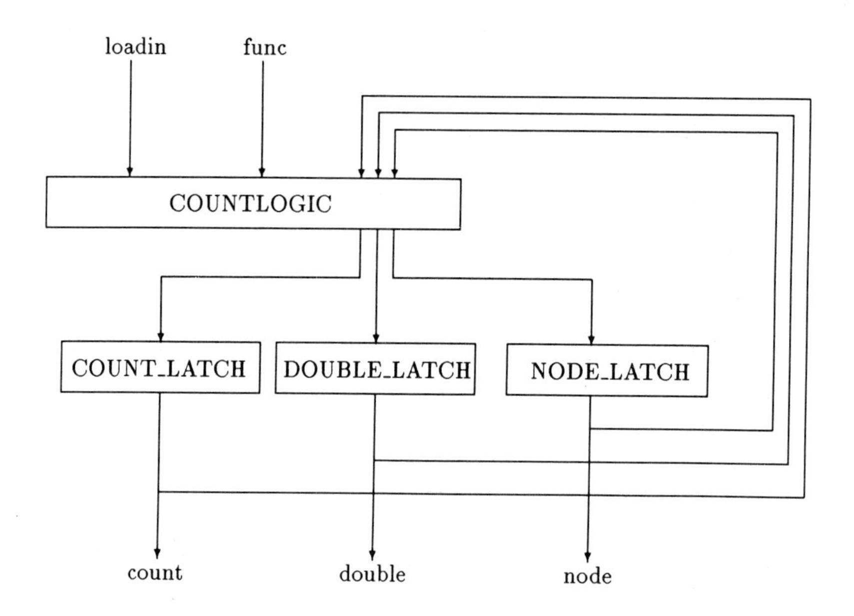

A state of the high level design is a triple `(count,double,node)` where `count` is the 6-bit word stored in the register `COUNT_LATCH`, `double` is the boolean value stored in `DOUBLE_LATCH` and `node` the 2-bit word stored in `NODE_LATCH`. If the high level design is in state `(count,double,node)` and `loadin` and `count` are being input, then the next state is `(count',double',node')` if

```
COUNTLOGIC((count,double,node),loadin,func) = (count',double',node')
```

The actual definition of `COUNTLOGIC` is

```
COUNTLOGIC((count,double,node),loadin,func) =
 let twice = EL 0 (BITS2 func) in
 (MULTIPLEX(INCLOGIC(count,INCCON node),loadin,MPLXCON node),
  twice,
  NEXTNODE(node,func,double))
```

where the functions `MULTIPLEX`, `INCLOGIC`, `INCCON`, `MPLXCON` and `NEXTNODE` are specified by the following HOL definitions:

```
MULTIPLEX(incout,loadin,mplxsel) =
 (mplxsel => incout | loadin)

INCLOGIC(count,noinc) =
 let countval = VAL6 count in
 (noinc => count | ((countval=63) => WORD6 0 | WORD6(countval+1)))

INCCON(node) = (VAL2 node = 0)

MPLXCON(node) = ~(VAL2 node = 3)

NEXTNODE(node,func,double) =
 let funcnum  = VAL2 func in
 let nodenum  = VAL2 node in
 let fetchnode = WORD2 0   in
 let inc1node  = WORD2 1   in
 let inc2node  = WORD2 2   in
 let loadnode  = WORD2 3   in
 ((nodenum=0) => ((funcnum=0) => fetchnode |
                  (funcnum=1) => loadnode | inc1node) |
  (nodenum=1) => (double => inc2node | fetchnode) | fetchnode)
```

We outline the proof that the host machine is correctly implemented by the high level design. This can be expressed very simply by

```
NEXT = COUNTLOGIC
```

To verify the high level design we prove that `NEXT = COUNTLOGIC`. To do this it is sufficient, by extensionality, to prove

```
!count double node loadin func.
  NEXT((count,double,node),loadin,func) =
  COUNTLOGIC((count,double,node),loadin,func)
```

This is easily proved by expanding out the definitions and then doing case analysis on the terms `node`, `func` and `double`.

7 SPECIFICATION OF THE CIRCUIT

The counter circuit schematic diagram below is reproduced from Cullyer & Pygott [1985].

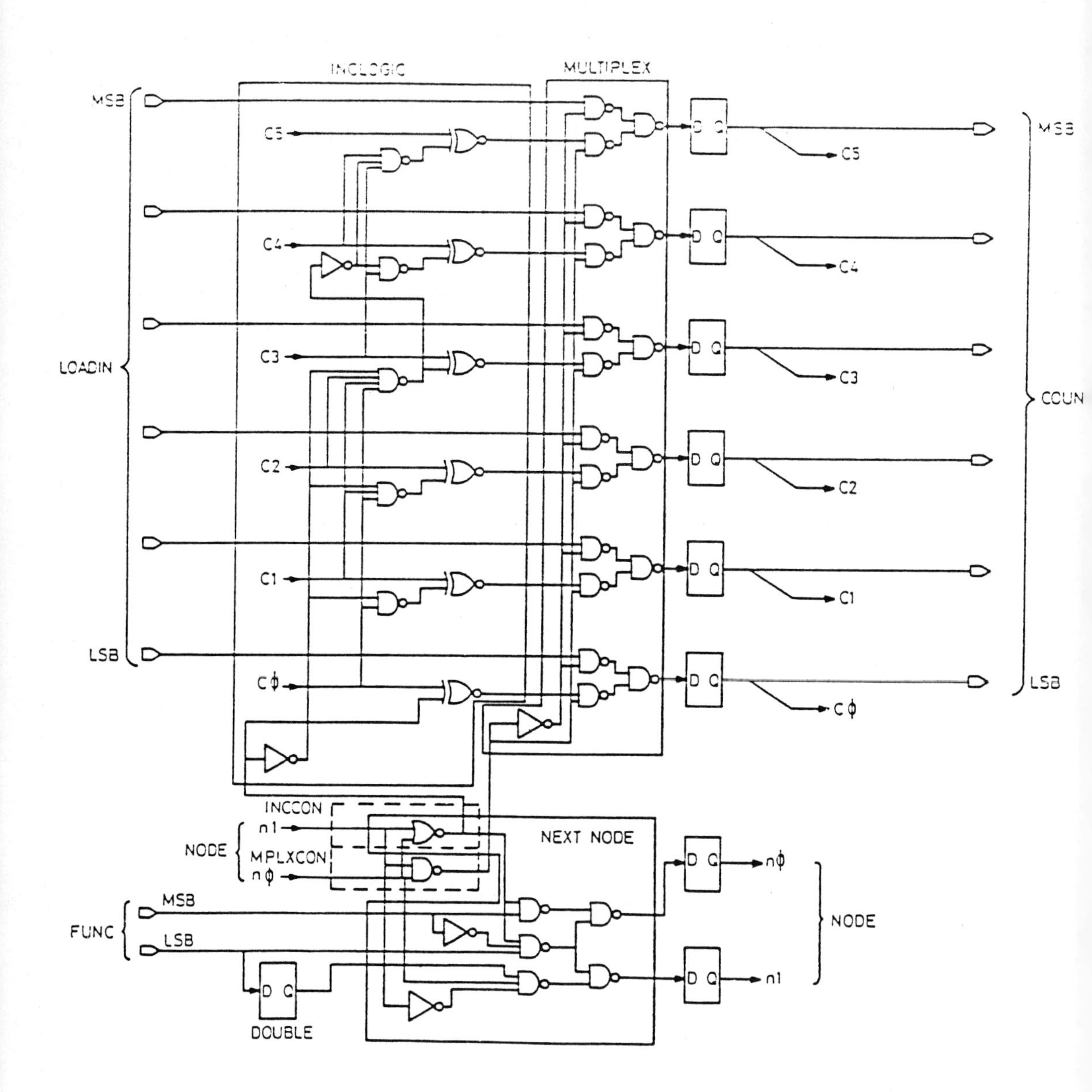
INCLOGIC
MULTIPLEX
MSB
LOADIN
LSB
C5
C4
C3
C2
C1
Cϕ
COUN
INCCON
NODE
n1
MPLXCON
nϕ
NEXT NODE
FUNC
DOUBLE
D Q

We prove correct two representations of this circuit. The first one (which we call the logical specification) is a direct translation into logic of the circuit diagram. The second one is a translation of the ELLA text given in Cullyer & Pygott [1985]. Since ELLA is a language specifically designed for hardware specification, one might think it better suited to this task than pure logic. However, as the example below shows, this is debatable. One could argue that the 'relational' style of representing structure that is used in HOL is a more direct reflection of schematic diagrams than the functional style of ELLA. Furthermore, Camilleri has shown that the relational style can be transformed into the functional style automatically (Camilleri [1988]). As we illustrate below, an ELLA-like functional style can be implemented in HOL if desired. (This is necessary if one wants to reason about ELLA specifications, as the ELLA system does not provide any proof tools, so cannot be used by itself for formal verification. It is, however, interfaced to many CAD tools that are not available in HOL.)

7.1 The logical specification

The combinatorial behaviour of the circuit `COUNTLOGIC` is determined by the behaviour of its five component parts, `INCCON`, `MPLXCON`, `MULTIPLEX`, `INCLOGIC` and `NEXTNODE`, whose behavioural specifications are given in Section 6. For each of these parts there is an implementation in terms of gates, as shown in the circuit diagram.

To describe the circuit, we first define gate constants `INV`, `NAND`, and so on, as logical relations between inputs and outputs:

```
NOR(in1,in2,out) = (out = ~(in1 \/ in2))

NAND(in1,in2,out) = (out = ~(in1 /\ in2))

NAND3(in1,in2,in3,out) = (out = ~(in1 /\ in2 /\ in3))

NAND4(in1,in2,in3,in4,out) = (out = ~(in1 /\ in2 /\ in3 /\ in4))

INV(in,out) = (out = ~in)

XNOR(in1,in2,out) = (out = (in1 = in2))
```

Then, for example, we can define the predicate `INCLOGIC_IMP` (the implementation of `INCLOGIC`) in terms of these gate constants. We let `c0`,...,`c5` represent the boolean input values to the circuit, as in the circuit diagram, and `b` the boolean value sent to `INCLOGIC` by `INCCON`. The variables `c0'`,...,`c5'` represent the boolean values output by `INCLOGIC_IMP` (and sent on to `MULTIPLEX`); `c0` and `c0'` are the least significant bits. `INCLOGIC_IMP` holds of `c0`,...,`c5`, `b` and `c0'`,...,`c5'`. We then introduce variables `b'`, `x1`, x2, x3, x3', x4 and x5 which, respectively and from bottom to top in the circuit, name the internal lines emanating from the inverter; the first NAND gate; the second NAND gate; the third NAND gate; the second inverter; the fourth NAND gate; and the fifth NAND gate. In the description of INCLOGIC_IMP, local lines are 'hidden' by existential quantification. A rationale of this technique of description is given in other papers (Gordon [1986], Camilleri *et al.* [1987]).

```
INCLOGIC_IMP(c0,c1,c2,c3,c4,c5,b,c0',c1',c2',c3',c4',c5') =
 ?b' x1 x2 x3 x3' x4 x5.
  INV(b,b')                  /\
  INV(x3,x3')                /\
  NAND(b',c0,x1)             /\
  NAND3(b',c0,c1,x2)         /\
  NAND4(b',c0,c1,c2,x3)      /\
  NAND(x3',c3,x4)            /\
  NAND3(x3', c3,c4,x5)       /\
  XNOR(c0,b,c0')             /\
  XNOR(c1,x1,c1')            /\
  XNOR(c2,x2,c2')            /\
  XNOR(c3,x3,c3')            /\
  XNOR(c4,x4,c4')            /\
  XNOR(c5,x5,c5')
```

The other four components are treated similarly; INCLOGIC is the most complex of the five. The analogous descriptions are called INCCON_IMP, MPLXCON_IMP, MULTIPLEX_IMP and NEXTNODE_IMP. Further details of the others appear in Cohn & Gordon [1986].

Finally, we relate the definition of the whole combinatorial part of the counter to the whole implementation. In Section 6, the high level design was specified using the function COUNTLOGIC, defined by

```
COUNTLOGIC((count,double,node),loadin,func) =
 let twice = EL 0 (BITS2 func) in
 (MULTIPLEX(INCLOGIC(count,INCCON node),loadin,MPLXCON node),
  twice,
  NEXTNODE(node,func,double))
```

The predicate COUNTLOGIC_IMP, defined below, formalizes the implementation of the combinatorial part of the high level design. It takes as arguments all of the boolean inputs to the circuit: c0,...,c5, and l0,...,l5, and n0 and n1 as in the circuit diagram; as well as f0 and f1, jointly called FUNC in the circuit; and double, extracted from FUNC in the schematic diagram on page 82. It also takes the various outputs as arguments: c0',...,c5', corresponding to c0',...,c5', n0' and n1' corresponding to n0 and n1, and double' corresponding to double.

As we have already done for the components, we describe the whole circuit COUNTLOGIC_IMP by giving names to the internal lines, and existentially quantifying over these names; in this case, we need the names b1 for the boolean output of MPLXCON, b2 for the boolean output of INCCON, and d0,...,d5 for the boolean outputs of INCLOGIC (with d0 the least significant bit). The description of the circuit is

```
COUNTLOGIC_IMP(c0,c1,c2,c3,c4,c5,n0,n1,l0,l1,l2,l3,l4,l5,f0,f1,double,
               c0',c1',c2',c3',c4',c5',n0',n1',double') =
?b1 b2 d0 d1 d2 d3 d4 d5.
  INCCON_IMP(n0,n1,b2)                                     /\
  MPLXCON_IMP(n0,n1,b1)                                    /\
  NEXTNODE_IMP(n0,n1,f0,f1,double,n0',n1')                 /\
  INCLOGIC_IMP(c0,c1,c2,c3,c4,c5,b2,d0,d1,d2,d3,d4,d5)    /\
  MULTIPLEX_IMP(d0,d1,d2,d3,d4,d5,l0,l1,l2,l3,l4,l5,b1,
               c0',c1',c2',c3',c4',c5')                    /\
  (double' = f0)
```

where INCLOGIC_IMP and the four other '_IMP' predicates are defined in terms of gate constants, as explained above.

7.2 The ELLA specification

In Cullyer & Pygott [1985] the circuit is specified in the hardware description language ELLA. The specification is easily translated into higher order logic, and gives an alternative formalization to the specification INCLOGIC_IMP in the previous section. For example, the ELLA definition of INCLOGIC is called INCCIRC in Cullyer & Pygott [1985]:

```
FN INCCIRC = (word6 count, bool:  noinc) -> word6
BEGIN LET noincbar = INV noinc.
      LET ic1 = XNOR(count[1], noinc).
      LET ic2 = XNOR(count[2],NAND2(noincbar, count[1])).
      LET ic3 = XNOR(count[3],NAND3(noincbar,count[1],count[2])).
      LET carry4bar = NAND4(noincbar,count[1],count[2],count[3]).
      LET ic4 = XNOR(count[4], carrybar).
      LET ic5 = XNOR(count[5],NAND2(carry4, count[4])).
      LET ic6 = XNOR(count[6],NAND3(carry4, count[4], count[5])).
      OUTPUT(ic1, ic2, ic3, ic4, ic5, ic6)
END.
```

To produce an equivalent, we define new gate constants which are functional rather than relational (in lower case, to distinguish):

```
nor(in1,in2) = ~(in1 \/ in2)


nand(in1,in2) = ~(in1 /\ in2)
```

```
nand3(in1,in2,in3) = ~(in1 /\ in2 /\ in3)

nand4(in1,in2,in3,in4) = ~(in1 /\ in2 /\ in3 /\ in4)

inv in = ~in

xnor(in1,in2) = (in1 = in2)
```

(We use `nand` rather than `nand2`, to be consistent with `NAND`).

In the translation of the definition of `INCCIRC` to HOL, the ELLA phrase '`count[5]`', for example, becomes the HOL term '`BIT6 count 5`'. We use the HOL `let-in` construct for ELLA's repeated `LET` construction, so that the ELLA keyword `OUTPUT` is not necessary. A coercion, `V6`, is added to correct the type of the result, and the normal bit order is used. The HOL `t:ty` notation replaces ELLA's `ty:t`. The HOL definition of `INCCIRC` is thus

```
INCCIRC count noinc =
let noincbar = inv noinc in
let ic1 = xnor(BIT6 count 1,noinc) in
let ic2 = xnor(BIT6 count 2,nand(noincbar,BIT6 count 1)) in
let ic3 = xnor(BIT6 count 3,nand3(noincbar,BIT6 count 1,BIT6 count 2)) in
let carry4bar = nand4(noincbar,BIT6 count 1,
                      BIT6 count 2, BIT6 count 3) in
let ic4 = xnor(BIT6 count 4, carry4bar) in
let carry4 = inv carry4bar in
let ic5 = xnor(BIT6 count 5,nand(carry4,BIT6 count 4)) in
let ic6 = xnor(BIT6 count 6, nand3(carry4,BIT6 count 4,BIT6 count 5)) in
V6(ic6,ic5,ic4,ic3,ic2,ic1)
```

The other four components have ELLA definitions which can be translated in a similar way; further details of the translation process and the other components can be found in Cohn & Gordon [1986].

8 VERIFICATION OF THE CIRCUIT

We now verify that both the ELLA and the purely logical representation of the circuit meet the specifications given in the previous section.

8.1 Verification of the logical specification

Our first goal is to relate the predicates such as `INCLOGIC_IMP`, from Section 7.1, which together represent the circuit, to the functions such as `INCLOGIC`, which specify the components of the high level design. The definition of `INCLOGIC` is

```
INCLOGIC(count,noinc) =
 let countval = VAL6 count
 in (noinc => count | ((countval=63)=>WORD6 0 | WORD6(countval+1)))
```

`INCLOGIC` takes a 6-bit word, `count`, represented for `INCLOGIC_IMP` by the booleans `c0`,...,`c5`, and a boolean, `noinc`, and returns a 6-bit word, represented for `INCLOGIC_IMP` by the booleans `c0'`,...,`c5'`. To relate these two rather different representations, it is convenient to introduce an intermediate level: a *relational specification*, which is a predicate we shall call `INCLOGIC_SPEC`:

```
INCLOGIC_SPEC(c0,c1,c2,c3,c4,c5,b,c0',c1',c2',c3',c4',c5') =
 (c5',c4',c3',c2',c1',c0') =
  (b => (c5,c4,c3,c2,c1,c0) |
        ((c5,c4,c3,c2,c1,c0) = (T,T,T,T,T,T)) => (F,F,F,F,F,F) |
        TUPLE6(BITS6(V6(c5,c4,c3,c2,c1,c0)+1)))
```

`INCLOGIC_SPEC`, which has the same type as `INCLOGIC_IMP`, holds of its (thirteen) arguments if the outputs, `c0'`,...,`c5'`, are equal to a certain function of the inputs, `b` and `c0`,...,`c5`. That function, on inspection, is really just `INCLOGIC` phrased in terms of separate boolean values. We then show that

```
INCLOGIC_SPEC(c0,c1,c2,c3,c4,c5,b,c0',c1',c2',c3',c4',c5') =
 (V6(c5',c4',c3',c2',c1',c0') =
  INCLOGIC(V6(c5,c4,c3,c2,c1,c0),b))
```

and that

```
INCLOGIC_SPEC(c0,c1,c2,c3,c4,c5,b,c0',c1',c2',c3',c4',c5') =
INCLOGIC_IMP (c0,c1,c2,c3,c4,c5,b,c0',c1',c2',c3',c4',c5')
```

and thereby relate `INCLOGIC` to `INCLOGIC_IMP`. The first goal is proved by unfolding the definitions, doing case analysis on the boolean `b`, and then case analysis on whether `(c5,c4,c3,c2,c1,c0) = (T,T,T,T,T,T)`. The second goal is proved similarly, but with a further analysis of the sixty-four possible combinations of the booleans `c0`, `c1`, `c2`, `c3`, `c4` and `c5`. Analogous results are proved for `INCCON`, `MPLXCON`, `MULTIPLEX` and `NEXTNODE`.

These proofs require some simple facts (axioms and lemmas) about bit strings and boolean lists, such as

```
~(V[b0;b1;b2;b3;b4;b5] = 63) ==>
(V(BITS6(WORD6(V[b0;b1;b2;b3;b4;b5]+1))) =
(V[b0;b1;b2;b3;b4;b5]+1))
```

To tie these results together, we make an intermediate level specification of COUNTLOGIC, just as we have done for each of its components:

```
COUNTLOGIC_SPEC(c0,c1,c2,c3,c4,c5,n0,n1,l0,l1,l2,l3,l4,l5,f0,f1,double,
                c0',c1',c2',c3',c4',c5',n0',n1',double') =
?b1 b2 d0 d1 d2 d3 d4 d5.
  INCCON_SPEC(n0,n1,b2)                                      /\
  MPLXCON_SPEC(n0,n1,b1)                                     /\
  NEXTNODE_SPEC(n0,n1,f0,f1,double,n0',n1')                  /\
  INCLOGIC_SPEC(c0,c1,c2,c3,c4,c5,b2,d0,d1,d2,d3,d4,d5) /\
  MULTIPLEX_SPEC(d0,d1,d2,d3,d4,d5,l0,l1,l2,l3,l4,l5,b1,
                 c0',c1',c2',c3',c4',c5')                    /\
  (double' = f0)
```

This definition of COUNTLOGIC_SPEC is very like that of COUNTLOGIC_IMP (in Section 7.1) except that it is in terms of the intermediate level specifications of the components rather than the circuit level definitions. We show that

```
COUNTLOGIC_SPEC(c0,c1,c2,c3,c4,c5,n0,n1,l0,l1,l2,l3,l4,l5,f0,f1,double,
                c0',c1',c2',c3',c4',c5',n0',n1',double') =
 ((V6(c5',c4',c3',c2',c1',c0'),double',V2(n1',n0')) =
  COUNTLOGIC((V6(c5,c4,c3,c2,c1,c0),double,V2(n1,n0)),
             V6(l5,l4,l3,l2,l1,l0),
             V2(f1,f0)))
```

and

```
COUNTLOGIC_SPEC(c0,c1,c2,c3,c4,c5,n0,n1,l0,l1,l2,l3,l4,l5,f0,f1,double,
                c0',c1',c2',c3',c4',c5',n0',n1',double') =
 COUNTLOGIC_IMP(c0,c1,c2,c3,c4,c5,n0,n1,l0,l1,l2,l3,l4,l5,f0,f1,double,
                c0',c1',c2',c3',c4',c5',n0',n1',double')
```

so that by transitivity

```
COUNTLOGIC_IMP (c0,c1,c2,c3,c4,c5,n0,n1,l0,l1,l2,l3,l4,l5,f0,f1,double,
                c0',c1',c2',c3',c4',c5',n0',n1',double') =
 ((V6(c5',c4',c3',c2',c1',c0'),double',V2(n1',n0')) =
  COUNTLOGIC((V6(c5,c4,c3,c2,c1,c0),double,V2(n1,n0)),
             V6(l5,l4,l3,l2,l1,l0),
             V2(f1,f0)))
```

The first goal is proved by unfolding definitions and using the five theorems (for the five component parts) stating that their definitions are related to their specifications. The second goal is proved similarly, using the theorems equating the specifications and implementations.

8.2 Verification of the ELLA specification

Given the HOL version of the ELLA definition, we wish to prove, for example, that

```
INCLOGIC_IMP(c0,c1,c2,c3,c4,c5,b,c0',c1',c2',c3',c4',c5') =
(V6(c5',c4',c3',c2',c1',c0') = INCCIRC(V6(c5,c4,c3,c2,c1,c0)))
```

The proof is by unfolding definitions and doing a case analysis on the boolean term `b`. Some simple lemmas about bit strings and boolean lists are again required. The statements and proofs for the other four component are similar, with case analysis on appropriate boolean terms, such as `f0`, `f1`, `n0`, and `n1` as necessary. The ELLA definitions given in Cullyer & Pygott [1985] for `MULTIPLEX`, `INCCON` and `NEXTNODE` are combined into a single definition (as these three components overlap in the circuit diagram), so the equivalence statements must select out the appropriate parts. Details are given in Cohn & Gordon [1986].

9 CORRECTNESS OF THE COUNTER IMPLEMENTATION

To complete the proof of the counter, we must tie together the results about the combinatorial behaviour of `COUNTLOGIC` and the results about the time behaviour (correctness and termination) of `NEXT`. We have already shown that `COUNTLOGIC` and `NEXT` are the same, in Section 6. We have also shown, in Section 5, that

```
(!t.state_sig(t+1) = NEXT(state_sig t,loadin_sig t,func_sig t)) /\
    (NODE(state_sig t)=#00)                                     /\
    NEXTTIME (t,t') (\x. NODE(state_sig x)=#00)
    ==>
    (COUNT(state_sig t') =
     COUNTER(COUNT(state_sig t),loadin_sig(t+1),func_sig t))
```

and

```
(!t.state_sig(t+1) = NEXT(state_sig t,loadin_sig t,func_sig t)) /\
   (NODE(state_sig t) = #00)
    ==>
   ?t'.NEXTTIME(t,t')(\x.NODE(state_sig x)=#00)
```

so we simply substitute to get statements of the partial correctness and termination of `COUNTLOGIC`:

```
(!t.state_sig(t+1) = COUNTLOGIC(state_sig t,loadin_sig t,func_sig t)) /\
    (NODE(state_sig t)=#00) /\
    NEXTTIME (t,t') (\x. NODE(state_sig x)=#00)
    ==>
    (COUNT(state_sig t') =
     COUNTER(COUNT(state_sig t),loadin_sig(t+1),func_sig t))
```

and

```
(!t.state_sig(t+1) = COUNTLOGIC(state_sig t,loadin_sig t,func_sig t)) /\
    (NODE(state_sig t) = #00)
    ==>
    ?t'.NEXTTIME(t,t')(\x.NODE(state_sig x)=#00)
```

To relate COUNTLOGIC to the temporal behaviour of the circuit, we first define COUNT-LOGIC_DEV (for device) which is like COUNTLOGIC but takes *signals* as arguments:

```
COUNTLOGIC_DEV
(c0_sig,c1_sig,c2_sig,c3_sig,c4_sig,c5_sig,n0_sig,n1_sig,
 l0_sig,l1_sig,l2_sig,l3_sig,l4_sig,l5_sig,f0_sig,f1_sig,double_sig,
 c0'_sig,c1'_sig,c2'_sig,c3'_sig,c4'_sig,c5'_sig,
 n0'_sig,n1'_sig,double'_sig) =
!t:num. COUNTLOGIC_IMP
 (c0_sig t,c1_sig t,c2_sig t,c3_sig t,c4_sig t,c5_sig t,
  n0_sig t,n1_sig t,
  l0_sig t,l1_sig t,l2_sig t,l3_sig t,l4_sig t,l5_sig t,
  f0_sig t,f1_sig t,double_sig t,
  c0'_sig t,c1'_sig t,c2'_sig t,c3'_sig t,c4'_sig t,c5'_sig t,
  n0'_sig t,n1'_sig t,double'_sig t)
```

We then introduce the definition of a LATCH:

```
LATCH(in,out)  =  !t. out(t+1) = in t
```

This enables us to give a temporal description of the entire circuit. To do this we define one more predicate, CIRCUIT_IMP, to represent the entire circuit diagram in HOL:

```
CIRCUIT_IMP
(((c5_sig,c4_sig,c3_sig,c2_sig,c1_sig,c0_sig),
 double_sig,
 (n1_sig,n0_sig)),
(l5_sig,l4_sig,l3_sig,l2_sig,l1_sig,l0_sig),
(f1_sig,f0_sig)) =
?c0'_sig c1'_sig c2'_sig c3'_sig c4'_sig c5'_sig double'_sig
 n0'_sig n1'_sig.
 (COUNTLOGIC_DEV
  (c0_sig,c1_sig,c2_sig,c3_sig,c4_sig,c5_sig,n0_sig,n1_sig,
   l0_sig,l1_sig,l2_sig,l3_sig,l4_sig,l5_sig,f0_sig,f1_sig,double_sig,
   c0'_sig,c1'_sig,c2'_sig,c3'_sig,c4'_sig,c5'_sig,
   n0'_sig,n1'_sig,double'_sig)          /\
```

```
    LATCH(c5'_sig,c5_sig)                    /\
    LATCH(c4'_sig,c4_sig)                    /\
    LATCH(c3'_sig,c3_sig)                    /\
    LATCH(c2'_sig,c2_sig)                    /\
    LATCH(c1'_sig,c1_sig)                    /\
    LATCH(c0'_sig,c0_sig)                    /\
    LATCH(n1'_sig,n1_sig)                    /\
    LATCH(n0'_sig,n0_sig)                    /\
    LATCH(double'_sig,double_sig))
```

Using the definition of COUNTLOGIC_DEV and the result from Section 8 relating COUNTLOGIC_IMP to COUNTLOGIC, we can prove the following lemma, relating CIRCUIT_IMP to COUNTLOGIC:

```
  CIRCUIT_IMP(((c5_sig,c4_sig,c3_sig,c2_sig,c1_sig,c0_sig),
               double_sig,
               (n1_sig,n0_sig)),
              (l5_sig,l4_sig,l3_sig,l2_sig,l1_sig,l0_sig),
              (f1_sig,f0_sig)) =
   !t.(V6(c5_sig(t+1),c4_sig(t+1),c3_sig(t+1),
          c2_sig(t+1),c1_sig(t+1),c0_sig(t+1)),
       double_sig (t+1),
       V2(n1_sig(t+1),n0_sig(t+1))) =
       COUNTLOGIC((V6(c5_sig t,c4_sig t,c3_sig t,
                      c2_sig t,c1_sig t,c0_sig t_),
                   double_sig t,
                   V2(n1_sig t,n0_sig t)),
                  V6(l5_sig t,l4_sig t,l3_sig t,l2_sig t,l1_sig t,l0_sig t),
                  V2(f1_sig t,f0_sig t))
```

We then use this lemma in combination with the initial result of this section (the partial correctness and termination of COUNTLOGIC) to prove the partial correctness and the terminiation of the circuit:

```
  CIRCUIT_IMP
   (((c5_sig,c4_sig,c3_sig,c2_sig,c1_sig,c0_sig),
     double_sig,
     (n1_sig,n0_sig)),
    (l5_sig,l4_sig,l3_sig,l2_sig,l1_sig,l0_sig),
    (f1_sig,f0_sig)) /\
  (V2(n1_sig t,n0_sig t) = #00) /\
  (NEXTTIME (t,t') (\x. V2(n1_sig x,n0_sig x) = #00))
  ==>
  (V6(c5_sig t',c4_sig t',c3_sig t',c2_sig t',c1_sig t',c0_sig t') =
    COUNTER(V6(c5_sig t,c4_sig t,c3_sig t,c2_sig t,c1_sig t,c0_sig t),
            V6(l5_sig(t+1),l4_sig(t+1),l3_sig(t+1),
               l2_sig(t+1),l1_sig(t+1),l0_sig(t+1)),
            V2(f1_sig t,f0_sig t)))
```

and

```
CIRCUIT_IMP(((c5_sig,c4_sig,c3_sig,c2_sig,c1_sig,c0_sig),
             double_sig,
             n1_sig,n0_sig),
            (l5_sig,l4_sig,l3_sig,l2_sig,l1_sig,l0_sig),
            f1_sig,f0_sig) /\
(V2(n1_sig t,n0_sig t) = #00)
==>
(?t'. NEXTTIME(t,t')(\x. V2(n1_sig x,n0_sig x) =#00))
```

Using the predefined constants VS2 and VS6, the two theorems above can be simplified to

```
CIRCUIT_IMP((count_sigs,double_sig,node_sigs),loadin_sigs,func_sigs) /\
 (VS2 node_sigs t = #00) /\
 (NEXTTIME (t,t') (\x. VS2 node_sigs x = #00))
 ==>
 (VS6 count_sigs t' =
  COUNTER(VS6 count_sigs t,
          VS6 loadin_sigs (t+1),
          VS2 func_sigs t))
```

and

```
CIRCUIT_IMP((count_sigs,double_sig,node_sigs),loadin_sigs,func_sigs) /\
(VS2 node_sigs t = #00)
==>
(?t'. NEXTTIME(t,t')(\x. VS2 node_sigs x = #00))
```

These theorems are easily combined to prove the total correctness of the circuit, namely

```
CIRCUIT_IMP((count_sigs,double_sig,node_sigs),loadin_sigs,func_sigs) /\
(VS2 node_sigs t = #00)
==>
(?t'. NEXTTIME(t,t')(\x. VS2 node_sigs x = #00) /\
       (VS6 count_sigs t' =
        COUNTER(VS6 count_sigs t,
                VS6 loadin_sigs (t+1),
                VS2 func_sigs t)))
```

This completes the verification of the counter.

10 CONCLUSIONS

Although the counter described in this report is quite simple, its formal verification may appear to the reader to be depressingly long and intricate, and it might seem that the verification of much more complex devices would be impractical. We believe, however, that real hardware *can* be formally verified using existing techniques. Our optimism is based on the success of the LCF methodology in partially automating the production of long proofs: succinct and general meta-language procedures can generate long and detailed chains of inferences. Proofs of the sort described in this chapter consist largely in unfolding definitions, performing obvious case analyses, and doing simple rewritings. None of these operations requires much cleverness, and producing them is really simpler than it may appear. With a little experience, HOL proofs can be produced quite rapidly. In fact, more of the time and effort go into writing, correcting and typing in the definitions than into producing the proofs, even though (measured in number of inference steps or in CPU time used) the proofs are large objects. In the present proof, for example, the formal specifications (i)–(iv) below had to be invented and added to the RSRE definitions before the problem statement was complete and the proof could be undertaken:

(i) The apparatus for relating different timescales (`NEXTTIME`, etc.);
(ii) The relational specifications of the circuit (`COUNTLOGIC_SPEC`);
(iii) The structural description of the circuit (`CIRCUIT_IMP`);
(iv) The temporal description of the circuit (`CIRCUIT_IMP`).

Furthermore, the correctness criteria had to be formulated for the host machine, the high level design, the circuit, and the complete counter implementation. Given all of that, generating the proofs was straightforward.

The subsequent proofs of correctness properties of the Viper microprocessor to the register transfer level (Cohn [1987, 1989]) also support our optimism about the methodology. These proofs are orders of magnitude larger and more complex than the counter proof. Nonetheless, they share the property of having been relatively simple to generate by machine, once all the definitions were in place. This latter, again, was the major creative phase of the project.

REFERENCES

Andrews [1986]
P. B. Andrews (1986), *An Introduction to Mathematical Logic and Type Theory: to Truth through Proof*, Academic Press Inc., Harcourt Brace Jovanovich, New York.

Camilleri [1988]
A. C. Camilleri (1988) Ph.D. Thesis, University of Cambridge. (Dr. Camilleri now at: Hewlett-Packard Laboratories, Filton Road, Stoke Gifford, Bristol BS12 6QZ, England.)

Camilleri *et al.* [1987]
A. Camilleri, M. Gordon and T. Melham (1987) 'Hardware Verification using Higher-Order Logic', in *Proceedings of the IFIP WG 10.2 Working Conference: From H.D.L. Descriptions to Guaranteed Correct Circuit Designs, Grenoble, September 1986*, ed. D. Borrione, North-Holland, Amsterdam.

Cohn [1987]
A. Cohn, 'A proof of correctness of the Viper microprocessor: The first level', in *VLSI Specification, Verification and Synthesis*, ed. G. Birtwistle & P.A. Subrahmanyam, Kluwer, Dortrecht; also University of Cambridge, Computer Laboratory, Tech. Report No. 104.

Cohn [1989]
A. Cohn (1989) 'Correctness properties of the Viper block model: The second level', *Current Trends in Hardware Verification and Automated Theorem Proving*, ed. G. Birtwistle and P. A. Subrahmanyam, Springer-Verlag, 1989; also University of Cambridge, Computer Laboratory, Tech. Report No. 104.

Cohn & Gordon [1986]
A. Cohn and M. Gordon (1986) 'A mechanized proof of correctness of a simple counter', University of Cambridge, Computer Laboratory, Tech. Report No. 94.

Cullyer [1985]
W. J. Cullyer (1985) 'Viper Microprocessor: formal specification', RSRE Report 85013.

Cullyer [1986]
W. J. Cullyer (1986) 'Viper — correspondence between the specification and the "major state machine" ', RSRE report No. 86004.

Cullyer [1987]
W. J. Cullyer (1987) 'Implementing safety-critical systems: the Viper microprocessor', in *VLSI Specification, Verification and Synthesis*, ed. G. Birtwistle & P.A. Subrahmanyam, Kluwer, Dortrecht.

Cullyer & Pygott [1985]
W. J. Cullyer and C. H. Pygott (1985) 'Hardware proofs using LCF_LSM and ELLA', RSRE Memo. 3832.

Gordon [1983]
M. Gordon (1983) 'Proving a computer correct', University of Cambridge, Computer Laboratory, Tech. Report No. 42.

Gordon [1985]
M. Gordon (1985) 'HOL: a machine oriented formulation of higher-order logic', University of Cambridge, Computer Laboratory, Tech. Report No. 68.

Gordon [1986]
M. Gordon (1986) 'Why higher-order logic is a good formalism for specifying and verifying hardware', in *Formal Aspects of VLSI Design*, ed. G. Milne & P. A. Subrahmanyam, North-Holland, Amsterdam.

Gordon [1987]
M. Gordon (1987) 'HOL: a proof generating system for higher-order logic', in *VLSI Specification, Verification and Synthesis*, ed. G. Birtwistle & P.A. Subrahmanyam, Kluwer, Dortrecht; also earlier version in University of Cambridge, Computer Laboratory, Tech. Report No. 103.

Gordon *et al.* [1979]
M. Gordon, R. Milner and C. P. Wadsworth (1979) Edinburgh LCF, Lecture Notes in Computer Science No. 78, Springer-Verlag.

Hanna & Daeche [1985]
F. K. Hanna and N. Daeche (1985) 'Specification and verification using higher-order logic', in *Proceedings of the Seventh International Conference on Computer Hardware Design Languages, Tokyo, 1985.*

Halpern *et al.* [1983]
J. Halpern, Z. Manna and B. Moszkowski (1983) 'A hardware semantics based on temporal intervals', in *Proceedings of the Tenth International Colloquium on Automata, Languages and Programming, Barcelona, Spain, 1983.*

Paulson [1987]
L. C. Paulson (1987) *Logic and Computation: Interactive Proof with Cambridge LCF*, Cambridge University Press.

Pygott [1986]
C. H. Pygott (1986) 'Viper: the electronic block model', RSRE Report. No. 86006.

3 A formal model for the hierarchical design of synchronous and systolic algorithms

KEVIN McEVOY*

ABSTRACT A model of computation for the design of synchronous and systolic algorithms is presented. The model is hierarchically structured, and so can express the development of an algorithm through many levels of abstraction. The syntax of the model is based on the directed graph, and the synchronous semantics are state-transitional. A semantic representation of ripple-carries is included. The cells available in the data structure of a computation graph are defined by a graph signature. In order to develop two-level pipelining in the model, we need to express serial functions as primitives, and so a data structure may include history-sensitive functions.

A typical step in a hierarchical design is the substitution of a single data element by a string of data elements so as to refine an algorithm to a lower level of abstraction. Such a refinement is formalised through the definition of parallel and serial homomorphisms of data structures.

Central to recent work on synchronous algorithms has been the work of H. T. Kung and others on systolic design. The Retiming Lemma of Leiserson & Saxe [1981] has become an important optimisation tool in the automation of systolic design (for example, in the elimination of ripple-carries). This lemma and the Cut Theorem (Kung & Lam [1984]) are proved in the formal model.

The use of these tools is demonstrated in a design for the matrix multiplication algorithm presented in H. T. Kung [1984].

1 INTRODUCTION

A synchronous algorithm is a parallel algorithm in which processor communication is synchronised by a clock so that the processors communicate regularly in lock-step. The pattern and direction of communcation are defined by the algorithm, but the

* The author wishes to thank Prof P. M. Dew for his advice during the preparation of this chapter.

times at which communication takes place are controlled by the clock; when the clock 'ticks' precisely one data item must be transmitted along every communication channel, and communication cannot take place at any other time. A common notation for presenting such algorithms is the signal flow graph, of which Figure 3.1 is an example (taken from Kung & Lin [1984]). Rectangles represent registers (or latches). This diagrammatic representation is clear and intuitive, and in this chapter it is developed into a model of computation for the design of algorithms of arbitrary complexity over any data structure. That the array computes the particular one-dimensional convolution

$$y_{i+3} = w_0 x_i + w_1 x_{i+1} + w_2 x_{i+2} + w_3 x_{i+3}, \quad i = 0, 1, 2, \ldots,$$

is usually shown by constructing snapshots of the data as it is clocked step by step through the diagram. This method is formalised as a state transition semantics. The use of denotational semantics in similar models has been examined in Kahn [1974], Gordon [1980], Brookes [1984] and others.

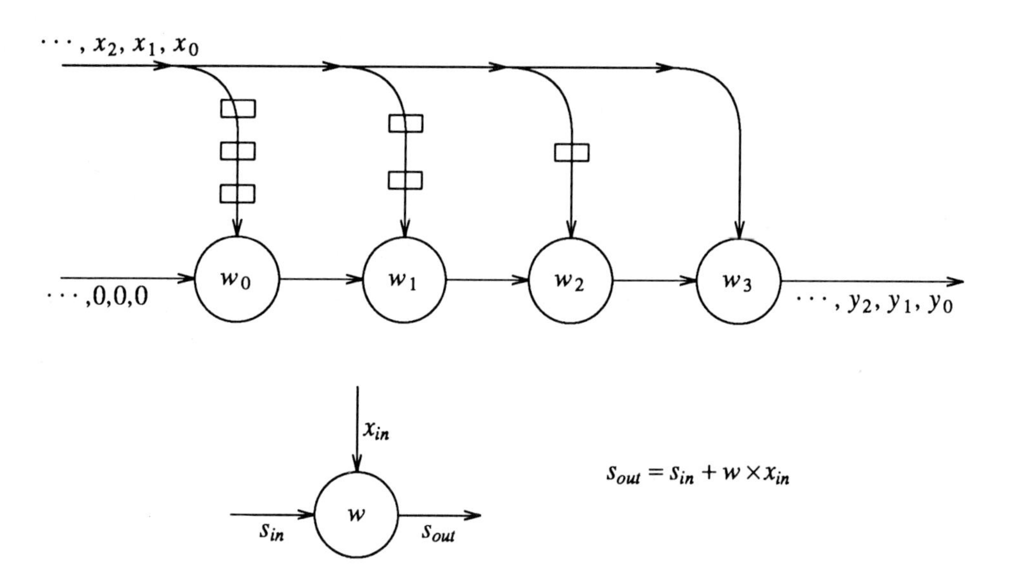

Figure 3.1. A finite impulse response filter in signal flow graph notation

The model will support hierarchical design. A very simple design refinement of the filter would be the substitution of the inner product nodes by the graph shown in Figure 3.2. This would allow pipelining of the inner product, and so reduce the period of the computation. However, a more interesting refinement is the substitution of the carry-chain adder shown in Figure 3.3 for the plus processor in Figure 3.2. The plus processor computes on natural numbers whereas the carry-chain adder computes on booleans, and the formal verification of such a substitution is one of the main concerns of this chapter. The following principle is made precise; if Figure 3.3 is shown to compute addition then substitution of Figures 3.2 and 3.3 for the inner product nodes in Figure 3.1 gives a correct design for the filter array at a lower level of abstraction.

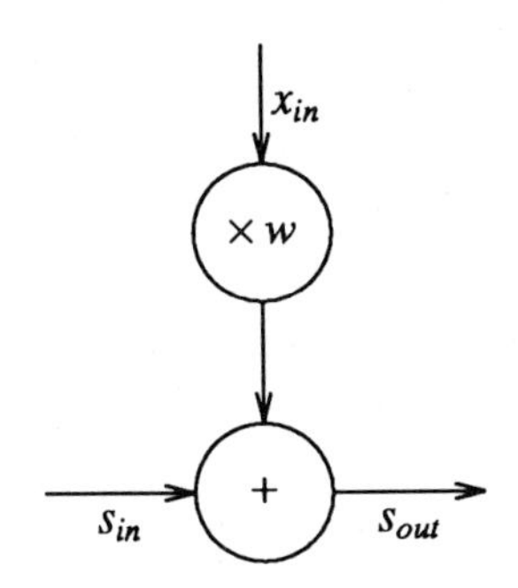

Figure 3.2. Inner product graph

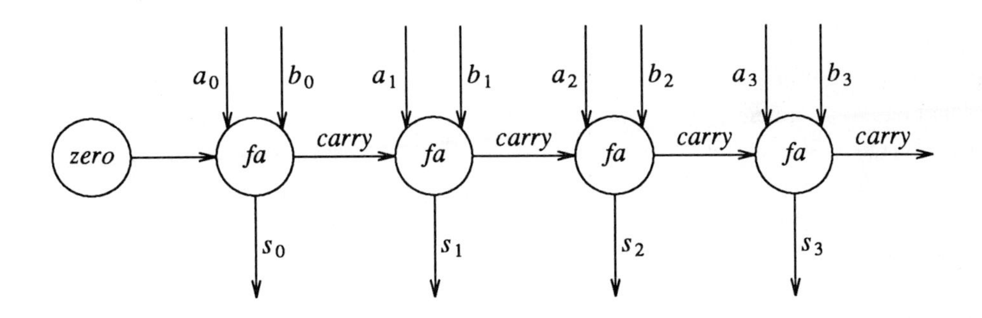

Figure 3.3. A Carry-chain adder

The syntax of the computation graph model is constructed from three types of primitive: *cells* (to evaluate the primitive functions in the data structure); *channels* (for communication of data elements); and *delay* (to provide storage and timing). A state-transition semantics is used to construct a functional description of a computation graph from the functional semantics of its cells. This functional semantics for the graph provides a formal basis for the verification of its substitution for a cell at a higher level of abstraction.

The model is powerful enough to express computations over arbitrary data structures, and in particular computations over more than one data set (e.g., the palindrome recogniser in Leiserson & Saxe [1981]). This capability is particularly useful in expressing the communication of control signals such as resets, and is realised through the representation of a data structure by a graphical presentation of a many-sorted algebra.

Section 2 contains a precise syntax for a computation graph, and in Section 3 its computational behaviour is defined in terms of a state-transition semantics. Section 4 is a short section showing how the formalism can describe the behaviour of a computation graph as a stream processor (this work is not used in the following sections). In Section 5 algorithm refinement within the model is described in terms of both graph substitution, and homomorphic refinement of data structures (parallel and serial). Section 6 contains formal proofs of the retiming results of Leiserson and Saxe and Kung and Lam (these results do not use the hierarchical definition of computation graph, and so can be read in isolation from Section 5). Section 7 is a worked example.

Computations will be carried out on *strings* of inputs, and we have used the following, fairly standard, notation. S^* is a set of all strings (or words) of elements of S (and S^+ is the set of all non-empty strings). We have used the stack operators *top*, *pop* and *push* on strings: for any $s \in S$ and $w \in S^+$, $push(s, w)$ adds s onto w, $top(w)$ is the last element pushed onto w, and $pop(w)$ is the remainder after having removed $top(w)$ from w. We also want to iterate *pop* so that $pop(w, 0) = w$ and $pop(w, n+1) = pop(pop(w, n))$. *proj* is the projection function so that $proj_k(w) = top(pop(w, m-k))$, where $m = |w|$, the length of the string w. $\hat{}$ is the concatenation operator and $\emptyset$ is the empty string.

The terms defined in the following definition describe common methods of constructing functions on strings from other functions.

Definition 1.1 Let X, Y be sets.

(i) Let $f : X \to Y$. $g : X^* \to Y^*$ is *defined by pointwise application of* f if $g(\emptyset) = \emptyset$ and $g(push(x, \overline{x})) = push(f(x), g(\overline{x}))$.
(ii) Let $\alpha : X^+ \to Y$. $\beta : X^+ \to Y^+$ is *defined by stacking the values of* α if for $|\overline{x}| = 1$, $\beta(\overline{x}) = push(\alpha(\overline{x}), \emptyset)$, and otherwise $\beta(\overline{x}) = push(\alpha(\overline{x}), \beta(pop(\overline{x})))$.
(iii) Let $a : X \to Y^+$. $b : X^* \to Y^*$ is *defined by concatenating the values of* a if $b(\emptyset) = \emptyset$, and for $\overline{x} \in X^+$, $b(push(x, \overline{x})) = a(x)\hat{}b(\overline{x})$.

2 SYNTAX OF THE MODEL

The communication pattern of the system of cells and channels is presented as a directed graph. The position of the edges (or channels) is independent of inputs, and so in terms of Cook's dichotomy of models of parallel computation (Cook [1981]) the model is of fixed structure. In the context of synchronous and systolic design this restriction is not seen as a disadvantage – systolic designs are regular and simple, and not reconfigurable. Input and output are achieved through two particular nodes (or cells), v_{in} and v_{out}, with the convention that the channels incident on v_{in} and v_{out} are the only channels which can transmit input and output data.

Definition 2.1

(i) A *digraph*, D, is a structure $\langle V, E, tl, hd\rangle$ where V and E are disjoint finite sets, $hd : E \to V$, and $tl : E \to V$. The members of V and E are called *cells* and *channels* respectively, tl is the *tail function* and hd the *head function.*

(ii) A *communication graph*, D, is a structure $\langle V, v_{in}, v_{out}, E, tl, hd\rangle$ satisfying the following conditions:

(a) $\langle V, E, tl, hd\rangle$ is a digraph;

(b) $v_{in}, v_{out} \in V$;

(c) for no channel $e \in E$ does $hd(e) = v_{in}$ or $tl(e) = v_{out}$.

(iii) We will need a notation for the channels incident into and incident from a given cell v in a communication graph D. Define $HEADS_D(v) = \{e \in E \mid hd(e) = v\}$, and $TAILS_D(v) = \{e \in E \mid tl(e) = v\}$. Also, the input channels and the output channels are given names: $D_{in} = TAILS_D(v_{in})$ and $D_{out} = HEADS_D(v_{out})$.

(iv) We will also need a notation for those cells which are to represent processors (rather than v_{in} and v_{out} which have no semantic importance in themselves). We call these cells the *interior* of the communcation graph and define $V_{interior} = V - \{v_{in}, v_{out}\}$.

The cells in the communcation graph must be chosen from a given set of cells available to the algorithm designer – this is part of the data structure of the algorithm. The syntax of the data structure of an algorithm is presented as a graph signature, where a graph signature is a graph theoretic presentation of the usual definition of signature (see, for example, Goguen *et al.* [1977]) but generalised to allow the direct treatment of many-valued functions. A graph signature is a formalism for the templates used by Masuzawa *et al.* [1983]. The signature of a data structure must contain information on the number of data sets and functions, and the types of these functions; in a graph signature the information on the type of the function will be represented as a communcation graph with only one interior cell – this cell represents the function itself, and each of the argument and value places has an associated channel. A computation graph is to compute on data elements of more than one type or sort, and so we introduce a *sort function*, st, to strongly type the channels of the graph.

Definition 2.2 An S-*sorted primitive computation graph* is a structure $\langle D, st \rangle$ satisfying the following conditions: the interior of the communication graph D has only one cell; every channel in D is either incident from v_{in} or incident into v_{out} but not both; S is a set (whose members are called sorts); and $st : E \rightarrow S$, where E is the channel set for D. A *graph signature* is a structure $\langle S, \Sigma \rangle$ where Σ is a set of S-sorted primitive computation graphs. (See Figure 3.4).

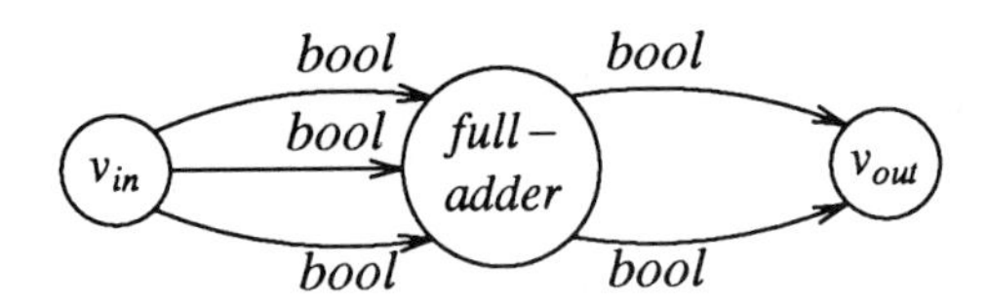

Figure 3.4. A $\{bool\}$-sorted primitive computation graph

A simple computation graph might be formed by connecting together the output channels from some primitive computation graphs to the input channels to some other primitive computation graphs. However our aim is to support the *structured* design of synchronous algorithms; a complex problem is not sensibly solved by connecting together a very large number of primitive computation graphs – one reason (of many) being that this approach introduces too much scope for error. A complex problem should be solved by connecting together a small number of computation graphs; the behaviour of these computation graphs will necessarily be only slightly less complicated than the original problem, and so these graphs must be further refined in the same way. If for some purpose it is necessary to describe the complete solution of the problem in terms of primitive computation graphs, then this task should be carried out by an automatic compiler from the top-level computation graph together with the (recursive) definition of the cells.

One form of refinement in hierarchical design is the replacement of a functional definition of a primitive operation by an algorithmic description. In the syntax of this model such a refinement is the replacement of a cell of a communication graph by another communication graph. In order to support such refinements computation graphs are defined recursively. At the bottom level in the recursion are primitive computation graphs. High level computation graphs are defined by using a *graph function*, gf, to replace a cell v of a communication graph by another computation graph $gf(v)$, and so embed communication graphs. However this substitution can only be carried out if for every input channel in v there is a corresponding input channel in $gf(v)$ of the same sort, and for every output channel from v there is a corresponding output channel in $gf(v)$ of the same sort. Mathematically this condition is stated as the existence of a *connection function* cin_v, which for every input channel e to v associates a corresponding input channel $cin_v(e)$ in $gf(v)$; the one-to-one nature of this association is expressed as the condition that cin_v is a bijection. Similarly a function $cout_v$ must exist, and these two functions must exist for every cell in the interior of the communication graph.

A computation graph is essentially a well-formed term over a graph signature.

Definition 2.3 The set $\mathbf{G}(\Gamma)$ of *computation graphs* over a graph signature $\Gamma = \langle S, \Sigma \rangle$ is inductively defined as follows.

Basis. The members of Σ are computation graphs over Γ.

Induction. A structure $G = \langle D, gf, cn, st, d \rangle$ which satisfies the following conditions is a computation graph.

(i) $D = \langle V, v_{in}, v_{out}, E, tl, hd \rangle$ is a communication graph.
(ii) $gf : V_{interior} \to \mathbf{G}(\Gamma)$ is such that for every $v \in V_{interior}$, $gf(v)$ has already been defined to be a computation graph.
(iii) cn is a pair $\langle cin, cout \rangle$ of families $\langle cin_v \mid v \in V_{interior} \rangle$ and $\langle cout_v \mid v \in V_{interior} \rangle$ of bijections. Given $v \in V_{interior}$, $cin_v : HEADS_D(v) \to gf(v)_{in}$ and $cout_v : TAILS_D(v) \to gf(v)_{out}$.
(iv) $st : E \to S$. Given $v \in V$, if st_v is the sort function of $gf(v)$ then*
  (a) for every $e \in HEADS_D(v)$, $st(e) = st_v \circ cin_v(e)$; and
  (b) for every $f \in TAILS_D(v)$, $st(f) = st_v \circ cout_v(f)$.
(v) $d : E \to \mathbf{N}$.

A signal flow graph can be represented as a picture of a communication graph (but note that a picture has geometric properties of position and distance which are not well-defined on a graph). Circles represent cells and arrows channels. To improve presentation of a picture v_{in} and v_{out} may each be represented at more than one place. Each channel is represented once only. The computation graph shown in Figure 3.5 defines the FIR filter of Figure 3.1. The semantics of this computation graph are defined in the next section, but the intended interpretations of $PLUS$ and $MULT$ are the standard addition and multiplication functions; $ZERO$ outputs the constant zero, and W_i outputs a constant weight w_i.

---

* The symbol $\circ$ is used for composition of functions.

By its nature a synchronous algorithm cannot be purely functional; there must be some element of state for the storage of data values between clock cycles. One method for introducing state into such a model such as this is to have an implicit association of one delay with every node or cell. However, we chose to model delay as a number $d(e)$ associated with every channel, with the interpretation that any data value reaching the tail of a channel e will leave the head of e exactly $d(e)$ clock cycles later. We offer the following reasons to support this approach: it provides the apparatus necessary to support the retiming transformations of Leiserson and Saxe (see Section 6); the clean functional semantics of the cells provides a simpler basis for the semantic verification of the graph for node substitution on which hierarchical design is based; the data structure retains a separate identity from the control structure (as in the relationship between an expression and the assignment statement in a programming language). (In S. Y. Kung [1984] an alternative semantics for the delay function is presented – the value of the function is interpreted as defining a number of storage registers on the channel and these registers communicate data through a hand-shaking protocol.)

(a) The $\{num\}$-sorted signature, where for every channel e, $st(e) = num$.

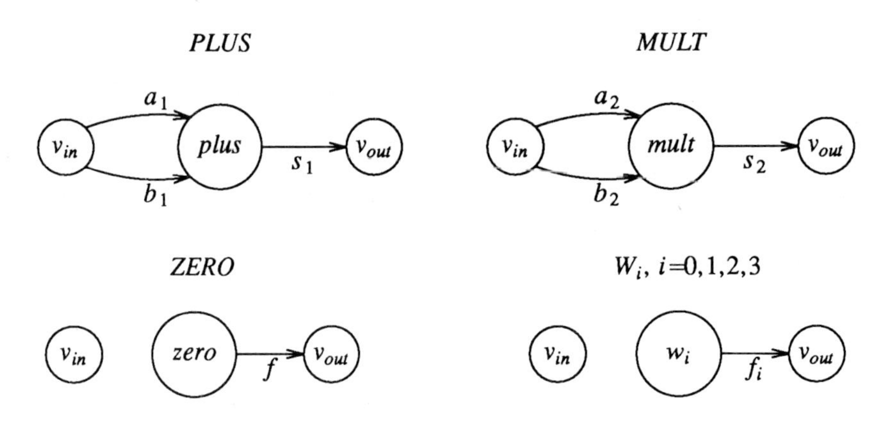

Figure 3.5. Computation graph for the FIR filter

(b) $IP_i = \langle D_i, gf_i, cn^i, st_i, d_i \rangle$.

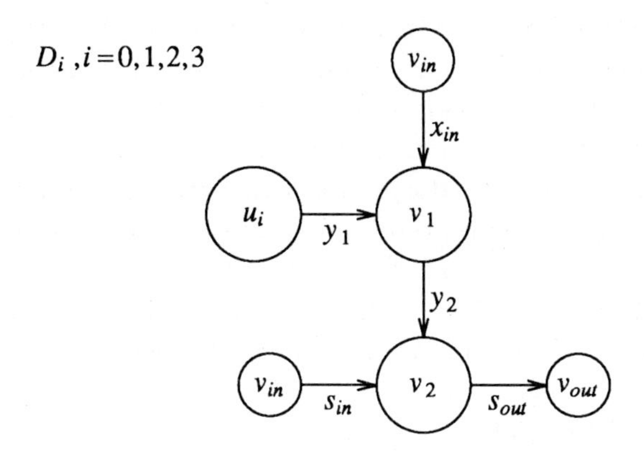

$$gf_i(v_1) = MULT \qquad gf_i(v_2) = PLUS \qquad gf_i(u_i) = W_i$$
$$cin^i_{v_1}(y_1) = a_2 \qquad cin^i_{v_1}(x_{in}) = b_2 \qquad cout^i_{v_1}(y_2) = s_2$$
$$cin^i_{v_2}(s_{in}) = a_1 \qquad cin^i_{v_2}(y_2) = b_1 \qquad cout^i_{v_2}(s_{out}) = s_1$$
$$cout^i_{u_i}(y_1) = f_i$$

For every channel e, $d_i(e) = 0$ and $st_i(e) = num$.

(c) $G = \langle D, gf, cn, st, d \rangle$.

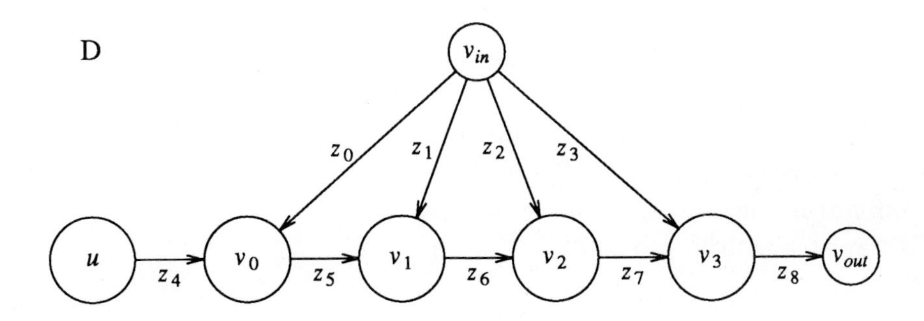

$$gf(u) = ZERO \qquad gf(v_i) = IP_i, \quad 1 = 0,1,2,3$$
$$cin_{v_j}(z_j) = x_{in} \quad cin_{v_j}(z_{j+4}) = s_{in} \quad cout_{v_j}(z_{j+5}) = s_{out} \quad j = 0,1,2,3$$
$$cout_u(z_4) = f$$
$$d(z_j) = 3 - j, \quad j = 0,1,2,3 \qquad d(z_k) = 0, \quad k = 4,5,6,7,8$$
$$st(z_m) = num, \quad m = 0,1,\ldots,8$$

Figure 3.5. (continued) Computation graph for the FIR filter

The introduction of non-zero delays introduces pipelining, and this can be used to reduce the period of a computation and increase throughput. Also, non-zero delays (i.e., storage elements) are essential for the contruction of meaningful cycles in communication graphs. Nevertheless, designers do use *ripple-carries* (paths with zero delay) to distribute communications, as they allow timing details to be kept to a minimum, and these ripple-carries are permitted in this model. Leiserson and Saxe (see Section 6) have shown that these ripple-carries can then be removed in a later (possibly automatic) stage in the design process. Indeed timing is one aspect of synchronous design where automation at an algorithmic level is possible (see Leiserson, Rose & Saxe [1983]); the model described here can support this automation.

3 SYNCHRONOUS SEMANTICS FOR THE MODEL

The computational behaviour of a computation graph G might be described informally as follows. Inputs are placed on the tails of the channels incident from v_{in}. Each such data value placed on the tail of a channel e will appear at the head of e $d(e)$ clock cycles later. At this point the data value becomes an input, or an argument, for the function associated with the cell $hd(e)$. The value of this function is defined by the semantics of the computation graph $gf \circ hd(e)$, and this value must be calculated recursively. These values are placed on the channels incident from $hd(e)$, and are pumped on through the graph until those values reaching the head of the channels incident into v_{out} are output. In each clock cycle precisely one data value flows through each delay element. At the bottom level in the recursion the semantics of a primitive computation graph is a defined part of the data structure.

The purpose of this section is to make the description in the previous paragraph precise – as a mathematical function. Two computation graphs which have the same such functional semantics can be substituted for each other, and in particular a primitive computation graph can be *implemented* at a lower level of abstraction by a non-primitive computation graph, which computes on a more basic data structure. In this way an algorithm is refined.

The basis in the inductive definition of the computational behaviour of a computation graph is to define the semantics of a primitive computation graph G. This is given as part of the data structure, and is not computed – it sets the level of abstraction of the algorithm. It is a function Ω_G from inputs to outputs, where an input is a well-typed assignment of values to the channels which are incident from v_{in}, and an output is a well-typed assignment of values to the channels which are incident into v_{out}. However, before making the definition we consider our declared aim, which is that primitive computation graphs and non-primitive computation graphs can be substituted for each other when their computational behaviour is described by the same function.

Now consider the computation graph in Figure 3.6 which is intended to be interpreted as a serial adder. In order to support hierarchical design, the model should be able to define, at a higher level of abstraction, a primitive computation graph with the same computational behaviour as the serial adder. As the carry bit is stored between clock cycles this computation graph has state, and so its behaviour cannot be represented as a simple function from inputs to outputs; the current output of a serial adder depends upon the previous inputs as well as the current input. In order to be able to describe such behaviour the semantics of a primitive computation graph is defined as a function from strings of inputs to outputs. (Indeed, in practice the production of pipelined arithmetic units has required the development of two-level pipelining – see Kung & Lam [1984] – where the functional units of the data structure are themselves pipelined and so history-sensitive.)

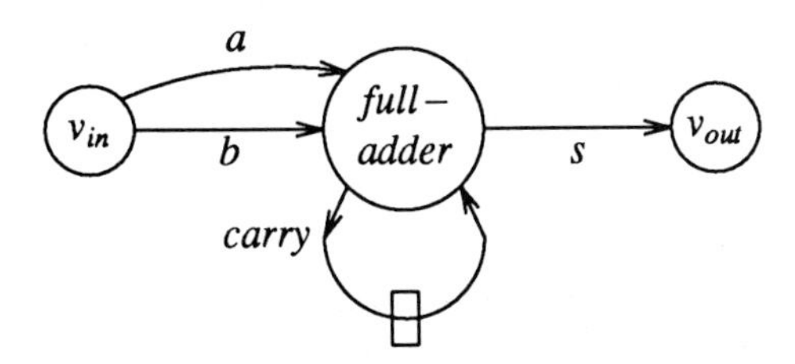

Figure 3.6. A serial adder

Definition 3.1 Let G be a computation graph over the graph signature $\langle S, \Sigma \rangle$, with communication graph D and sort function st. Let E be the channel set of D.

(i) A *sort interpretation* is a function $\Theta : S \to A$, where A is a set of sets, which are called the carrier sets of the interpretation.

(ii) Given $E' \subseteq E$, a function $g : E' \to \cup A$ is said to be *well-typed* if for every $e \in E'$, $g(e) \in \Theta \circ st(e)$.

(iii) A computation will be described as a sequence of states, where a state is an assignment of values to the storage present in the graph. Define

$$STATES_{\langle G,\Theta \rangle} = \{\tau \mid \tau : E \to \cup A \text{ and } \tau \text{ is well-typed}\}.$$

(iv) Inputs and outputs for G must be defined.

$$IN_{\langle G,\Theta \rangle} = \{\tau \mid \tau : D_{in} \to \cup A \text{ and } \tau \text{ is well-typed}\}$$

and

$$OUT_{\langle G,\Theta \rangle} = \{\tau \mid \tau : D_{out} \to \cup A \text{ and } \tau \text{ is well-typed}\}.$$

Definition 3.2 A *processor interpretation* of a graph signature $\Gamma = \langle S, \Sigma \rangle$ is a pair $\langle \Theta, \Phi \rangle$ such that Θ is a sort interpretation for S, and Φ interprets the primitive computation graphs in Σ as follows. For each $\sigma \in \Sigma$,

$$\Phi(\sigma) : IN^{+}_{\langle G,\Theta \rangle} \rightarrow OUT_{\langle G,\Theta \rangle}.$$

The pair $\langle \Gamma, \langle \Theta, \Phi \rangle \rangle$ is called an *interpreted signature.*

Note: Two special cases might need further explanation. (1) If $D_{in} = \emptyset$ then $\Phi(\sigma)$ is a function with no arguments, which is a constant, and so in this case $\Phi(\sigma) \in OUT_{\langle G,\Theta \rangle}$. (2) If $D_{out} = \emptyset$ then $\Phi(\sigma)$ is the empty function and so $\Phi(\sigma) = \emptyset$.

Definition 3.3 A *processor* is a pair $\langle G, \Psi \rangle$ where G is a computation graph, and Ψ is a processor interpretation of the graph signature of G.

Example 3.4 Consider a processor interpretation $\langle \Theta, \Phi \rangle$ for the graph signature of Figure 3.5. $\Theta(num) = \mathbf{N}$, the set of natural numbers. $((\Phi(ZERO))(I))(f) = 0$ and $((\Phi(W_i))(I))(f_i) = x_i$, where $x_i \in \mathbf{N}$, $i = 0, 1, 2, 3$.

$$((\Phi(PLUS))(I))(s_1) = top(I)(a_1) + top(I)(b_1)$$

where $+$ is the standard addition function (and *top* returns the current input). $\Phi(MULT)$ has a similar definition.

In the above example the output of the function $\Phi(PLUS)$ is independent of inputs during previous clock cycles. Delay, or storage, was defined as one of the primitives of the model, and in many cases one of the aims of hierarchical design is that the data structure at the bottom level is *delay-free.*

Definition 3.5 A processor interpretation $\langle \Theta, \Phi \rangle$ of a graph signature $\langle S, \Sigma \rangle$ is *algebraic* (or *delay-free*) if every $\sigma \in \Sigma$ satisfies the following condition. For every $I, J \in IN^{+}_{\langle \sigma,\Theta \rangle}$, if $top(I) = top(J)$ then $(\Phi(\sigma))(I) = (\Phi(\sigma))(J)$.

The synchronous semantics of the model are based on the assumption that at the end of each clock cycle there is a flow of one data element along every channel in the graph. (This pumping around of data is particularly relevant to the systolic model of computation.) Therefore at the end of each clock cycle every cell v must assign a value to the channels in $TAILS(v)$. This leaves no facility for diverging computations within cells and so all functions in the function set of a processor interpretation must be total. (A total function may contain an error value in its range, but convergence with the output of an error value is fundamentally different behaviour to that of divergence.)

We must now mathematically describe the pumping of data through a processor P, and from this derive a functional semantics

$$\Omega_P : IN_P^+ \to OUT_P.$$

(Given $P = \langle G, \langle \Theta, \Phi \rangle \rangle$ we now write IN_P for $IN_{\langle G,\Theta \rangle}$. Similarly we write $STATES_P$ and OUT_P.) In order to follow the pumping of the data through the graph we, unfortunately, need to define several auxiliary functions. The *head state function* HS and the *tail state function* TS calculate the data values to be found at the head and tail of every channel in the graph at any point in the computation. Ω_P is just the function HS with domain restricted to the output channels. The value of HS on a given channel e is the value of TS on e $d(e)$ clock cycles previously. (See part (2a) of the following definition.) The computation is pumped along by the input I, and so the length of a computation is determined by $|I|$, the length of I. So '$d(e)$ clock cycles previously' is interpreted as $pop(I, d(e))$. The value of TS must be defined inductively, and by cases. Initially, when $I = \emptyset$, the values held in these delays, or registers, are not determined, and so could be anything. (See part (2bi) of the following definition.) If $I \neq \emptyset$ there are two cases. The value of TS on the input channels is defined by the input I. (See part (2b$ii\alpha$).) The value of TS on any other channel e is determined by the processor $\langle gf \circ tl(e), \Psi \rangle$. (See part (2b$ii\beta$).) The function $\Omega_{\langle gf \circ tl(e), \Psi \rangle}$ must be evaluated on its arguments (this function has, by induction, already been defined). But what are the arguments to be passed to this function? They are the values of HS (both current and previous values) on the channels which are incident into $tl(e)$. These values are available, but unfortunately they are not in the correct form to be presented to $\Omega_{\langle gf \circ tl(e), \Psi \rangle}$ and so auxiliary functions $PRED$ and *pred* must be defined to collect together the inputs into $tl(e)$ and transform them into inputs to $gf \circ tl(e)$ (using the function *cin*). This definition uses some notation on strings which was defined in the introduction (e.g., *pop* and *stacking the values of*); note that $pop(\emptyset) = \emptyset$.

Definition 3.6 For any processor $P = \langle G, \Psi \rangle$ define a synchronous computation function

$$\Omega_P : IN_P^+ \to OUT_P$$

by recursion on the definition of G as follows. Let $\Gamma = \langle S, \Sigma \rangle$ be the signature of G with processor interpretation $\Psi = \langle \Theta, \Phi \rangle$. $\Omega_P(I)$ is defined to be the restriction of $HS(I)$ to domain G_{out}.

(1) *Basis.* If $G \in \Sigma$ then $\Omega_P = \Phi(G)$.
(2) *Induction.* Otherwise let $G = \langle D, gf, cn, st, d \rangle$ and $D = \langle V, v_{in}, v_{out}, E, tl, hd \rangle$. As well as $TS : IN_P^* \to STATES_P$ and $HS : IN_P^+ \to STATES_P$ define families $\langle PRED_v \mid v \in V_{interior} \rangle$ and $\langle pred_v \mid v \in V_{interior} \rangle$ of auxiliary functions such

that for every $v \in V_{interior}$, $PRED_v : IN_P^+ \rightarrow IN_{\langle gf(v),\Psi\rangle}$ and $pred_v : IN_P^+ \rightarrow IN_{\langle gf(v),\Psi\rangle}^+$.

(2a) $HS(I)$ is defined by $(HS(I))(e) = (TS \circ pop(I, d(e)))(e)$, for any $e \in E$.

(2b) $TS(I)$ is defined by recursion on $|I|$.

(2bi) *Basis.* For any $e \in E$, $(TS(\emptyset))(e) \in \Theta \circ st(e)$ (a non-deterministic choice).

(2bii) *Induction.*

(2b$ii\alpha$) If $e \in D_{in}$ then $(TS(I))(e) = (top(I))(e)$.

(2b$ii\beta$) Otherwise $e \in TAILS_D(v)$ for some $v \in V_{interior}$ and in this case define

$$(TS(I))(e) = (\Omega_{\langle gf(v),\Psi\rangle} \circ pred_v(I))(cout_v(e))$$

where for each $e' \in gf(v)_{in}$,

$$(PRED_v(I))(e') = (HS(I))(cin_v^{-1}(e'))$$

and $pred_v$ is defined by stacking the values of $PRED_v$.

This is a recursive definition which describes a computation as a sequence of states as defined by HS. There is a mutual recursion in the definitions of HS and TS and therefore it is necessary to check that this recursive definition is well-founded. The linc

$$(TS(I))(e) = (\Omega_{\langle gf(v),\Psi\rangle} \circ pred_v(I))(cout_v(e))$$

defines $TS(I)$ in terms of $HS(I)$ (as Ω is a restriction of HS), but here $gf(v)$ is strictly lower than G in the depth ordering implicit in the inductive definition of computation graph. The other line in which the mutual recursion is introduced is

$$(HS(I))(e) = (TS \circ pop(I, d(e)))(e).$$

There is no problem here as long as $d(e) \neq 0$, as then the recursion is well-founded on the ordering given by $|I|$. If $d(e) = 0$ then the situation is more problematic, and the recursion must be defined along the ripple-carry, starting at the beginning of the ripple-carry – i.e., along the *Rank* of e where this is defined to be the length of the longest path which has head $tl(e)$ and in which every channel has delay zero. (See Definition 5.8) This leaves us with the problem where there is no such longest path – i.e., the graph contains a cycle in which every channel has delay zero. In such a graph the recursive definition of Ω is *not* well-founded; but this is natural and proper as such a graph models a clocked circuit which contains a cycle with no latches or registers, and such a circuit will exhibit race conditions.

The phrase $(TS(\emptyset))(e) \in \Theta \circ st(e)$ introduces non-determinism into Definition 3.6. This reflects the fact that at the start of a computation one has no control over what values are stored in the registers. It follows that the function $\Omega_P(I)$ is non-deterministic during its initialisation period. This non-determinism could be eliminated by introducing another parameter (an initial state) into the functionality of Ω; however, it would then be the case that the equality of Ω_P and Ω_Q would require the equality of the underlying computation graphs, and this condition would be too strong for our purposes (in particular the computational equivalence of a primitive computation graph with a non-primitive computation graph).

Definition 3.7 Let $P = \langle G, \Psi \rangle$ be a processor with communication graph D. The *initialisation period of P with input $I \in IN_P^+$* is the least $n \in \mathbf{N}$ such that $1 \leq n \leq |I|$ and

$$(\Omega_P \circ pop(I,\ |I| - n))(e)$$

is uniquely determined for every $e \in D_{out}$ (if such an n exists).

We can now define computational equivalence of processors. In order to simplify the work in Section 5, computational equivalence is given the strongest, and simplest, of definitions – that of identical computational behaviour. In practice this may not be the most useful of definitions, and in Section 6 we give a weaker definition which allows equivalent processors to have different initialisation periods and latencies. A weaker definition should also allow isomorphism (rather than identity) of the underlying communication graphs.

Definition 3.8 The processors $P = \langle G, \Psi \rangle$ and $Q = \langle H, \Psi' \rangle$ are said to be *computationally equivalent* if $\Omega_P = \Omega_Q$.

4 COMPUTATIONS ON STRINGS AND STREAMS

In this short section we discuss briefly how Ω_P can be used as a basis for the definition of more general forms of computational behaviour on computation graphs. The function Ω_P defines the result of a computation to be the output during the last clock cycle. However, in the majority of synchronous algorithms every output is of interest, and so, as a computation progresses through time, it is natural to describe the computational behaviour of a synchronous processor as a monotone function on strings.

Definition 4.1 For any processor P define a function $\Omega_P^{string} : IN_P^+ \to OUT_P^+$ by stacking the values of Ω_P.

The phrase 'defined by stacking the values of' was defined in the introduction.

Furthermore, in many applications (such as signal processing) the input to a computation is a stream (a sequence indexed by $\mathbf{N}$) rather than a string. We can describe such infinite computations on a processor by using computations on strings as a basis to define a continuous function on streams.

Definition 4.2 For any set S, let S^∞ be the set of all countably infinite words on S, and let $\sqsubseteq$ be the standard complete partial ordering on $S^\infty \cup S^*$ defined by $\overline{s} \sqsubseteq \overline{t}$ if and only if $\overline{s}$ is an initial segment of $\overline{t}$. Let $\sqcup$ denote the least upper bound on this ordering (when it exists).

Definition 4.3 For any processor P define a function $\Omega_P^{stream} : IN_P^\infty \to OUT_P^\infty$ by

$$\Omega_P^{stream}(J) = \sqcup\{\Omega_P^{string}(I) \mid I \sqsubseteq J \;\&\; I \in IN_P^+\}.$$

(The existence of the least upper bound for every $J \in IN_P^\infty$ is a consequence of the monotonicity of Ω_P^{string} and the completeness of the partial order.)

5 ALGORITHM REFINEMENT

A refinement step in a hierarchical design of an algorithm usually consists of a homomorphic refinement of the data structure. There are two possibilities: a refinement of the data elements or a refinement of the functions on that data.

The refinement of a function in the data structure implies its replacement by an algorithm which computes that function. In this model such a refinement is the substitution of a primitive computation graph by another computation graph which may have a different signature. Using Definition 3.8 we can now make precise the statement that the computation graph of Figure 3.5(b) is computationally equivalent (under the interpretation of Example 3.4) to an inner product primitive computation graph; this allows us to verify that the substitution of Figure 3.5(b) for the inner product primitive computation graph in any processor is indeed a valid refinement. Such a substitution we shall call a function refinement.

Definition 5.1 Let $\Gamma = \langle S, \Sigma \rangle$ and $\Gamma' = \langle S', \Sigma' \rangle$ be graph signatures with processor interpretations Ψ and Ψ'. If $\eta : \Sigma \to \mathbf{G}(\Gamma')$ is such that for every $\sigma \in \Sigma$, $\langle \sigma, \Psi \rangle$ is computationally equivalent to $\langle \eta(\sigma), \Psi' \rangle$, then η is said to be a *function-refinement* from $\langle \Gamma, \Psi \rangle$ into $\langle \Gamma', \Psi' \rangle$ and $FunRef_\eta : \mathbf{G}(\Gamma) \to \mathbf{G}(\Gamma')$ is defined by recursion on the definition of G as follows. If $G \in \Sigma$ then $FunRef_\eta(G) = \eta(G)$. Otherwise G is of the form $\langle D, gf, cn, st, d \rangle$ and $FunRef_\eta(G) = \langle D, gf', cn, st, d \rangle$ where $gf'(v) = FunRef_\eta \circ gf(v)$.

Lemma 5.2 *If η is a function-refinement from $\langle \Gamma, \Psi \rangle$ into $\langle \Gamma', \Psi' \rangle$ then for any $G \in \mathbf{G}(\Gamma)$, the processor $\langle FunRef_\eta(G), \Psi' \rangle$ is computationally equivalent to $\langle G, \Psi \rangle$.*

Proof (by induction on the definition of G). If G is primitive then there is nothing to prove. Otherwise $G = \langle D, gf, cn, st, d\rangle$ and $FunRef_\eta(G) = \langle D, gf', cn, st, d\rangle$ where for all $v \in V_{interior}$ (the cell set), $gf'(v) = FunRef_\eta(gf(v))$. In the induction step the inductive hypothesis gives us that that $\langle gf(v), \Psi\rangle$ and $\langle gf'(v), \Psi'\rangle$ are computationally equivalent, and so the definitions of $\Omega_{\langle G,\Psi\rangle}$ and $\Omega_{\langle FunRef_\eta(G),\Psi'\rangle}$ are identical.

The first stage in the refinement of the data elements is the refinement of the signature. Each data element is to be expressed at a lower level of abstraction by a string (or word) of data elements and this string can be transmitted either in parallel or in series. Let the length of the string be n. If the data is to be transmitted in parallel then the essential refinement of the graph is the replacement of each channel by n parallel channels. In a serial refinement the delay on each channel is increased by a factor of n.

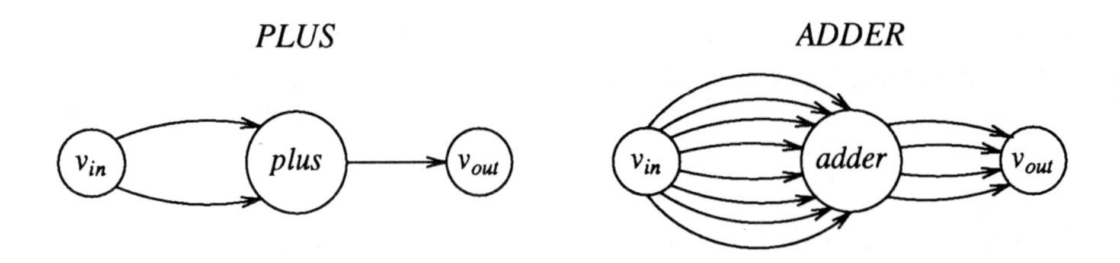

Figure 3.7. A simple parallel refinement

Figure 3.7 shows a simple parallel refinement of the plus cell of Figure 3.5(a) to a primitive computation graph for a (specification of) parallel adder on four-bit words. Section 7 contains examples of the refinements defined in this section.

Definition 5.3 Given a graph signature $\Gamma = \langle S, \Sigma\rangle$, a sort set S', and a function $\lambda : S \to (S')^+$, define the *parallel λ-refinement* of Γ to be the S'-sorted signature $\langle S', \{ParRef_\lambda(\sigma) \mid \sigma \in \Sigma\}\rangle$ defined as follows. Let $\sigma = \langle D, st\rangle \in \Sigma$ and $D = \langle V, v_{in}, v_{out}, E, tl, hd\rangle$. Then $ParRef_\lambda(\sigma) = \langle\langle V, v_{in}, v_{out}, E', tl', hd'\rangle, st'\rangle$, where

$$E' = \{e_k \mid e \in E \;\&\; 1 \leq k \leq |\lambda(st(e))|\},$$

and for all $e_k \in E'$, $tl'(e_k) = tl(e)$, $hd'(e_k) = hd(e)$, and $st'(e_k) = proj_k \circ \lambda \circ st(e)$.

If Figure 3.7(a) is interpreted as addition on numbers, and Figure 3.7(b) is interpreted as a functional description of a parallel adder on booleans, then we need to be able to express the fact that Figures 3.7(a) and 3.7(b) are in some way computationally equivalent. The mathematical tool for the expression of semantic equivalence is the morphism. The following notation allows the definition of a homomorphism over a signature refinement to be stated more concisely.

Definition 5.4 If $\Theta : S \to A$ is a sort interpretation and $w \in S^+$ then define $\vec{\Theta}(w)$ to be the set product of A defined by $\vec{\Theta}(s_1 s_2 ... s_n) = \Theta(s_1) \times \Theta(s_2) \times \ldots \times \Theta(s_n)$.

Definition 5.5 Let $\langle \Gamma, \langle \Theta, \Phi \rangle \rangle$ and $\langle \Gamma', \langle \Theta', \Phi' \rangle \rangle$ be interpreted signatures such that Γ' is the parallel λ-refinement of $\Gamma = \langle S, \Sigma \rangle$ for a given function λ. Let the sets of carrier sets of the interpretations be A and A'. A *parallel λ-homomorphism* is a family $\langle h_s \mid s \in S \rangle$ of functions $h_s : \Theta(s) \to \vec{\Theta}'(\lambda(s))$ which satisfy the following conditions for every $\sigma \in \Sigma$. Let D be the communication graph of G and st the sort function. Let $e \in D_{out}$, $s = st(e)$ and $1 \leq k \leq |\lambda(s)|$. Then, for every $I \in IN_P^+$,

$$((\Phi' \circ ParRef_\lambda(\sigma))(par_h(I)))(e_k) = proj_k \circ h_s \circ ((\Phi(\sigma))(I))(e),$$

where $P = \langle \sigma, \langle \Theta, \Phi \rangle \rangle$, $Q = \langle ParRef_\lambda(\sigma), \langle \Theta', \Phi' \rangle \rangle$, and $par_h : IN_P^* \to IN_Q^*$ is defined as follows. First define $PAR_h : IN_P \to IN_Q$ by, for $J \in IN_P$, $f \in D_{in}$, $t = st(f)$, $1 \leq m \leq |\lambda(t)|$,

$$(PAR_h(J))(f_m) = proj_m \circ h_t \circ J(f).$$

Then par_h is defined by pointwise application of PAR_h. (The phrase 'defined by pointwise application of' was defined in the introduction.) If such a homomorphism exists then $\langle \Gamma', \langle \Theta', \Phi' \rangle \rangle$ is said to be a *parallel data refinement of* $\langle \Gamma, \langle \Theta, \Phi \rangle \rangle$.

We can now express the equivalence of Figures 3.7(a) and 3.7(b), but we need to know that this semantic equivalence is retained when arbitrary computation graphs are built from these two signatures. The domain of the function $ParRef_\lambda$ can easily be extended to all computation graphs over S-sorted signatures.

Definition 5.6 If Γ' is the parallel λ-refinement of the signature Γ, define the function $ParRef_\lambda : \mathbf{G}(\Gamma) \to \mathbf{G}(\Gamma')$ by recursion on the definition of G as follows. If G is primitive then $ParRef_\lambda(G)$ was defined in Definition 5.3. Otherwise let $G = \langle D, gf, cn, st, d \rangle$ and define $ParRef_\lambda(G) = \langle D', gf', CN, st', d' \rangle$ as follows. Define D' and st' as in Definition 5.3. $d'(e_k) = d(e)$ for all $e_k \in E'$. For all $v \in V_{interior}$, $gf'(v) = ParRef_\lambda \circ gf(v)$. Also, for all $v \in V_{interior}$, define $CIN_v : HEADS_{D'}(v) \to gf'(v)_{in}$ by $CIN_v(e_k) = (cin_v(e))_k$, and $COUT_v : TAILS_{D'}(v) \to gf'(v)_{out}$ by $COUT_v(e_k) = (cout_v(e))_k$.

$ParRef_\lambda(G)$ is called the *parallel λ-refinement of G*.

Definition 5.7 Let $P = \langle G, \Psi \rangle$ and $P' = \langle G', \Psi' \rangle$ be processors with signatures Γ and Γ' such that G' is a parallel refinement of G and the interpreted signature $\langle \Gamma', \Psi' \rangle$ is a parallel data refinement of $\langle \Gamma, \Psi \rangle$. Then P' is said to be a *parallel processor refinement of* P. The following lemma is proved by induction on the definition of Ω. In the proof we need to make explicit the orderings on which the inductions are made, and so we state the following definition.

Definition 5.8 For any computation graph $G = \langle D, gf, cn, st, d \rangle$ with

$$D = \langle V, v_{in}, v_{out}, E, tl, hd \rangle$$

define $Rank_G : V \to \mathbf{N}$ by

$$Rank_G(v) = \max\{0\} \cup \{|P| : P \in Paths_D \ \&\ hd^*(P) = v \ \&\ d^*(P) = 0\}.$$

That is, the rank of v is the length of the longest path which ends at v and in which every channel has delay zero, or zero if no such path exists (see Definition 6.4). Also define $Erank_G : E \to \mathbf{N}$ by $Erank_G(e) = Rank_G \circ tl(e)$.

Lemma 5.9 If $Q = \langle H, \Psi' \rangle$ is a parallel processor refinement of $P = \langle G, \Psi \rangle$ then Q is computationally a homomorphic refinement of P in the sense of Definition 5.5. That is, let D be the communication graph of G and st the sort function. Let $\langle h_s \mid s \in S \rangle$ be the parallel λ-homomorphism defining the data refinement, and let $e \in D_{out}$, $s = st(e)$ and $1 \leq k \leq |\lambda(s)|$. Then, for every $I \in IN_P^+$,

$$proj_k \circ h_s \circ (\Omega_P(I))(e) \subseteq (\Omega_Q \circ par_h(I))(e_k).$$

par_h was defined in Definition 5.5.

Proof (by induction on the definition of Ω_P). Let $\langle S, \Sigma \rangle$ be the signature of G. (1) If $G \in \Sigma$ then $\Omega_P = \Phi(G)$ and $\Omega_Q = \Phi' \circ ParRef_\lambda(G)$, and so the result follows from Definition 5.5. (2) Otherwise let $G = \langle D, gf, cn, st, d \rangle$, $H = \langle D', gf', CN, st', wt \rangle$ and $D = \langle V, v_{in}, v_{out}, E, tl, hd \rangle$. Given $I \in IN_P^*$, $e \in E$, $s = st(e)$ and $1 \leq k \leq |\lambda(s)|$, we prove by induction on $|I|$ that

(A) $proj_k \circ h_s \circ (TS_P(I))(e) \subseteq (TS_Q \circ par_h(I))(e_k)$; and

(B) $proj_k \circ h_s \circ (HS_P(I))(e) \subseteq (HS_Q \circ par_h(I))(e_k)$, or $I = \emptyset$.

The basis case is $I = \emptyset$. By Definition 3.6,

$$\begin{aligned} proj_k \circ h_s \circ (TS_P(\emptyset))(e) &= proj_k \circ h_s \circ \Theta \circ st(e) \\ &\subseteq \Theta' \circ proj_k \circ \lambda \circ st(e) \quad \text{by definition of } h_s \\ &= \Theta' \circ st(e_k) \\ &= (TS_Q(\emptyset))(e_k) \\ &= (TS_Q \circ par_h(\emptyset))(e_k). \end{aligned}$$

The inductive step on $|I|$ is proved in (I) and (II). Assume $I \neq \emptyset$.

(I) If $d(e) \neq 0$ then

$$\begin{aligned} proj_k \circ h_s \circ (HS_P(I))(e) &= proj_k \circ h_s \circ (TS_P \circ pop(I, d(e)))(e) \\ &= (TS_Q \circ par_h \circ pop(I, d(e)))(e_k) \quad \text{by induction} \\ &= (TS_Q \circ pop(par_h(I), d(e)))(e_k) \\ &\qquad \text{by pointwise definition of } par_h \\ &= (HS_Q \circ par_h(I))(e_k) \quad \text{as } d(e) = wt(e_k) \end{aligned}$$

(II) Otherwise the proof is by induction on $Erank_G(e)$.

(IIa) Let $Erank_H(e) = 0$.

(i) If $e \in D_{in}$ then

$$\begin{aligned} proj_k \circ h_s \circ (TS_P(I))(e) &= proj_k \circ h_s \circ (top(I))(e) \\ &= (PAR_h \circ top(I))(e_k) \\ &= (top \circ par_h(I))(e_k) \\ &\qquad \text{by pointwise definition of } par_h \\ &= (TS_Q \circ par_h(I))(e_k) \end{aligned}$$

(ii) Otherwise let $v \in V_{interior}$ be such that $e \in TAILS_D(v)$ and (assuming that $HEADS(v) \neq \emptyset$) let $f \in (gf(v))_{in}$. As $Erank_H(e) = 0$, $wt(cin_v^{-1}(f)) \neq 0$ and so, by (I),

$$\begin{aligned} (PAR_h^v \circ PRED_v^P(I))(f_k) &= proj_k \circ h_s \circ (PRED_v^P(I))(f) \\ &\qquad \text{by definition of } PAR_h \\ &= proj_k \circ h_s \circ (HS_P(I))(cin_v^{-1}(f)) \quad \text{by definition of } PRED \\ &\subseteq (HS_Q \circ par_h(I))((cin_v^{-1}(f))_k) \qquad (\dagger) \\ &= (HS_Q \circ par_h(I))(CIN_v^{-1}(f_k)) \\ &= (PRED_v^Q \circ par_h(I))(f_k) \quad \text{by definition of } PRED \end{aligned}$$

Therefore, by induction on $|I|$,

$$par_h^v \circ pred_v^P(I) \subseteq pred_v^Q \circ par_h(I)$$

and so

$$\begin{aligned} &proj_k \circ h_s \circ (TS_P(I))(e) \\ &\quad = proj_k \circ h_s \circ (\Omega_{\langle gf(v), \Psi \rangle} \circ pred_v^P(I))(cout_v(e)) \\ &\quad \subseteq (\Omega_{\langle gf'(v), \Psi' \rangle} \circ par_h^v \circ pred_v^P(I))((cout_v(e))_k) \quad \text{by induction} \\ &\quad \subseteq (\Omega_{\langle gf'(v), \Psi' \rangle} \circ pred_v^Q \circ par_h(I))(COUT_v(e_k)) \\ &\quad = (TS_Q \circ par_h(I))(e_k) \end{aligned}$$

Given $d(e) = 0$,

$$\begin{aligned} proj_k \circ h_s \circ (HS_P(I))(e) &= proj_k \circ h_s \circ (TS_P(I))(e) \quad \text{as } d(e) = 0 \\ &\subseteq (TS_Q \circ par_h(I))(e_k) \quad \text{by (i) or (ii)} \\ &= (HS_Q \circ par_h(I))(e_k) \quad \text{as } wt(e) = 0 \end{aligned}$$

(IIb) The induction step on $Erank_H(e)$ is similar to the basis. (A) is proved as in (ii) above; if $d \circ cin_v^{-1}(f) = 0$ then $Erank \circ cin_v^{-1}(f) < Erank(e)$ and so (†) follows by induction. (B) is proved as above.

In a serial refinement a data element is refined to a string of data elements and the elements of the string are transmitted along the same channel over a number of clock cycles. So in this case the signature is not changed, only the interpretation and the delays.

One advantage of serial refinements as compared to parallel refinements is that individual elements of a given data set may be refined to strings of different lengths. For example, whereas the given parallel adder can add only four-bit words, a serial adder is able to add any two words of the same length. However, the exploitation of this extra power requires a greater sophistication than we present here. In the case of a serial adder the two words being added must be of the same length and so the representation of a given number depends upon the number it is being added to; for example, the addition of eight and three is represented by $\langle 1,0,0,0\rangle \oplus \langle 0,0,1,1\rangle$ and the addition of one and three by $\langle 0,1\rangle \oplus \langle 1,1\rangle$. We shall assume that in a serial refinement of a data set all elements are represented by strings of the same length. Also, to keep timing considerations simple we shall assume that the lengths of refinement vectors of elements from different data sets are the same.

Definition 5.10 Let $\langle \Theta, \Phi\rangle$ and $\langle \Theta', \Phi'\rangle$ be processor interpretations for the signature $\Gamma = \langle S, \Sigma\rangle$, and let the sets of carrier sets be A and A'. Given $n \in \mathbf{N}$, $n \neq 0$, a *serial homomorphism of length n* is a family $\langle h_s \mid s \in S\rangle$ of functions $h_s : \Theta(s) \to \vec{\Theta}'(s^n)$ which satisfy the following conditions for every $\sigma \in \Sigma$ (where s^n is a string of s's of length n). Let D be the communication graph of G and st the sort function. Let $e \in D_{out}$, $s = st(e)$ and $1 \leq k \leq n$.

(1) Then for every $I \in IN_P^+$,

$$((\Phi'(\sigma))(pop(ser_h(I), n-k)))(e) = proj_k \circ h_s \circ ((\Phi(\sigma))(I))(e),$$

where $P = \langle \sigma, \langle \Theta, \Phi\rangle\rangle$, $Q = \langle \sigma, \langle \Theta', \Phi'\rangle\rangle$, and $ser_h : IN_P^+ \to IN_Q^+$ is defined as follows. First define $SER_h : IN_P \to IN_Q^+$. For $J \in IN_P$, $f \in D_{in}$, $t = st(f)$,

$SER_h(J)$ is a string of length n such that for $1 \leq m \leq n$,

$$(proj_m \circ SER_h(J))(f) = proj_m \circ h_t \circ J(f).$$

Then ser_h is defined by concatenating the values of SER_h. (The phrase 'defined by concatenating the values of' was defined in the introduction.)

(2) If such a homomorphism exists then $\langle \Gamma, \langle \Theta', \Phi' \rangle \rangle$ is said to be a *serial data refinement of* $\langle \Gamma, \langle \Theta, \Phi \rangle \rangle$ *of length n.*

Definition 5.11 For any computation graph G and $n \in \mathbf{N}$, $n \neq 0$, define the computation graph $SerRef_n(G)$ by recursion on the definition of G as follows. If G is primitive then $SerRef_n(G) = G$; otherwise $G = \langle \langle V, v_{in}, v_{out}, E, tl, hd \rangle, gf, cn, st, d \rangle$ and in this case set $SerRef_n(G) = \langle \langle V, v_{in}, v_{out}, E, tl, hd \rangle, gf', cn, st, wt \rangle$ where $wt : E \to \mathbf{N}$ is defined by $wt(e) = n \times d(e)$, and for $v \in V_{interior}$, $gf'(v) = SerRef_n \circ gf(v)$.

$SerRef_n(G)$ is called the *serial refinement of G of length n.*

Definition 5.12 Let $P = \langle G, \Psi \rangle$ and $P' = \langle G', \Psi' \rangle$ be processors with signature Γ such that G' is the serial refinement of G of length n and $\langle \Gamma, \Psi' \rangle$ is a serial data refinement of $\langle \Gamma, \Psi \rangle$ of length n. Then P' is said to be a *serial refinement of P.*

Lemma 5.13 If the processor $Q = \langle H, \Psi' \rangle$ is a serial refinement of length n of the processor $P = \langle G, \Psi \rangle$ then Q is computationally a homomorphic refinement of P in the sense of Definition 5.10. That is, let D be the communication graph of G and st the sort function. Let $e \in D_{out}$, $s = st(e)$ and $1 \leq k \leq n$. Then, for every $I \in IN_P^+$,

$$proj_k \circ h_s \circ (\Omega_P(I))(e) \subseteq (\Omega_Q \circ pop(ser_h(I), n-k))(e).$$

ser_h was defined in Definition 5.10.

Proof (by induction on the definition of Ω_P). Let $\langle S, \Sigma \rangle$ be the signature of G. (1) If $G \in \Sigma$ then $\Omega_P = \Phi(G)$ and $\Omega_Q = \Phi'(G)$, and so the result follows from Definition 5.10. (2) Otherwise let $G = \langle D, gf, cn, st, d \rangle$ and $H = \langle D, gf', cn, st, wt \rangle$ and $D = \langle V, v_{in}, v_{out}, E, tl, hd \rangle$. Given $I \in IN_P^*$, $e \in E$ and $s = st(e)$ we prove by induction on $|I|$ that, for $1 \leq k \leq n$,

(A) $proj_k \circ h_s \circ (TS_P(I))(e) \subseteq (TS_Q \circ pop(ser_h(I), n-k))(e)$; and

(B) $proj_k \circ h_s \circ (HS_P(I))(e) \subseteq (HS_Q \circ pop(ser_h(I), n-k))(e)$, or $I = \emptyset$.

The basis case is $I = \emptyset$.

$$\begin{aligned} proj_k \circ h_s \circ (TS_P(\emptyset))(e) &= proj_k \circ h_s \circ \Theta \circ st(e) \\ &\subseteq \Theta' \circ st(e) \quad \text{by definition of } h_s \\ &= (TS_Q(\emptyset))(e) \\ &= (TS_Q \circ pop(ser_h(\emptyset), n-k)(e). \end{aligned}$$

The inductive step on $|I|$ is proved in (I) and (II). Assume $I \neq \emptyset$.

(I) If $d(e) \neq 0$ then

$$\begin{aligned} proj_k \circ h_s \circ (HS_P(I))(e) &= proj_k \circ h_s \circ (TS_P \circ pop(I, d(e)))(e) \\ &\subseteq (TS_Q \circ pop(ser_h \circ pop(I, d(e)), n-k))(e) \quad \text{by induction} \\ &= (TS_Q \circ pop(pop(ser_h(I), n \times d(e)), n-k))(e) \\ &\quad \text{by definition of } ser_h \\ &= (TS_Q \circ pop(pop(ser_h(I), n-k), wt(e)))(e) \\ &= (HS_Q \circ pop(ser_h(I), n-k))(e) \end{aligned}$$

(IIa) Let $Erank_H(e) = 0$.

(i) If $e \in D_{in}$ then

$$\begin{aligned} proj_k \circ h_s \circ (TS_P(I))(e) &= proj_k \circ h_s \circ (top(I))(e) \\ &= (proj_k \circ SER_h \circ top(I))(e) \quad \text{by definition of } SER_h \\ &= (top \circ pop(SER_h \circ top(I), n-k))(e) \\ &\quad \text{by definition of } proj_k \\ &= (top \circ pop(ser_h(I), n-k))(e) \quad \text{as } |SER_h(top(I))| = n \\ &= (TS_Q \circ pop(ser_h(I), n-k))(e) \end{aligned}$$

(ii) Otherwise let $v \in V_{interior}$ be such that $e \in TAILS_D(v)$ and let $f \in (gf(v))_{in}$.

$$\begin{aligned} (proj_k \circ SER_h^{gf(v)} \circ PRED_v^G(I))(f) & \\ &= proj_k \circ h_t \circ (PRED_v^P(I))(f) \\ &= proj_k \circ h_t \circ (HS_P(I))(cin_v^{-1}(f)) \\ &\subseteq (HS_Q \circ pop(ser_h(I), n-k))(cin_v^{-1}(f)) \quad \text{by(I)} \\ &= (PRED_v^Q \circ pop(ser_h(I), n-k))(f) \end{aligned}$$

Therefore,

$$\begin{aligned} top \circ pop(SER_h^{gf(v)} \circ PRED_v^P(I), n-k) & \\ &= proj_k \circ SER_h^{gf(v)} \circ PRED_v^P(I) \\ &\subseteq PRED_v^Q \circ pop(ser_h(I), n-k) \end{aligned}$$

So, by induction on $|I|$,

$$pop(ser_h^{gf(v)} \circ pred_v^P(I), n-k) \subseteq pred_v^Q \circ pop(ser_h(I), n-k)$$

It follows that

$$\begin{aligned}
&proj_k \circ h_t \circ (TS_P(I))(e) \\
&\quad = proj_k \circ h_t \circ (\Omega_{\langle gf(v),\Psi\rangle} \circ pred_v^P(I))(cn_v^{out}(e)) \\
&\quad \subseteq (\Omega_{\langle gf'(v),\Psi'\rangle} \circ pop(ser_h^v \circ pred_v^P(I), n-k)(cn_v^{out}(e) \\
&\qquad \text{by induction} \\
&\quad \subseteq (\Omega_{\langle gf'(v),\Psi'\rangle} \circ pred_v^H \circ pop(ser_h(I), n-k))(cn_v^{out}(e)) \\
&\quad = (TS_Q \circ pop(ser_h(I), n-k))(e)
\end{aligned}$$

The rest of the proof follows as before.

6 RETIMING

A serial refinement increases the delay in a computation graph. However, if the communication graph is acyclic then it is possible to remove some or all of this increase in the delay by retiming. A retiming of a synchronous processor is a change in the delay function which does not alter the computational behaviour of the processor. Figure 3.8 shows a very simple retiming; for more interesting examples the reader is referred to the literature (e.g., Leiserson & Saxe [1981], Leiserson & Saxe [1983], Leiserson, Rose & Saxe [1983], Kung & Lam [1984], S. Y. Kung [1984], McEvoy & Dew [1988]). Retiming was introduced in Leiserson & Saxe [1981] as a tool for pipelining ripple-carry circuits; or more particularly, for producing systolic designs from semi-systolic designs. The automation of the Retiming Lemma is an important step in the development of systolic design, as the designer is able to make liberal use of broadcasts and ripple-carries, and so is relieved of much of the burdensome detail of checking that the right data is in the right place at the right time.

A retiming of a processor may drastically change its behaviour under initialisation. We have noted that as the initial values held in the registers defined by the delay function are not defined, the function Ω_P is non-deterministic. If Ω_P is sensitive to these initial values then some branches of the computation may vanish completely under a retiming. Consider an interpretation of the graphs of Figure 3.8 for which the cell v outputs the value input on channel a if there was ever a 1 input on channel e; otherwise v outputs the value input on channel b. Then whereas Figure 3.8(a) will always output the stream of values input on channel b, Figure 3.8(b) could output the stream of values input on channel a (depending on the value originally held in the register on channel e). Nevertheless, it is not unreasonable to suggest that these two processors are in some sense computationally equivalent, and in this section a relation $\Omega_P \approx \Omega_Q$ which is weaker than equality is considered. This relation can express the computational equivalence of Figures 3.8(a) and 3.8(b). However transitivity is lost, and so $\approx$ is not an equivalence relation. The relations $\approx$ and $=$ agree on deterministic functions.

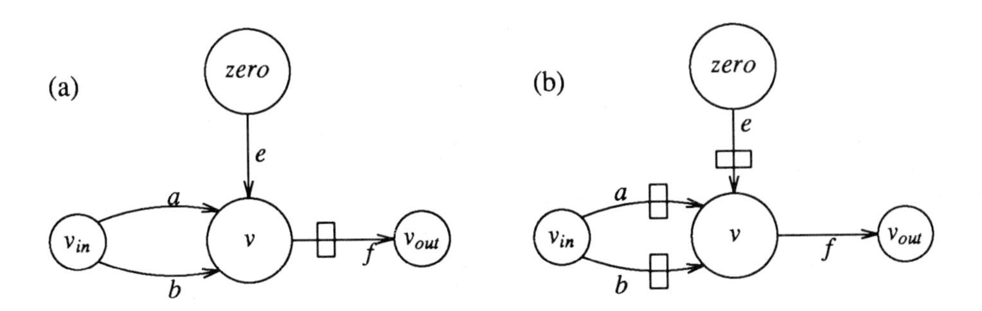

Figure 3.8. A simple retiming

Definition 6.1 Define a relation $\approx$ on non-deterministic total functions as follows. Given sets A, B and non-deterministic total functions $f, g : A \to B$, define $f \approx g$ if and only if for every $a \in A$, there exists $b \in B$ for which $f(a) = b$ and $g(a) = b$.

The latency of a processor is the number of clock cycles required for inputs to propagate through to v_{out}. A retiming of a processor may change the latency, and this behaviour is reflected in finite shifts of the output.

Definition 6.2 Let $G = \langle D, gf, cn, st, d\rangle$ be a non-primitive computation graph. A *retiming* of G is a non-primitive computation graph $H = \langle D, gf, cn, st, wt\rangle$ satisfying the condition that there exist $m, n \in \mathbf{N}$ such that for any processor interpretation Ψ of the signature of G, and any $I \in IN^{+}_{\langle G,\Psi\rangle}$ for which $|I| > \max\{m, n\}$,

$$\Omega_{\langle G,\Psi\rangle} \circ pop(I, m) \approx \Omega_{\langle H,\Psi\rangle} \circ pop(I, n).$$

Lemma 6.3 (Retiming Lemma, Leiserson and Saxe) *Let $G = \langle D, gf, cn, st, d\rangle$ be a non-primitive computation graph with $D = \langle V, v_{in}, v_{out}, E, tl, hd\rangle$. Given a function*

$$lag : V \to \mathbf{Z}$$

define $wt : E \to \mathbf{Z}$ by

$$wt(e) = d(e) + lag \circ hd(e) - lag \circ tl(e).$$

If, for all $e \in E$, $wt(e) \geq 0$ then $H = \langle D, gf, cn, st, wt\rangle$ is a retiming of G. Such a retiming is called a lag-defined *retiming of G.*

Proof (by induction on $|I|$). Given a processor interpretation Ψ of the signature of G, let $P = \langle G, \Psi\rangle$ and $Q = \langle H, \Psi\rangle$. Let $k = \min\{lag(v) \mid v \in V\}$, and define $\alpha : E \to \mathbf{N}$ by $\alpha(e) = \max\{lag(v_{in}) - k, lag \circ hd(e) - k\}$. Given $e \in E$ and $I \in IN^{*}_{Q}$, we prove that

(A) $(TS_Q \circ pop(I, lag(v_{in}) - k))(e) \approx (TS_P \circ pop(I, lag \circ tl(e) - k))(e)$; and
(B) either $|I| \leq \alpha(e)$ or

$$(HS_Q \circ pop(I, lag(v_{in}) - k))(e) \approx (HS_P \circ pop(I, lag \circ hd(e) - k))(e).$$

The lemma follows from (B), the retiming constants being $lag(v_{in}) - k$ and $lag(v_{out}) - k$. The basis is the case $|I| \leq \alpha(e)$ which is clear from the definition of $TS(\emptyset)(e)$ and $\approx$. For the inductive step assume that $|I| \geq \alpha(e)$.

(I) If $wt(e) \neq 0$ then

$$\begin{aligned}
&(HS_Q \circ pop(I, lag(v_{in}) - k))(e) \\
&\quad = (TS_Q \circ pop(pop(I, lag(v_{in}) - k), wt(e)))(e) \\
&\quad = (TS_Q \circ pop(pop(I, wt(e)), lag(v_{in}) - k))(e) \\
&\quad \approx (TS_P \circ pop(pop(I, wt(e)), lag \circ tl(e) - k))(e) \quad \text{by induction} \\
&\quad = (TS_P \circ pop(I, wt(e) + lag \circ tl(e) - k))(e) \\
&\quad = (TS_P \circ pop(I, d(e) + lag \circ hd(e) - k))(e) \\
&\quad = (HS_P \circ pop(I, lag \circ hd(e) - k))(e).
\end{aligned}$$

(II) Otherwise the proof is by induction on $Erank_H(e)$.
(IIa) If $Erank_H(e) = 0$ then either
(i) $e \in D_{in}$ and so (A) follows from $tl(e) = v_{in}$; or otherwise
(ii) let $v \in V_{interior}$ be such that $e \in TAILS_D(v)$ and (assuming that $HEADS(v) \neq \emptyset$) let $f \in gf(v)_{in}$. As $Erank_H(e) = 0$, $wt(cin_v^{-1}(f)) \neq 0$ and so by (I),

$$\begin{aligned}
&(PRED_v^Q \circ pop(I, lag(v_{in}) - k))(f) \\
&\quad = (HS_Q \circ pop(I, lag(v_{in}) - k))(cin_v^{-1}(f)) \\
&\quad \approx (HS_P \circ pop(I, lag \circ hd \circ cin_v^{-1}(f) - k))(cin_v^{-1}(f)) \quad (\dagger) \\
&\quad = (PRED_v^P \circ pop(I, lag(v) - k))(f).
\end{aligned}$$

By induction on $|I|$,

$$pred_v^Q \circ pop(I, lag(v_{in}) - k) = pred_v^P \circ pop(I, lag(v) - k).$$

Therefore,

$$\begin{aligned}
&(TS_Q \circ pop(I, lag(v_{in}) - k))(e) \\
&\quad = (\Omega_{\langle gf(v), \Psi \rangle} \circ pred_v^Q \circ pop(I, lag(v_{in}) - k))(cout_v(e)) \\
&\quad \approx (\Omega_{\langle gf(v), \Psi \rangle} \circ pred_v^P \circ pop(I, lag(v) - k)))(cout_v(e)) \\
&\quad = (TS_P \circ pop(I, lag \circ tl(e) - k))(e).
\end{aligned}$$

Given $wt(e) = 0$, (B) is now proved as in (I) except that where induction on $|I|$ was used, either (i) or (ii) is now used.

(IIb) The induction step on $Erank_H(e)$ is very similar to the basis. The induction hypothesis is used at (†).

Brookes [1984] has noted that the lemma can be generalised by allowing negative delays and so omitting the hypothesis that $wt(e) \geq 0$ for all $e \in E$. Given that the underlying communication graph is directed, the only reasonable semantics for $d(e) = -n$ for $n \in \mathbf{N}$ is that data appears on $hd(e)$ n clock cycles before it appears on $tl(e)$. Such behaviour is unrealistic and so is not permitted here. Also, Brookes allows cycles in which every channel has zero delay. However, although retiming can eliminate ripple-carries, the following proposition shows that it cannot be used to introduce delay into such cycles. Before stating the proposition we introduce some graph theoretic notation.

Definition 6.4 A path in a communication graph D is a string of channels for which the head of any edge in the path is the tail of the next channel in the path. $Paths_D \subset E^+$ is the set of all paths in D. For any such path P define

(a) $tl^*(P)$ to be the tail of the first channel in the path, and $hd^*(P)$ to be the head of the last channel in the path,
(b) P is a *cycle* if $tl^*(P) = hd^*(P)$,
(c) P is *simple* if it does not strictly contain any cycles (so P may be both a cycle and simple),
(d) If $d : E \to \mathbf{N}$ is a delay function for D then define $d^* : Paths_D \to \mathbf{N}$ by

$$d^*(P) = \sum_{i=1}^{|P|} d \circ proj_i(P).$$

Proposition 6.5 Let

$$H = \langle D, gf, cn, st, wt \rangle$$

be a lag-defined retiming of

$$G = \langle D, gf, cn, st, d \rangle$$

defined by the function lag. Then, for every $P \in Paths_D$,

$$wt^*(P) = d^*(P) + lag \circ hd^*(P) - lag \circ tl^*(P).$$

Proof

$$
\begin{aligned}
wt^*(P) &= \sum_{i=1}^{|P|} wt \circ proj_i(P) \\
&= \sum_{i=1}^{|P|} [d \circ proj_i(P) + lag \circ hd \circ proj_i(P) - lag \circ tl \circ proj_i(P)] \\
&= \sum_{i=1}^{|P|} d \circ proj_i(P) + lag \circ hd \circ proj_{|P|}(P) - lag \circ tl \circ proj_1(P) \\
&= d^*(P) + lag \circ hd^*(P) - lag \circ tl^*(P).
\end{aligned}
$$

Leiserson, Rose & Saxe [1983] have developed the Retiming Lemma as an important tool within a systolic design methodology. Its most important use has been in the elimination of ripple-carries, as expressed in the following lemma.

Lemma 6.6 *Let $G = \langle D, gf, cn, st, d\rangle$ be a non-primitive computation graph. A necessary and sufficient condition for the existence of a lag-defined retiming of G in which every channel has non-zero delay is that every cycle C in D satisfies*

$$d^*(C) \geq |C|.$$

Proof Let $D = \{V, v_{in}, v_{out}, E, tl, hd\}$. First assume that such a lag-defined retiming exists, and let $lag : V \to \mathbf{Z}$ define such a retiming $G = \langle D, gf, cn, st, wt\rangle$ of G. Let C be a cycle in D. Then as $wt \geq 1$ for every $e \in E$, $wt^*(C) \geq |C|$ and so by the previous proposition, $d^*(C) = wt^*(C) \geq |C|$.

To prove the converse assume the condition on cycles holds, and define $lag : V \to \mathbf{Z}$ by

$$lag(v) = \begin{cases} 0 & \text{if } HEADS(v) = \emptyset, \\ \max\{|P| - d^*(P) : P \in Paths_D \ \& \ hd^*(P) = v\} & \text{otherwise.} \end{cases}$$

Assuming that lag is total (to be proved below), define $wt : E \to \mathbf{Z}$ by $wt(e) = d(e) + lag \circ hd(e) - lag \circ tl(e)$. We show that, for all $e \in E$, $wt(e) \geq 1$ and so by the Retiming Lemma $\langle D, gf, cn, st, wt\rangle$ is a retiming of G. If $HEADS \circ tl(e) = \emptyset$ then $wt(e) = d(e) + lag \circ hd(e) \geq d(e) + (1 - d(e)) = 1$. Otherwise let P be a path such that $lag \circ tl(e) = |P| - d^*(P)$ and $hd^*(P) = tl(e)$. Then the path $push(e, P)$ witnesses that $lag \circ hd(e) \geq (|P| + 1) - (d^*(P) + d(e)) = lag \circ tl(e) + 1 - d(e)$. Therefore $wt(e) \geq 1$. It remains to show that lag is a total function. We prove that if a path P is not simple then there exists another path R with $hd^*(R) = hd^*(P)$,

$|R| < |P|$ and $|R| - d^*(R) \geq |P| - d^*(P)$. So if a cell v satisfies $HEADS(v) \neq \emptyset$ then $lag(v) = \max\{|P| - d^*(P) : P \in Paths_D \;\&\; hd^*(P) = v \;\&\; P \text{ is simple}\}$, and as there are only finitely many simple paths in any graph this maximum must exist. Let P be a path which is not simple. Then for some i, j, $1 \leq i < j < n$, one of three cases holds, and define a path R and a cycle C as follows.

(i) $hd(e_i) = hd(e_j)$; let $R = \langle e_n, \ldots, e_{j+1}, e_i, \ldots, e_1 \rangle$ and $C = \langle e_j, \ldots, e_{i+1} \rangle$.
(ii) $hd(e_i) = hd(e_n)$; let $R = \langle e_i, \ldots, e_1 \rangle$ and $C = \langle e_n, \ldots, e_{i+1} \rangle$.
(iii) $tl(e_1) = tl(e_j)$; let $R = \langle e_n, \ldots, e_j \rangle$ and $C = \langle e_{j-1}, \ldots, e_1 \rangle$.

Then

$$|P| - d^*(P) = (|R| + |C|) - (d^*(R) + d^*(C)) = (|R| - d^*(R)) + (|C| - d^*(C)).$$

Now as C is a cycle, by assumption $|C| - d^*(C) \leq 0$, and therefore

$$|R| - d^*(R) \geq |P| - d^*(P).$$

If a computation graph G does contain a cycle C for which $d^*(C) < |C|$ then it is still possible to eliminate ripple-carries by first interleaving the algorithm. A suitable factor $k \in \mathbf{N}$ is chosen so that multiplying the delay of each channel in G by k satisfies the condition on cycles. This method is less satisfactory as the latency of the computation is increased k-fold, and because of the interleaving of the algorithm the resulting computation graph is no longer a simple retiming. However, the interleaving ensures that the total number of clock cycles required to complete k computations remains the same, and so if the absence of ripple carries allows a shorter clock cycle then the computation time will decrease when averaged over k clock cycles.

Definition 6.7

(i) Given a non-primitive computation graph $G = \langle D, gf, cn, st, d \rangle$ and $k \in \mathbf{N}$, $k \neq 0$, the *k-fold interleave of* G is the computation graph $\langle D, gf, cn, st, wt \rangle$ where $wt(e) = k \times d(e)$ for all e in the channel set of D.
(ii) Given a set X and $k \in \mathbf{N}$, $k \neq 0$, define $filter_k : X^* \rightarrow X^*$ as follows. $filter_k(\emptyset) = \emptyset$. If $\overline{x} \in X^+$ and $|\overline{x}| = mk + n$ for $m, n \in \mathbf{N}$, $1 \leq n \leq k$ then $|filter_k(x)| = m + 1$ and for $j \in \mathbf{N}$, $1 \leq j \leq m + 1$, $proj_j \circ filter_k(\overline{x}) = proj_{(j-1)k+n}(\overline{x})$.

Lemma 6.8 Given a non-primitive computation graph $G = \langle D, gf, cn, st, d \rangle$ and $k \in \mathbf{N}$, $k \neq 0$, let $H = \langle D, gf, cn, st, wt \rangle$ be the k-fold interleave of G. Given a processor interpretation Ψ of the signature of G, let $P = \langle G, \Psi \rangle$ and $Q = \langle H, \Psi \rangle$. Then, for any $I \in IN_Q^+$,

$$\Omega_Q(I) = \Omega_P \circ filter_k(I).$$

Proof The structure of the proof is the same as that of the Retiming Lemma. Given $I \in IN_Q^*$ and $e \in E$, we prove that

$$\begin{aligned}&(A)\ (TS_Q(I))(e) = (TS_P \circ filter_k(I))(e); \quad \text{and}\\ &(B)\ (HS_Q(I))(e) = (HS_P \circ filter_k(I))(e), \quad \text{or } I = \emptyset.\end{aligned}$$

The basis case $I = \emptyset$ follows from the fact that $filter_k(\emptyset) = \emptyset$.

(I) If $wt(e) \neq 0$ then

$$\begin{aligned}(HS_Q(I))(e) &= (TS_Q \circ pop(I, wt(e)))(e)\\ &= (TS_P \circ filter_k \circ pop(I, wt(e)))(e) \quad \text{by induction}\\ &= (TS_P \circ pop(filter_k(I), d(e)))(e) \quad \text{as } wt(e) = k \times d(e)\\ &= (HS_P \circ filter_k(I))(e).\end{aligned}$$

(IIa) If $Erank_H(e) = 0$ then either

(i) $e \in D_{in}$ and so (A) follows from the fact that $top(I) = top \circ filter_k(I)$, or otherwise

(ii) let $v \in V_{interior}$ be such that $e \in TAILS_D(v)$ and (assuming that $HEADS(v) \neq \emptyset$) let $f \in gf(v)_{in}$.

$$\begin{aligned}(PRED_v^Q(I))(f) &= (HS_Q(I))(cin_v^{-1}(f))\\ &= (HS_P \circ filter_k(I))(cin_v^{-1}(f)) \quad \text{by } (I)\\ &= (PRED_v^P \circ filter_k(I))(f).\end{aligned}$$

So, $pred_v^Q(I) = (pred_v^P \circ filter_k(I))$, and therefore,

$$\begin{aligned}(TS_Q(I))(e) &= (\Omega_{\langle gf(v),\Psi\rangle} \circ pred_v^Q(I))(cout_v(e))\\ &= (\Omega_{\langle gf(v),\Psi\rangle} \circ pred_v^P \circ filter_k(I))(cout_v(e))\\ &= (TS_P \circ filter_k(I))(e).\end{aligned}$$

The rest of the proof follows as before.

Proposition 6.9 For every non-primitive computation graph, G, there exists $k \in \mathbf{N}$, $k \neq 0$, such that if wt is the delay function of the k-fold interleave of G then, for every cycle C in the communication graph, $wt^*(C) \geq |C|$.

Proof Choose the least k such that for each of the finitely many simple cycles, C, $k \times d^*(C) \geq |C|$ (where d is the delay function of G).

An application of the Retiming Lemma requires the assignment of a lag value to every cell in the communication graph, and so in complicated designs constructing the appropriate lag function can be laborious. A simpler method of retiming a computation graph is to use the Cut Theorem (Kung & Lam [1984]). An informal statement of the Cut Theorem is that given any minimal cutset for a computation graph and any integer k, if we add k registers to every channel in the cutset with a given direction and subtract k from every other channel in the cutset, then the result is a retiming of the graph (providing all delays are still positive).

Definition 6.10 Let $D = \langle V, v_{in}, v_{out}, E, tl, hd \rangle$ be a communication graph.

(i) The *undirected graph underlying* D is the digraph $\langle V, E_u, tl_u, hd_u \rangle$ where $E_u = E \cup \{e' \mid e \in E\}$, $tl_u(e) = tl(e)$, $tl_u(e') = hd(e)$, $hd_u(e) = hd(e)$ and $hd_u(e') = tl(e)$.
(ii) D is *connected* if for every $u, v \in V$, $u \neq v$, there is a path P in the underlying undirected graph such that $tl_u^*(P) = u$ and $hd_u^*(P) = v$. Otherwise D is *disconnected.*
(iii) Assume D is connected. $C \subseteq E$ is a *cutset for* D if $\langle V, v_{in}, v_{out}, E \backslash C, tl_c, hd_c \rangle$ is disconnected where tl_c and hd_c are tl and hd restricted to domain $E \backslash C$. A cutset C for D is *minimal* if no strict subset of C is a cutset for D.

Proposition 6.11 Let C be a minimal cutset for the connected communication graph $D = \langle V, v_{in}, v_{out}, E, tl, hd \rangle$ and let D_c be the resulting disconnected graph; that is, $D_c = \langle V, v_{in}, v_{out}, E \backslash C, tl_c, hd_c \rangle$ where tl_c and hd_c are tl and hd restricted to domain $E \backslash C$. Let V_{in} be the cell set of the connected component of D_c containing v_{in}; that is,

$$V_{in} = \{v_{in}\} \cup \{v \in V \mid \text{there is a path from } v_{in} \text{ to } v \text{ in the undirected graph underlying } D_c\}.$$

Given $e \in E$,

(i) If $e \notin C$ then $tl(e) \in V_{in} \Leftrightarrow hd(e) \in V_{in}$,
(ii) If $e \in C$ then $tl(e) \notin V_{in} \Leftrightarrow hd(e) \in V_{in}$.

Proof (i) Assume $tl(e) \in V_{in}$ and $tl(e) \neq v_{in}$. Let P be a path from v_{in} to $tl(e)$ in the undirected graph underlying D_c. Then as $e \notin C$ the path $push(e, P)$ witnesses that $hd(e) \in V_{in}$. The proof of the converse is symmetric. (ii) Assume $tl(e) \notin V_{in}$ and $hd(e) \neq v_{in}$. As C is minimal there is a simple path, P, from v_{in} to $hd(e)$ in the undirected graph underlying $\langle V, v_{in}, v_{out}, E \backslash (C \backslash \{e\}), tl_e, hd_e \rangle$. If P contains neither e nor e' then P witnesses that $hd(e) \in V_{in}$; otherwise, as P is simple $top(P) = e$ and so $pop(P)$ witnesses that $tl(e) \in V_{in}$ contradicting the assumption. Conversely, suppose if possible that both $tl(e), hd(e) \in V_{in}$ and let P and Q be paths from v_{in} to $tl(e)$ and $hd(e)$ in the undirected graph underlying D_c. Let $u \notin V_{in}$. As C is minimal let R be a simple path from $tl(e)$ to u in the undirected graph underlying $\langle V, v_{in}, v_{out}, E\backslash(C\backslash\{e\}), tl_e, hd_e\rangle$. If R contains neither e nor e' then $R\,\hat{}\,P$ contradicts that $u \notin V_{in}$; otherwise, as R is simple, $proj_1(R) = e$ and if R' is the remainder of R then $R'\,\hat{}\,Q$ contradicts that $u \notin V_{in}$.

Theorem 6.12 (Cut Theorem, Kung and Lam) *Let $G = \langle D, gf, cn, st, d\rangle$ be a non-primitive computation graph such that $D = \langle V, v_{in}, v_{out}, E, tl, hd\rangle$ is connected. Let C be a minimal cutset for D, and in the resulting disconnected graph let V_{in} be the connected component containing v_{in}. Given $k \in \mathbf{Z}$ define $wt : E \to \mathbf{Z}$ by*

$$wt(e) = \begin{cases} d(e) + k & \text{if } e \in C \text{ and } tl(e) \in V_{in}, \\ d(e) - k & \text{if } e \in C \text{ and } tl(e) \notin V_{in}, \\ d(e) & \text{if } e \notin C. \end{cases}$$

If, for all $e \in E$, $wt(e) \geq 0$ then $\langle D, gf, cn, st, wt\rangle$ is a retiming of G.

Proof Define $lag : V \to \mathbf{Z}$ by

$$lag(v) = \begin{cases} 0 & \text{if } v \in V_{in}, \\ k & \text{otherwise.} \end{cases}$$

If $e \notin C$ then by the previous proposition $lag \circ hd(e) = lag \circ tl(e)$. The proposition also gives (i) if $e \in C$ and $lag \circ tl(e) = 0$ then $lag \circ hd(e) = k$; and (ii) if $e \in C$ and $lag \circ tl(e) = k$ then $lag \circ hd(e) = 0$. So for all $e \in E$, $wt(e) = d(e) + lag \circ hd(e) - lag \circ tl(e)$ and the result follows from the Retiming Lemma.

The Cut Theorem could be stated for disconnected graphs by considering connected components.

7 A MATRIX MULTIPLICATION ALGORITHM

In this section we demonstrate the techniques described in the previous sections by presenting a hierarchical design for a matrix multiplication algorithm, given in H. T. Kung [1984]. Figure 3.9 illustrates the algorithm for matrices of order 8; we shall show the development of the design for matrices of order 2.

7.1 Functional Specification

(i) Signature: $\Gamma_1 = \langle S_1, \{MATMULT_1\}\rangle$, where $S_1 = \{sort_x, sort_w, sort_z\}$.
(ii) Computation Graph: $MATMULT_1 = \langle D_1, st_1\rangle$.

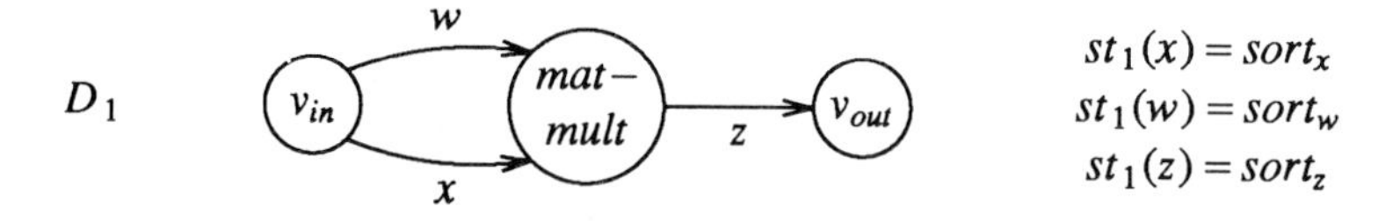

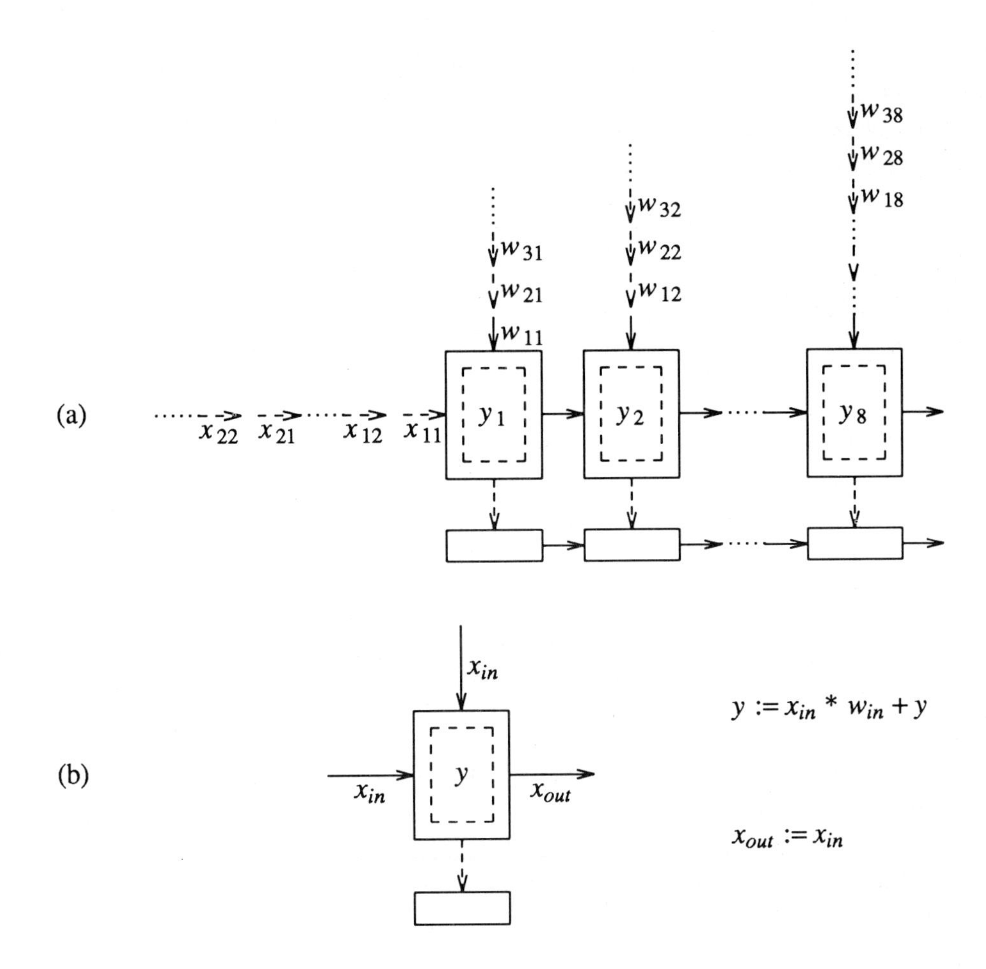

Figure 3.9. (a) Systolic array for matrix multiplication
(b) cell specification (taken from H. T. Kung [1984]).

(iii) Interpretation: $\Psi_1 = \langle \Theta_1, \Phi_1 \rangle$.

$$\Theta_1(sort_x) = \Theta_1(sort_w) = \Theta_1(sort_z) = M_{2,2},$$

where $M_{m,n}$ is the set of $m \times n$ matrices (with elements in **N**, say).

$$\Phi_1(MATMULT_1)(I)(z) = top(I)(x) \otimes top(I)(w),$$

where $\otimes$ denotes matrix multiplication.

7.2 Serial refinement

The matrix to be input on channel x is to be transmitted in serial as two row vectors.

(i) Signature: $\Gamma_2 = \Gamma_1$

(ii) Computation graph:

$$MATMULT_2 = SerRef(MATMULT_1, 2) = MATMULT_1$$

(see Lemma 5.13).

(iii) Interpretation: $\Psi_2 = \langle \Theta_2, \Phi_2 \rangle$.

$$\Theta_2(sort_x) = \Theta_2(sort_z) = M_{1,2}, \quad \Theta_2(sort_w) = M_{2,2}.$$

$$\Phi_2(MATMULT_2)(I)(z) = top(I)(x) \otimes top(I)(w).$$

(iv) Serial homomorphism h.

$$h_{sort_x}(m) = h_{sort_z}(m) = \langle row_1(m), row_2(m) \rangle,$$
$$h_{sort_w}(m) = \langle m, n \rangle,$$

where $row_i(m)$ is the ith row of the matrix $m, i = 1, 2$.

7.3 Parallel refinement

The matrix to be input on channel w is to be transmitted in parallel as two column vectors. Also, a reset line is introduced.

(i) Signature: $\Gamma_3 = \langle S_3, \{MATMULT_3\} \rangle$, where $S_3 = S_1 \cup \{reset\}$ and $\lambda : S_1 \to S_3^+$ is defined by

$$\lambda(sort_x) = \langle sort_x, reset \rangle,$$
$$\lambda(sort_w) = \langle sort_w, sort_w \rangle,$$
$$\lambda(sort_z) = \langle sort_z, sort_z \rangle.$$

(ii) Computation graph: $MATMULT_3 = \langle D_3, st_3 \rangle = ParRef_\lambda(MATMULT_2)$ (see Lemma 5.9).

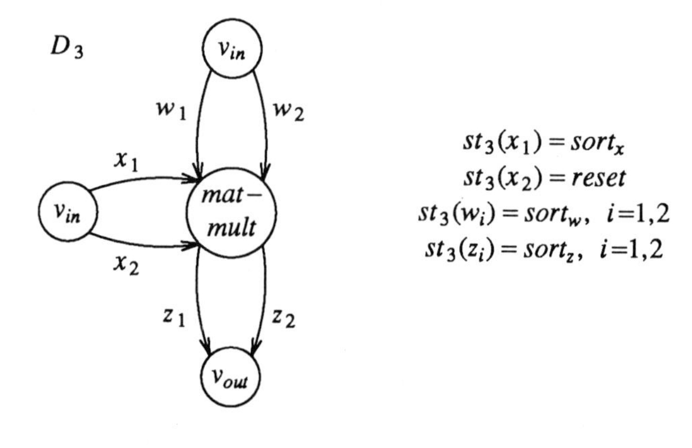

(iii) Interpretation:

$$\Psi_3 = \langle \Theta_3, \Phi_3 \rangle. \quad \Theta_3(sort_x) = M_{1,2}, \quad \Theta_3(sort_w) = M_{2,1},$$

$$\Theta_3(sort_z) = \mathbf{N}, \quad \Theta_3(reset) = \{true, false\}.$$

$$\Phi_3(MATMULT_3)(I)(z_i) = top(I)(x_1) \otimes top(I)(w_i), \quad 1 = 1,2.$$

(iv) Parallel homomorphism g.

$$g_{sort_w}(m) = \langle col_1(m), col_2(m) \rangle,$$
$$g_{sort_z}(m) = \langle proj_1(m), proj_2(m) \rangle,$$
$$g_{sort_x}(m) = \langle m, true \rangle,$$

where $col_i(m)$ is the ith column of the matrix m, and $proj_i(m)$ is the ith element of the vector $m, i = 1,2$.

7.4 Graph substitution

Pipelining of the algorithm is made possible by dividing the cell *matmult*. The following processor is computationally equivalent to the previous one (with x_2 relabelled as r_1).

(i) Signature: $\Gamma_4 = \langle S_3, \{VECMULT_1, VECMULT_2\} \rangle$.
(ii) Computation graph: $MATMULT_4 = \langle D_4, gf_4, cn, st_4, d_4 \rangle$.

$$gf_4(v_i) = VECMULT_i, \quad i = 1,2,$$

where

$$VECMULT_1 = \langle D_{41}, st_{41} \rangle$$

and

$$VECMULT_2 = \langle D_{42}, st_{42} \rangle.$$

$$cin_{v_i}(x_i) = x_{in}, \quad cout_{v_i}(z_i) = z, \quad i = 1,2.$$
$$cin_{v_i}(r_i) = r_{in}, \quad cout_{v_1}(x_2) = x_{out},$$
$$cin_{v_i}(w_i) = w, \quad cout_{v_1}(r_2) = r_{out}.$$

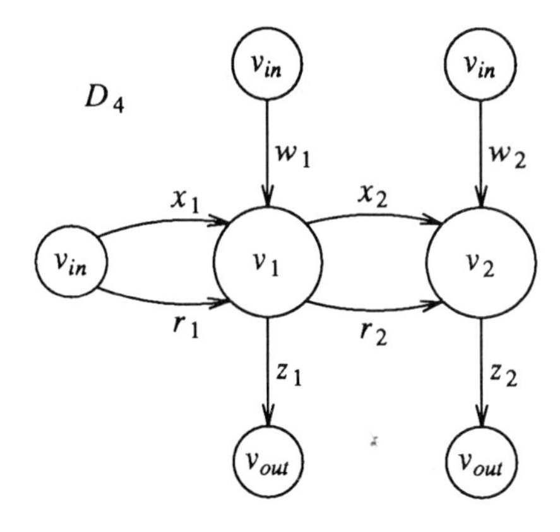

$$st_4(x_i) = sort_x, \ \ i=1,2$$
$$st_4(r_i) = reset, \ \ i=1,2$$
$$st_4(w_i) = sort_w, \ \ i=1,2$$
$$st_4(z_i) = sort_z, \ \ i=1,2$$

$$d_4(e) = 0, \ \text{all } e$$

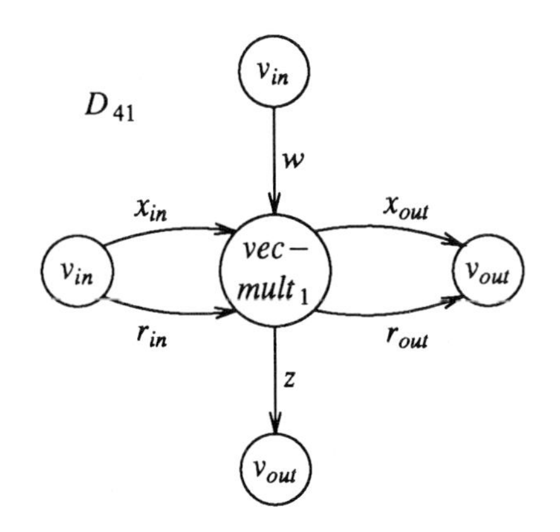

$$st_{41}(x_{in}) = st_{41}(x_{out}) = sort_x$$
$$st_{41}(r_{in}) = st_{41}(r_{out}) = reset$$
$$st_{41}(w) = sort_w$$
$$st_{41}(z) = sort_z$$

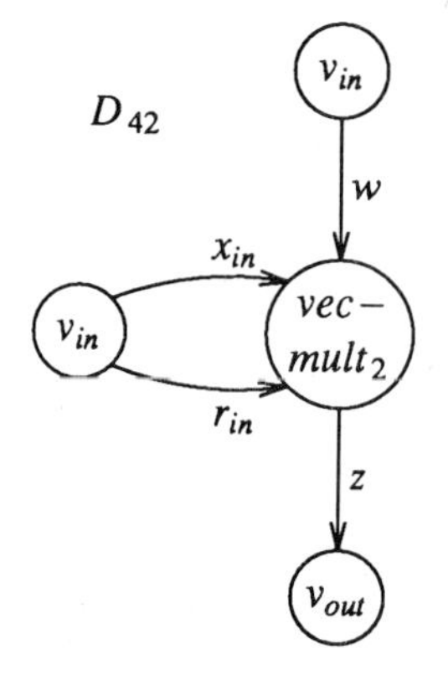

$$st_{42}(x) = sort_x$$
$$st_{42}(r) = reset$$
$$st_{42}(w) = sort_w$$
$$st_{42}(z) = sort_z$$

(iii) Interpretation:

$$\Psi_4 = \langle \Theta_4, \Phi_4 \rangle. \quad \Theta_4 = \Theta_3.$$

$$\Phi_4(VECMULT_i)(I)(z) = top(I)(x_{in}) \otimes top(I)(w), \quad i = 1, 2,$$
$$\Phi_4(VECMULT_1)(I)(r_{out}) = top(I)(r_{in}).$$
$$\Phi_4(VECMULT_1)(I)(x_{out}) = top(I)(x_{in}).$$

The pipelining of the algorithm will in fact be the last refinement, as the lack of delay will make the intervening serial refinement slightly simpler.

7.5 Serial refinement

The row and column vectors are to be transmitted in serial as pairs of numbers.

(i) Signature: $\Gamma_5 = \Gamma_4$.
(ii) Computation Graph:

$$MATMULT_5 = SerRef(MATMULT_4, 2) = MATMULT_4.$$

(iii) Interpretation: $\Psi_5 = \langle \Theta_5, \Phi_5 \rangle$.

$$\Theta_5(sort_x) = \Theta_5(sort_z) = \Theta_5(sort_w) = \mathbf{N}, \quad \Theta_5(reset) = \mathbf{B}.$$

$$\Phi_5(VECMULT_i)(I)(z) = \begin{cases} 0 & \text{if } top(I)(r_{in}) = true, \\ top(I)(x) \times top(I)(w) & \\ \quad + top \circ pop(I)(x) \times top \circ pop(I)(w) & \text{otherwise.} \end{cases}$$

$$\Phi_5(VECMULT_1)(I)(r_{out}) = top(I)(r_{in}),$$
$$\Phi_5(VECMULT_1)(I)(x_{out}) = top(I)(x_{in}).$$

(iv) Serial homomorphism h. $h_{sort_x} : M_{1,2} \to \mathbf{N}^2$ is defined by $h_{sort_x}(m) = \langle proj_1(m), proj_2(m) \rangle$. $h_{sort_w} : M_{2,1} \to \mathbf{N}^2$ is defined by

$$h_{sort_w}(m) = \langle proj_1(m), proj_2(m) \rangle.$$

$h_{sort_z} : \mathbf{N} \to \mathbf{N}^2$ is defined by $h_{sort_z}(n) = \langle 0, n \rangle$. $h_{reset} : \mathbf{B} \to \mathbf{B}^2$ is defined by $h_{reset}(b) = \langle true, false \rangle$.

7.6 Graph Substitution

Implementation of the inner product.

(i) Signature:

$$\Gamma_6 = \langle \{num, reset\}, \{MULT, ADD, ZERO, DUP_{reset}, DUP_{num}, MUX\} \rangle.$$

(ii) Computation Graph: $MATMULT_6 = FunRef_\eta(MATMULT_5)$ where $\eta : \{VECMULT_1, VECMULT_2\} \rightarrow \mathbf{G}(\Gamma_6)$ is defined by $\eta(VECMULT_i) = INPROD_i$, $i = 1, 2$ as follows.

$$INPROD_1 = \langle D_{61}, gf_{61}, cn^1, st_{61}, d_{61} \rangle, \quad \text{and}$$
$$INPROD_2 = \langle D_{62}, gf_{62}, cn^2, st_{62}, d_{62} \rangle.$$

$$gf_{61}(dup_n) = DUP_{num}, \quad gf_{61}(mult) = MULT,$$
$$gf_{61}(sum) = SUM, \quad gf_{61}(dup_r) = DUP_{reset}.$$

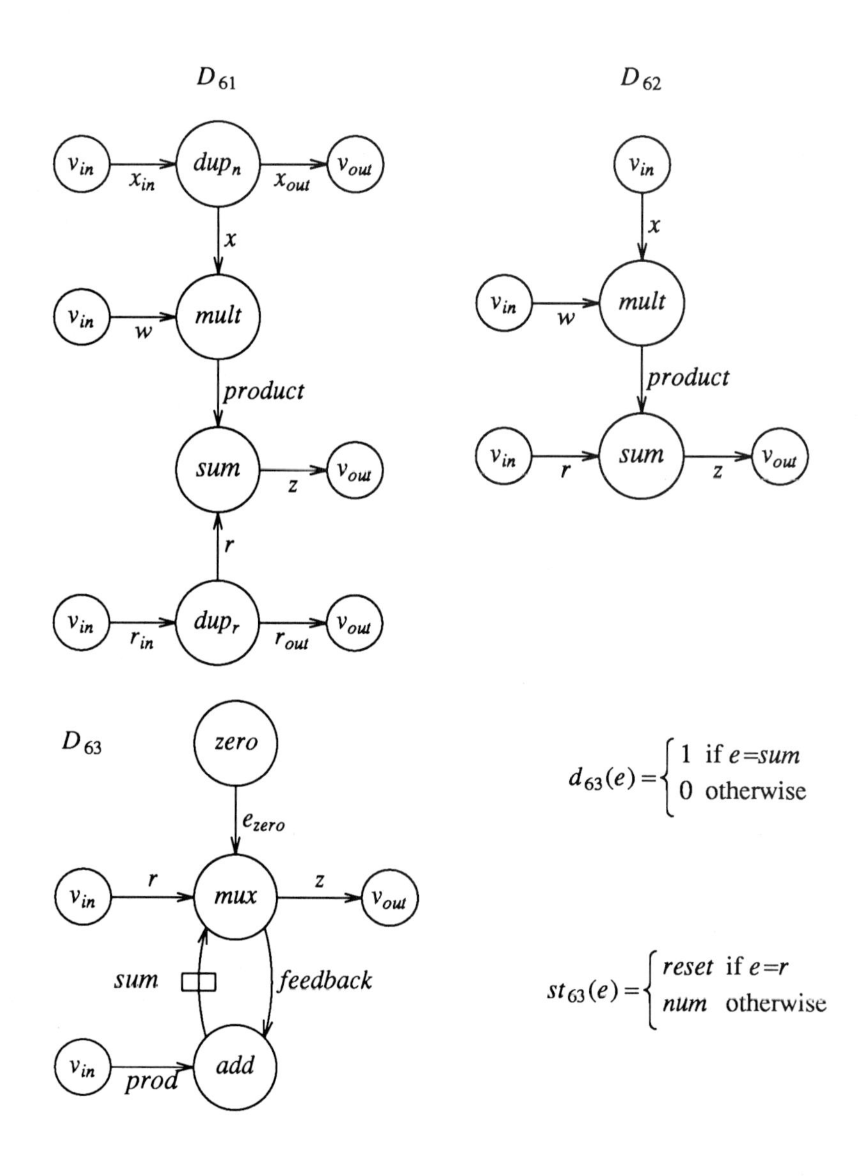

$$d_{61}(e) = 0, \quad \text{for all } e. \qquad st_{61}(e) = \begin{cases} reset & \text{if } e \in \{r_{in}, r, r_{out}\}, \\ num & \text{otherwise.} \end{cases}$$

$$cin^1_{dupn}(x_{in}) = e_1, \quad cout^1_{dupn}(x_{out}) = e_2, \quad cout^1_{dupn}(x) = e_3.$$

$$cin^1_{mult}(x) = x, \quad cin^1_{mult}(w) = w, \quad cout^1_{mult}(prod) = prod.$$

$$cin^1_{sum}(prod) = prod, \quad cin^1_{sum}(r) = r, \quad cout^1_{sum}(z) = z.$$

$$cin^1_{dupr}(r_{in}) = e_1, \quad cout^1_{dupr}(r_{out}) = e_2, \quad cout^1_{dupr}(r) = e_3.$$

$$gf_{62}(mult) = MULT, \quad gf_{62}(sum) = SUM.$$

$$d_{62}(e) = 0, \quad \text{for all } e. \qquad st_{62}(e) = \begin{cases} reset & \text{if } e = r \\ num & \text{otherwise.} \end{cases}$$

$$cin^2_{mult}(x) = x, \quad cin^2_{mult}(w) = w, \quad cout^2_{mult}(prod) = prod.$$

$$cin^2_{sum}(prod) = prod, \quad cin^2_{sum}(r) = r, \quad cout^2_{sum}(z) = z.$$

$$SUM = \langle D_{63}, gf_{63}, cn^3, st_{63}, d_{63} \rangle$$

$$gf_{63}(mux) = MUX, \quad gf_{63}(zero) = ZERO, \quad gf_{63}(add) = ADD.$$

$$cin^3_{mux}(r) = r, \quad cin^3_{mux}(sum) = e_1, \quad cin^3_{mux}(e_{zero}) = e_2.$$

$$cout^3_{mux}(z) = e_3, \quad cout^3_{mux}(feedback) = e_4, \quad cout^3_{zero}(e_{zero}) = out.$$

$$cin^3_{add}(prod) = x, \quad cin^3_{add}(feedback) = y, \quad cout^3_{add}(sum) = sum.$$

(iv) Interpretation: $\Psi_6 = \langle \Theta_6, \Phi_6 \rangle$. $\Theta_6(num) = \mathbf{N}$, $\Theta_6(reset) = \mathrm{B}$.

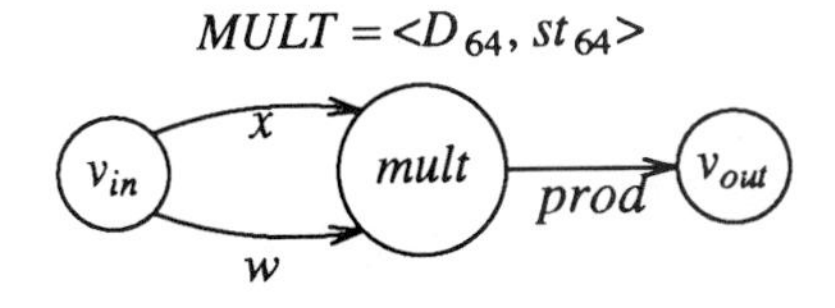

$st_{64}(e) = num$, all e

$\Phi_6(MULT)(I)(prod) = top\,(I)(x) \times top\,(I)(w)$

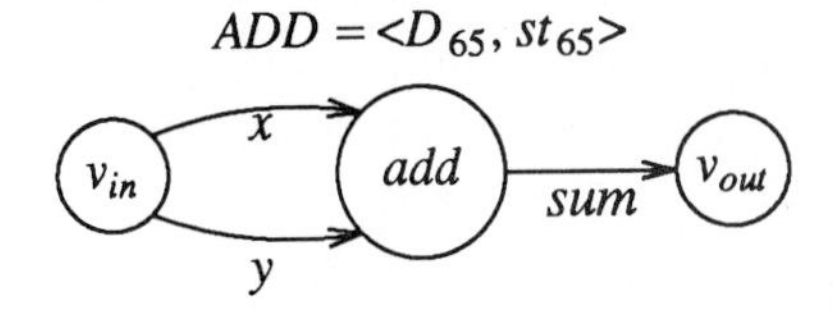

$st_{65}(e) = num$, all e

$\Phi_6(PLUS)(I)(sum) = top\,(I)(x) + top\,(I)(w)$

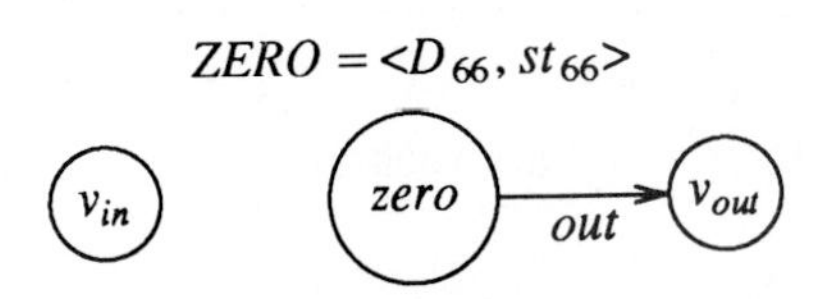

$st_{66}(out) = num$

$\Phi_6(PLUS)(I)(out) = 0$

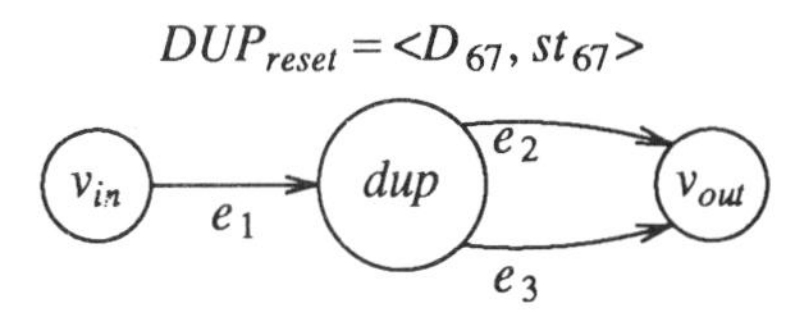

$$st_{67}(e) = reset, \text{ all } e$$
$$\Phi_6(DUP_{reset})(I)(e_i) = top\,(I)(e_1),\ i{=}2,3$$

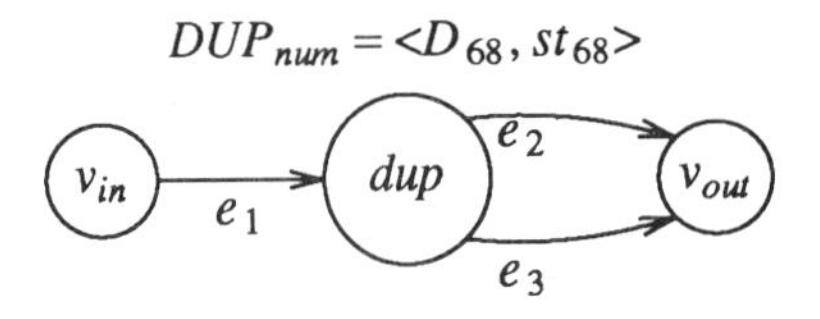

$$st_{68}(e) = num, \text{ all } e$$
$$\Phi_6(DUP_{num})(I)(e_i) = top\,(I)(e_1),\ i{=}2,3$$

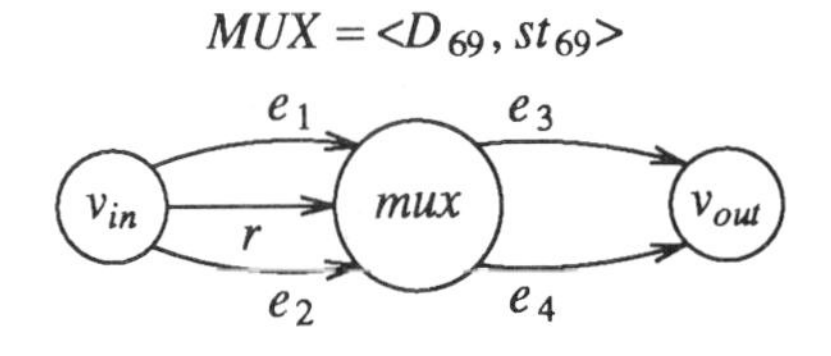

$$st_{69}(e_i) = num,\ i{=}1,2,3,4$$
$$st_{69}(r) = reset$$

$$\Phi_6(MUX)(I)(e_3) = \begin{cases} top(I)(e_1) & \text{if } top(I)(r) = true, \\ top(I)(e_2) & \text{otherwise.} \end{cases}$$

$$\Phi_6(MUX)(I)(e_4) = \begin{cases} top(I)(e_2) & \text{if } top(I)(r) = true, \\ top(I)(e_1) & \text{otherwise.} \end{cases}$$

7.7 Pipelining

Finally the algorithm is pipelined at the top level through a simple use of the cut theorem. The previous two sections have not altered the communication graph at the top level, and so it is still D_4. $\{w_2, x_2, r_2\}$ is a minimal cutset for the underlying undirected graph, and so we can add one delay to each of these channels. In H. T. Kung [1984] the algorithm is to be implemented on a data structure in which the functional cells themselves are not delay-free (in particular, the adders in the CMU Warp are pipelined). To achieve this implementation the necessary delay is introduced at the top level (using the cut theorem), and it is then transferred down the levels by absorbing it into the cell interpretations. This technique for two-level pipelining will be further discussed elsewhere.

REFERENCES

Brookes [1984]
S.D. Brookes (1984) 'Reasoning about synchronous systems', Technical Report CMU-CS-84-145, Department of Computer Science, Carnegie-Mellon University, Pittsburgh, Pennsylvania.

Cook [1981]
S.A. Cook (1981) 'Towards a complexity theory of synchronous parallel computation', *Enseignements Mathématique (2)*, **27**, 99–124.

Goguen *et al.* [1977]
J. A. Goguen, J. W. Thatcher, E. G. Wagner, and J. B. Wright (1977) 'Initial algebra semantics and continuous algebras' *J. ACM*, **24**, 68–95.

Gordon [1980]
M.Gordon (1980) 'The denotational semantics of sequential machines', *Information Processing Letters*, **10**, 1–3.

Kahn [1974]
G. Kahn (1974) 'The semantics of a simple language for parallel programming', 471–5 in *Information Processing '74, Proceedings IFIP Congress*, North-Holland, Amsterdam.

H. T. Kung [1984]
H. T. Kung (1984) 'The Warp Processor: a versatile systolic array for very high speed signal processing', Technical Report, Department of Computer Science, Carnegie-Mellon University, Pittsburgh, Pennsylvania.

Kung & Lam [1984]
H. T. Kung and M. S. Lam (1984) 'Wafer-scale integration and two-level pipelined implementations of systolic arrays', *J. Parallel and Dstributed Computing*, **1**, 32–63.

Kung & Lin [1984]
H. T. Kung and W. T. Lin (1984) 'An algebra for systolic computation', 141–60 in *Elliptic Problem Solvers 2, Proceedings, Conference, Monterey, Jan. 1983*, ed. G. Birkhoff & A. L. Schoenstadt, Academic Press.

S. Y. Kung [1984]
S. Y. Kung (1984) 'On supercomputing with systolic/wavefront array processors', *Proc. IEEE*, **72**, 867–84.

Leiserson, Rose & Saxe [1983]
C. E. Leiserson, F. M. Rose, and J. B. Saxe (1983) 'Optimising synchronous circuitry by retiming', 87–116 in *Third Caltech Conference on VLSI*, ed. R. Bryant, Springer Verlag.

Leiserson & Saxe [1981]
C. E. Leiserson and J. B. Saxe (1981) 'Optimising synchronous systems', 23–36 in *22nd Annual Symposium on Foundations of Computer Science, Nashville.*

Leiserson & Saxe [1983]
C. E. Leiserson and J. B. Saxe (1983) 'Optimising synchronous systems', *J. VLSI and Computer Systems*, **1**, 41–68.

Masuzawa *et al.* [1983]
T. Masuzawa, S. Nakauchi, K. Wada, K. Hagihara, and N. Tokura (1983) 'Systolic algorithm description language SADL and support system for systolic algorithm design', 20–4 in *Proceedings International Symposium on VLSI Technology, Systems and Applications.*

McEvoy & Dew [1988]
K. McEvoy and P. M. Dew (1988) 'The cut theorem: a tool for design of systolic algorithms', *Int. J. Computer Mathematics*, **25**, 203–33.

4 Verification of a Systolic Algorithm in Process Algebra

W. P. WEIJLAND

ABSTRACT

In designing VLSI-circuits it is very useful, if not necessary, to construct the specific circuit by placing simple components in regular configurations. Systolic systems are circuits built up from arrays of cells and therefore very suitable for formal analysis and induction methods. In the case of a palindrome recognizer a correctness proof is given using bisimulation semantics with asynchronous cooperation. The proof is carried out in the formal setting of the *Algebra of Communicating Processes* (see Bergstra & Klop [1986]), which provides us with an algebraical theory and a convenient proof system. An extensive introduction to this theory is included in this paper. The palindrome recognizer has also been studied by Hennessy [1986] in a setting of failure semantics with synchronous cooperation.

1 INTRODUCTION

In the current research on (hardware) verification one of the main goals is to find strong proof systems and tools to verify the designs of algorithms and architectures. For instance, in the development of integrated circuits the important stage of testing a prototype (to save the high costs of producing defective processors) can be dealt with much more efficiently, when a strong verification tool is available. Therefore, developing a verification theory has very high priority and is subject of study at many universities and scientific institutions.

However, working on detailed verification theories is not the only approach to this problem. Once having a basic theory, the development of case studies is of utmost importance to provide us with new ideas. Furthermore, one can focus on special design techniques, which turn out to fit conveniently in the theory. For example, because of the regular configuration of these circuits, *systolic arrays* are very suitable for formal analysis and induction methods. Indeed, systolic arrays have grown very popular in the last few years (see Huang & Lengauer [1987], Mead & Conway [1980] and Lipton & Lopresti [1985]).

In this chapter we will present a theory called *Algebra of Communicating Processes* (see Bergstra & Klop [1986]), which is an algebraic theory that provides us with a formal description of concurrent processes. Related to the work of Milner [1980], this theory has the aim of treating communicating processes in an *axiomatic* way, in order not to restrict ourselves unnecessarily to processes of one particular kind. Its language seems large enough to describe many kinds of processes and algorithms, while its equational theory turns out to be a strong tool for verification purposes. Some of the main theoretical results are presented in simple terms, to make it possible for the reader to understand how to work in ACP. Other references to formal systems for circuit description are that of CSP from Hoare [1985], and CIRCAL from Milne [1983].

Next, a simple description in ACP of a systolic algorithm for palindrome recognition (see Kung [1979]) will be presented. A systolic system can be looked at as a large integration of identical cells such that the behaviour of the total system strongly resembles the behaviour of the individual cells. In fact the total system behaves like one of its individual cells 'on a larger scale'. In designing VLSI-circuits it is very useful, if not necessary, to construct the specific circuit by placing simple components in regular configurations. Otherwise, one loses all intuition about the behaviour of the circuit that is eventually constructed. For this reason one may see systolic systems as a sort of regular subclass of VLSI-circuits.

Within the semantical setting of ACP, we will be able to prove *correctness* of the palindrome recognizer. Such a proof has already been presented by Hennessy [1986] using *synchronous* ('clocked') cooperation between cells. The development of a theory for synchronous cooperation is still the subject of ongoing research (see Weijland [1988]). In the following, however, we shall specify an *asynchronous* version of this algorithm.

This chapter is a revised version of the first part of Kossen & Weijland [1987a]. Most of the improvements only concern notational problems, although some of its formalism has been changed as well. General references to related work are for instance Hennessy[1986], Huang & Lengauer [1987], Kossen & Weijland [1987a,b], Kung [1979], Mead & Conway [1980], Mulder & Weijland [1987], Rem [1983] and Weijland [1987]).

2 THE ALGEBRA OF COMMUNICATING PROCESSES

The axiomatic framework in which we analyse the algorithm is ACP_τ, the *Algebra of Communicating Processes with silent steps*, as described in Bergstra & Klop [1985]. In this section, we give a brief review of ACP_τ.

Process algebra starts from a finite collection A of given objects, called atomic actions, atoms or steps. These actions are taken to be indivisible, usually have no duration and form the basic building blocks of our systems. The first two compositional operators we consider are $\cdot$, denoting sequential composition, and $+$ for alternative composition. If x and y are two processes, then $x \cdot y$ is the process that starts the execution of y after the completion of x, and $x + y$ is the process that chooses either x or y and executes the chosen process. Each time a choice is made, we choose from a set of alternatives. We do not specify whether the choice is made by the process itself, or by the environment. Axioms A1–5 in Table 4.1 below give the laws that $+$ and $\cdot$ obey. We leave out $\cdot$ and brackets as in regular algebra, so $xy + z$ means $(x \cdot y) + z$.

On intuitive grounds $x(y + z)$ and $xy + xz$ present different mechanisms (because the moment of choice is different), and therefore an axiom $x(y + z) = xy + xz$ is not included (see example below).

Example Playing Russian Roulette can best be described by the process $shot \cdot dead + shot \cdot alive$, instead of $shot \cdot (dead + alive)$, since in the second process after the shot, one can still choose between dying and surviving. This illustrates why $x(y + z) = xy + xz$ is not included as an axiom.

We have a special constant δ denoting deadlock, the acknowledgement of a process that it cannot do anything anymore, the absence of an alternative. Axioms A6,7 give the laws for δ. Together, the axioms A1–A7 are referred to as BPA, which stands for *Basic Process Algebra.*

Next, we have the parallel composition operator $\|$, called merge. The merge of processes x and y will interleave the actions of x and y, except for the communication actions. In $x \parallel y$, we can do a step from x or a step from y, or x and y both synchronously perform an action, which together with them makes up a new action, the communication action. This trichotomy is expressed in axiom CM1. Here, we use two auxiliary operators $⌊\!⌊$ (left-merge) and $|$ (communication merge). Thus, $x ⌊\!⌊ y$ is $x \parallel y$, but with the restriction that the first step comes from x, and $x \mid y$ is $x \parallel y$ with a communication step as the first step. Axioms CM2–9 give the laws for $⌊\!⌊$ and $|$. We assume the communication function is given on atomic actions and obeys laws C1–3.

Examples

$$a \parallel b = a ⌊\!⌊ b + b ⌊\!⌊ a + (a \mid b) = ab + ba + (a \mid b)$$

$$(ab) ⌊\!⌊ c = a(b \parallel c) = a(bc + cb + (b \mid c))$$

$$(ab) \mid (cd) = (a \mid c)(b \parallel d) = (a \mid c)(bd + db + (b \mid d)).$$

Finally, on the left-hand side of Table 4.1 we have the laws for the encapsulation operator ∂_H. Here H is a set of atoms, and ∂_H blocks actions from H by renaming them into δ. The operator ∂_H can be used to encapsulate a process, i.e., to block communications with the environment.

Example Suppose $H = \{b\}$, then $\partial_H(a \parallel b) = \partial_H(ab + ba + (a \mid b)) = a\delta + \delta a + (a \mid b)$ and, using axioms A6 and A7, we obtain: $= a\delta + \delta + (a \mid b) = a\delta + (a \mid b)$.

In all tables we have $a, b, c \in A \cup \{\delta\}$, x, y, z are arbitrary processes, and $H \subseteq A$.

Table 4.1. ACP $(a, b, c \in A \cup \{\delta\}, H \subseteq A)$.

$x + y = y + x$	A1	$x \parallel y = x \llcorner\!\llcorner y + y \llcorner\!\llcorner x + x \mid y$	CM1
$x + (y + z) = (x + y) + z$	A2	$a \llcorner\!\llcorner x = ax$	CM2
$x + x = x$	A3	$ax \llcorner\!\llcorner y = a(x \parallel y)$	CM3
$(x + y)z = xz + yz$	A4	$(x + y) \llcorner\!\llcorner z = x \llcorner\!\llcorner z + y \llcorner\!\llcorner z$	CM4
$(xy)z = x(yz)$	A5	$ax \mid b = (a \mid b)x$	CM5
$x + \delta = x$	A6	$a \mid bx = (a \mid b)x$	CM6
$\delta x = \delta$	A7	$ax \mid by = (a \mid b)(x \parallel y)$	CM7
		$(x + y) \mid z = x \mid z + y \mid z$	CM8
		$x \mid (y + z) = x \mid y + x \mid z$	CM9
$\partial_H(a) = a$ if $a \notin H$	D1		
$\partial_H(a) = \delta$ if $a \in H$	D2	$a \mid b = b \mid a$	C1
$\partial_H(x + y) = \partial_H(x) + \partial_H(y)$	D3	$(a \mid b) \mid c = a \mid (b \mid c)$	C2
$\partial_H(xy) = \partial_H(x) \cdot \partial_H(y)$	D4	$\delta \mid a = \delta$	C3

Next, we introduce laws for Milner's silent step, denoted by the new constant τ (see Milner [1980]). This constant can be looked at as an *internal* action, which cannot be seen from the outside. In fact, τ stands for zero or more machine steps and indicates that the machine is busy. Suppose we see atomic actions only *start*, not end. The silent step τ denotes an invisible action, so we do not see it start, or end. Now, it is clear that the processes a and at cannot be distinguished, since they both start with a and after a while they terminate. Thus, in general $x\tau = x$.

Since τ stands for *zero or more* internal machine steps, any process τx has a possibility of starting immediately with x. So, since τx has a summand x, we have $\tau x + x = \tau x$.

For similar reasons $a(\tau x + y)$ has a summand ax, since after this process has done a, and possibly some internal moves, it might do x without being able to choose y. So $a(\tau x + y) = a(\tau x + y) + ax$. In Table 4.2, the laws T1–3 are Milner's τ-laws, and TM1,2 and TC1–4 describe the interaction of τ and merge. Finally, τ_I is the abstraction operator, that renames atoms from I into τ.

Table 4.2. The silent step $\tau (a \in A \cup \{\delta\}, H, I \subseteq A)$.

$x\tau = x$	T1		
$\tau x + x = \tau x$	T2		
$a(\tau x + y) = a(\tau x + y) + ax$	T3		
$\tau \lfloor\!\lfloor x = \tau x$	TM1	$\partial_H(\tau) = \tau$	DT
$\tau x \lfloor\!\lfloor y = \tau(x \parallel y)$	TM2	$\tau_I(\tau) = \tau$	TI1
$\tau \mid x = \delta$	TC1	$\tau_I(a) = a$ if $a \notin I$	TI2
$x \mid \tau = \delta$	TC2	$\tau_I(a) = \tau$ if $a \in I$	TI3
$\tau x \mid y = x \mid y$	TC3	$\tau_I(x + y) = \tau_I(x) + \tau_I(y)$	TI4
$x \mid \tau y = x \mid y$	TC4	$\tau_I(xy) = \tau_I(x) \cdot \tau_I(y)$	TI5

The axioms of ACP in Table 4.1, together with the axioms in Table 4.2 above, form the system ACP_τ.

Definition The set BT of **basic terms** is inductively defined as follows:

(i) $\tau, \delta \in BT$;
(ii) if $t \in BT$, then $\tau t \in BT$;
(iii) if $t \in BT$ and $a \in A$, then $at \in BT$;
(iv) if $t, s \in BT$, then $t + s \in BT$.

Elimination Theorem (Bergstra & Klop [1985]) *Let t be a closed term over* ACP_τ. *Then there is a basic term s such that* $\mathrm{ACP}_\tau \vdash t = s$.

The elimination theorem allows us to use induction in proofs. The set of closed terms modulo derivability forms a model for ACP_τ (the initial algebra). However, most processes encountered in practice cannot be represented by a closed term, since they possibly contain infinite recursive calls. Therefore, most models of process algebra also contain infinite processes which are recursively specified. Let us first develop some terminology.

Definition

(i) Let t be a term over ACP_τ, and x a variable in t. Suppose that the abstraction operator τ_I does not occur in t. Then we say that an occurrence of x in t is **guarded** if t has a subterm of the form $a \cdot s$, with $a \in A_\delta$ (so $a \neq \tau$!) and this x occurs in s (i.e., each variable is 'preceded' by an atom).

(ii) A **recursive specification** over ACP_τ is a set of equations $\{x = t_x : x \in X\}$, with X a set of variables, and t_x a term over ACP_τ and variables only from X (for each $x \in X$).

(iii) A recursive specification $\{x = t_x : x \in X\}$ is **guarded** if no t_x contains an abstraction operator τ_I, and each occurrence of a variable in each t_x is guarded.

Notes

(i) The constant τ cannot be a guard, since the presence of a τ does not lead to unique solutions: for instance, the equation $x = \tau x$ has each process starting with a τ as a solution.

(ii) A definition of guardedness involving τ_I is very complicated, and therefore, we do not give such a definition here. The definition above suffices for our purposes.

Definition On ACP_τ, we can define a **projection operator** π_n, that cuts off a process after n atomic steps are executed, by the axioms in Table 4.3 ($n \geq 1, a \in A_\delta$, x, y are arbitrary processes).

Table 4.3. Projection ($n \geq 1, a \in A_\delta$).

$\pi_n(a) = a$	$\pi_n(\tau) = \tau$
$\pi_1(ax) = a$	$\pi_n(\tau x) = \tau \cdot \pi_n(x)$
$\pi_{n+1}(ax) = a \cdot \pi_n(x)$	
$\pi_n(x + y) = \pi_n(x) + \pi_n(y)$	

Remarks Because of the τ-laws, we must have that executing a τ does not increase depth. A process p is **finite** if it is equal to a closed term; otherwise p is **infinite**. Note that if p is finite, there is an n such that $\pi_n(p) = p$.

Projection Theorem (Baeten, Bergstra & Klop [1986]) *If the set of processes P forms a solution for a guarded recursive specification E, then $\pi_n(p)$ is equal to some closed ACP_τ-term for each $p \in P$ and $n \geq 1$, and this term does not depend on the particular solution P.*

The projection theorem leads us to formulate the following two principles, which together imply that each guarded recursive specification has a unique solution (that is determined by its finite projections).

The **Recursive Definition Principle (RDP)** is the assumption that each guarded recursive specification has at least one solution, and the **Recursive Specification Principle (RSP)** is the assumption that each guarded recursive specification has at most one solution. In applications in this chapter, we will assume RDP and RSP to be valid.

To give an example, if p is a solution of the guarded recursive specification $\{x = a \cdot x\}$, we find $\pi_n(p) = a^n$ for all $n \geq 1$, so we can put $p = a^\omega$. For more information, see Baeten, Bergstra & Klop [1987]. Abusing language, we also use the variables in a guarded recursive specification for the process that is its unique solution.

In Baeten, Bergstra & Klop [1987], a model is presented for ACP_τ, consisting of rooted, directed multigraphs, with edges labeled by elements of $A \cup \{\delta, \tau\}$, modulo a congruence relation called rooted $\tau\delta$-bisimulation (comparable to Milner's observational congruence, see Milner [1980]). In this model all axioms presented in this chapter hold, and also principles RDP and RSP hold.

The axioms of **Standard Concurrency** (displayed in Table 4.4) will also be used in the sequel. A proof that they hold for all closed terms can be found in Bergstra & Klop [1985].

Table 4.4. Standard concurrency ($a \in A_\delta$).

$$(x \mathbin{\lfloor\!\lfloor} y) \mathbin{\lfloor\!\lfloor} z = x \mathbin{\lfloor\!\lfloor} (y \parallel z)$$
$$(x \mid ay) \mathbin{\lfloor\!\lfloor} z = x \mid (ay \mathbin{\lfloor\!\lfloor} z)$$
$$x \mid y = y \mid x$$
$$x \parallel y = y \parallel x$$
$$x \mid (y \mid z) = (x \mid y) \mid z$$
$$x \parallel (y \parallel z) = (x \parallel y) \parallel z$$

As one can easily see encapsulation and abstraction cannot in general be distributed over $\parallel$ since in a merge processes may do a communication step and thus it is of great importance which comes first, the encapsulation (or abstraction) operator or the merge. Next *conditional axioms* will be presented to state conditions for distributing τ_I and ∂_H over $\parallel$.

Definition The **alphabet** of a process is the set of atomic actions that it can perform. So an alphabet is a subset of A. In order to define the alphabet function α on processes, we have the axioms in Table 4.5 (see Baeten, Bergstra & Klop [1986]).

Table 4.5. Alphabet ($a \in A, I \subseteq A$).

$\alpha(\delta) = \emptyset$	AB1
$\alpha(\tau) = \emptyset$	AB2
$\alpha(ax) = \{a\} \cup \alpha(x)$	AB3
$\alpha(\tau x) = \alpha(x)$	AB4
$\alpha(x + y) = \alpha(x) \cup \alpha(y)$	AB5
$\alpha(x) = \cup_{n \geq 1} \alpha(\pi_n(x))$	AB6
$\alpha(\tau_I(x)) = \alpha(x) - I$	AB7

Note that $\alpha(\delta) = \alpha(\tau) = \emptyset$ is necessary by axioms A6 and T1. The axioms AB6 and AB7 can be proved from AB1–5 for *closed* terms, but are needed here to define the alphabet on general processes. Now we can formulate the conditional axioms as is done in Table 4.6.

Table 4.6. Conditional axioms ($H, I \subseteq A$).

$\alpha(x) \mid (\alpha(y) \cap H) \subseteq H$	$\Rightarrow$	$\partial_H(x \parallel y) = \partial_H(x \parallel \partial_H(y))$	CA1
$\alpha(x) \mid (\alpha(y) \cap I) = \emptyset$	$\Rightarrow$	$\tau_I(x \parallel y) = \tau_I(x \parallel \tau_I(y))$	CA2
$\alpha(x) \cap H = \emptyset$	$\Rightarrow$	$\partial_H(x) = x$	CA3
$\alpha(x) \cap I = \emptyset$	$\Rightarrow$	$\tau_I(x) = x$	CA4
$H = J \cup K$	$\Rightarrow$	$\partial_H(x) = \partial_J \circ \partial_K(x)$	CA5
$I = J \cup K$	$\Rightarrow$	$\tau_I(x) = \tau_J \circ \tau_K(x)$	CA6
$H \cap I = \emptyset$	$\Rightarrow$	$\tau_I \circ \partial_H(x) = \partial_H \circ \tau_I(x)$	CA7

In Baeten, Bergstra & Klop [1986] the axioms CA1–7 have been proved to hold for all closed ACP_τ-terms. We will assume that they hold for all processes.

3 A PALINDROME RECOGNIZER

In the following we will describe a machine which is able to recognize palindromes from strings of input symbols, i.e., a machine that answers 'true' iff a given string of input symbols is equal to its reverse. Suppose S is a finite set of symbols from which the input strings are built up. We have a predicate *ispal* with strings of symbols as its domain which is true iff its argument is a palindrome. We write $|w|$ for the length of the string w. The actions of sending and receiving a symbol $d \in S$ along some channel are written as $s(d)$ and $r(d)$ respectively.

Now we can easily write down the specification of the palindrome recognizer PAL as is done in Table 4.7.

Table 4.7. A specification of the palindrome recognizer PAL.

$$\mathrm{PAL}(\varepsilon) = s(\mathrm{true}) \cdot \mathrm{PAL}(\varepsilon) + \Sigma_{x \in S} r(x) \cdot s(\mathrm{true}) \cdot \mathrm{PAL}(x)$$
$$\mathrm{PAL}(w) = \Sigma_{x \in S} r(x) \cdot s(\mathrm{ispal}(x \cdot w)) \cdot \mathrm{PAL}(x \cdot w) \qquad (|w| > 0)$$

The specification in Table 4.7 describes precisely our intuition about what a palindrome recognizer should do. In case it contains the empty string ε it will answer 'true' (since ε is a palindrome) and return in its original state, or it will receive a new value x and answer 'true' (since every one-word string is a palindrome) and enter the new state PAL(x). In case it contains a string w, however, it will receive a new symbol x, answer whether or not $x \cdot w$ is a palindrome and enter the state PAL($x \cdot w$).

Note that the machine PAL only *receives* input symbols. Since it is clear that a palindrome recognizer should not throw away any of its received information the machine described in Table 4.7 needs to be able to contain arbitrarily long strings of symbols. In practice, however, machines are of a finite size. So from a more practical point of view we should give a specification of a machine that works only on input strings with a limited length. This is done in Table 4.8 where a finite machine PAL_k is specified, working exactly like the previous palindrome recognizer but now with a limit to the length of its input. For reasons to be explained later this limit is put at length $2k$ instead of k.

We assume the machine PAL_k to have an input/output channel numbered $k+1$. So $s_{k+1}(d)$ and $r_{k+1}(d)$ will denote the actions of sending and receiving a symbol d. Furthermore, the fourth equation says that if PAL_k has reached its maximum capacity it will deadlock, for instance by announcing memory-overflow.

Table 4.8. A specification of PAL_k for arbitrary natural number k.

$$\mathrm{PAL}_0(w) = s_1(\mathrm{true}) \cdot \mathrm{PAL}_0(w) \qquad (0 \leq |w|)$$
$$\mathrm{PAL}_{k+1}(\varepsilon) = s_{k+2}(\mathrm{true}) \cdot \mathrm{PAL}_{k+1}(\varepsilon) + \Sigma_{x \in S} r_{k+2}(x) \cdot s_{k+2}(\mathrm{true}) \cdot \mathrm{PAL}_{k+1}(x)$$
$$\mathrm{PAL}_{k+1}(w) = \Sigma_{x \in S} r_{k+2}(x) \cdot s_{k+2}(\mathrm{ispal}(x \cdot w)) \cdot \mathrm{PAL}_{k+1}(x \cdot w) \quad (0 < |w| < 2(k+1))$$
$$\mathrm{PAL}_{k+1}(w) = \delta \qquad (2(k+1) \leq |w|)$$

We will now introduce an implementation of a palindrome recognizer. This particular implementation is a large chain of cells, each of which is itself a palindrome recognizer of size 2. We will prove that a concurrent merge of k such cells gives us exactly a palindrome recognizer of size $2k$, i.e., satifies the specification in Table 4.8.

Consider the cell pictured in Figure 4.1.

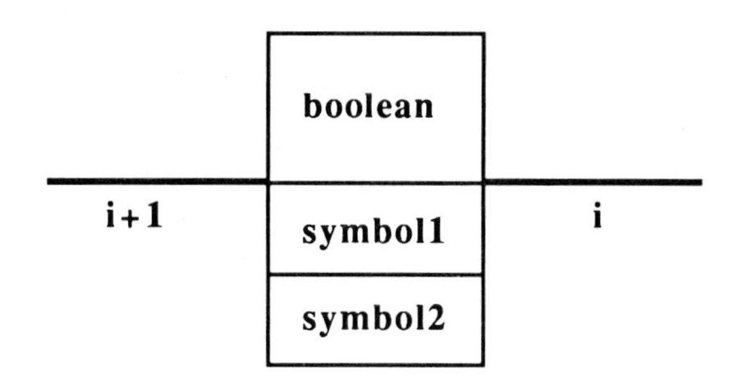

Figure 4.1. An individual cell, C_i, of the palindrome recognizer

The ith cell C_i has two communication channels i and $i+1$. Internally C_i has three storage locations, one for boolean values and two for symbols.

The cell C_i has three distinct states.

(1) In the initial state the cell carries no symbols, i.e., carries the empty word, and since the empty word is a palindrome it can always output the boolean value true to the left. If a symbol is input from the left it is stored in the location **symbol2**, then the boolean value true is output to the left since a word consisting of a single symbol always is a palindrome. The cell now is in state one.
(2) In state one a symbol is input from the left and a boolean from the right (in any order), and stored in the remaining locations **symbol1** and **boolean**. The cell is now in state two.
(3) In state two the cell contains two symbols **symbol1** and **symbol2** forming a word that is a palindrome iff **symbol1** = **symbol2**. Now a boolean value **b** is output to the left, which is calculated according to the formula

$$\mathbf{b} \Leftrightarrow \mathbf{boolean} \text{ and } (\mathbf{symbol1} = \mathbf{symbol2}).$$

Hence before deciding about its output the cell C_i *consults* messages received from its channels. Together with this boolean output the symbol in location **symbol1** is output to the right (in any order interleaved) making room for new input symbols. The cell is now in state one once more.

In the language of ACP_τ the behaviour of the cell C_i described above can be expressed by the equations shown in Table 4.9. The fourth equation defines a machine called TC which stands for *terminal cell.* Since TC never 'contains' any symbol (or always contains the empty string) it can always output a boolean value true and thus behaves like a palindrome recognizer of size zero since the empty string is a palindrome.

Table 4.9. Formal definition of the behaviour of an individual cell.

$$C_i = s_{i+1}(\text{true}) \cdot C_i + \Sigma_{x \in S} r_{i+1}(x) \cdot s_{i+1}(\text{true}) \cdot C'_i(x)$$
$$C'_i(x) = [\{\Sigma_{y \in S} r_{i+1}(y)\} \parallel \{\Sigma_{v \in \{\text{true,false}\}} r_i(v)\}] \cdot C''_i(x, y, v)$$
$$C''_i(x, y, v) = [s_{i+1}(x = y \text{ and } v) \parallel s_i(y)] \cdot C'_i(x)$$
$$\text{TC} = s_1(\text{true}) \cdot \text{TC}$$

Note that the second equation violates the scope rules of Σ since y and v are bound variables in the first term. We will nevertheless use this notation as a shorthand for the correct but much more complex term

$$\Sigma_{y \in S} r_{i+1}(y) \cdot [\Sigma_{v \in \{\text{true,false}\}} r_i(v) \cdot C''_i(x, y, v)] + \Sigma_{v \in \{\text{true,false}\}} r_i(v) \cdot [\Sigma_{y \in S} r_{i+1}(y) \cdot C''_i(x, y, v)].$$

We prefer not to introduce a formal notion here.

From the cells described above we now construct a larger machine by putting the cells in a chain and defining communications between connected cells. Consider the configuration as pictured in Figure 4.2. The cells C_{i-1} and C_i can communicate through channel i by the communication action $s_i(x) \mid r_i(x)$. Any separate action $s_i(x)$ or $r_i(x)$ will be encapsulated, except for $s_{k+1}(x)$ and $r_{k+1}(x)$, since there is no cell C_{k+1} to communicate with them. Hence these two actions can communicate with the outside world.

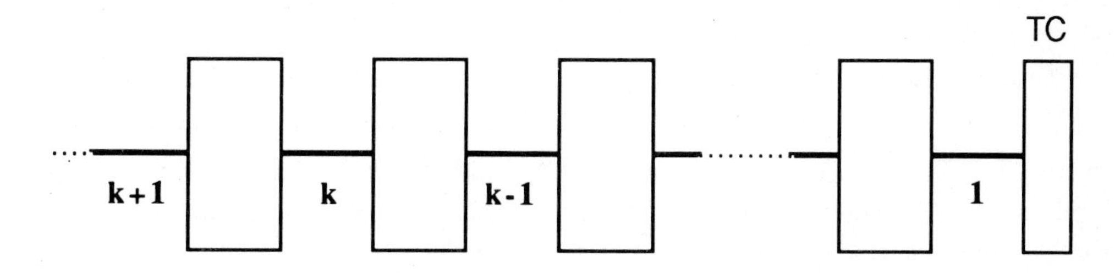

Figure 4.2. A chain configuration of k cells.

From now on we assume k to be fixed. We have the following *communication function* defined on atomic actions:

$$s_i(x) \mid r_i(x) = c_i(x) \qquad \text{for all } x \in S \text{ and } i < k+1;$$
$$a \mid b = \delta \qquad \text{for all other pairs of actions } a, b \in A.$$

The *encapsulation set* H_k of actions resulting in a deadlock is defined as

$$H_k = \{s_i(x), r_i(x) : x \in S \text{ and } i < k+1\}$$

The *abstraction set* I of internal communication actions is defined as

$$I = \{c_i(x) : x \in S \text{ and } i < \omega\}.$$

Note that none of the actions from I occur in the specification of Table 4.8. One can look at them as actions that are *invisible* or *hidden*, and cannot be influenced from outside.

The machine pictured in Figure 4.2 can be described algebraically as a communication merge $M(k)$ of k individual cells, i.e.,

$$M(k) = \tau_I \partial_{H_k}(C_k \parallel \ldots \parallel C_1 \parallel \mathrm{TC}).$$

In the following we will formally prove that $M(k)$ indeed is an implementation of the palindrome recognizer given in Table 4.8.

4 A FORMAL PROOF OF CORRECTNESS

Before turning to the formal proof itself let us first try an example to see how the machine works. Indeed this gives us some intuition about the practical behaviour of $M(k)$ which will be helpful later in this paragraph. The specific example given below was found in Hennessy[1986].

In Figure 4.3 four connected cells are pictured and we let the machine respond at the input string *abaabaab*. As we see, immediately after receiving a new input symbol the machine returns a boolean value at the leftmost channel, stating whether or not the string input so far is a palindrome. We get to the correctness theorem which will be proved by means of the equations of ACP_τ together with RSP, the Recursive Specification Principle, which says that if two processes satisfy the same guarded recursive specification then they are equal.

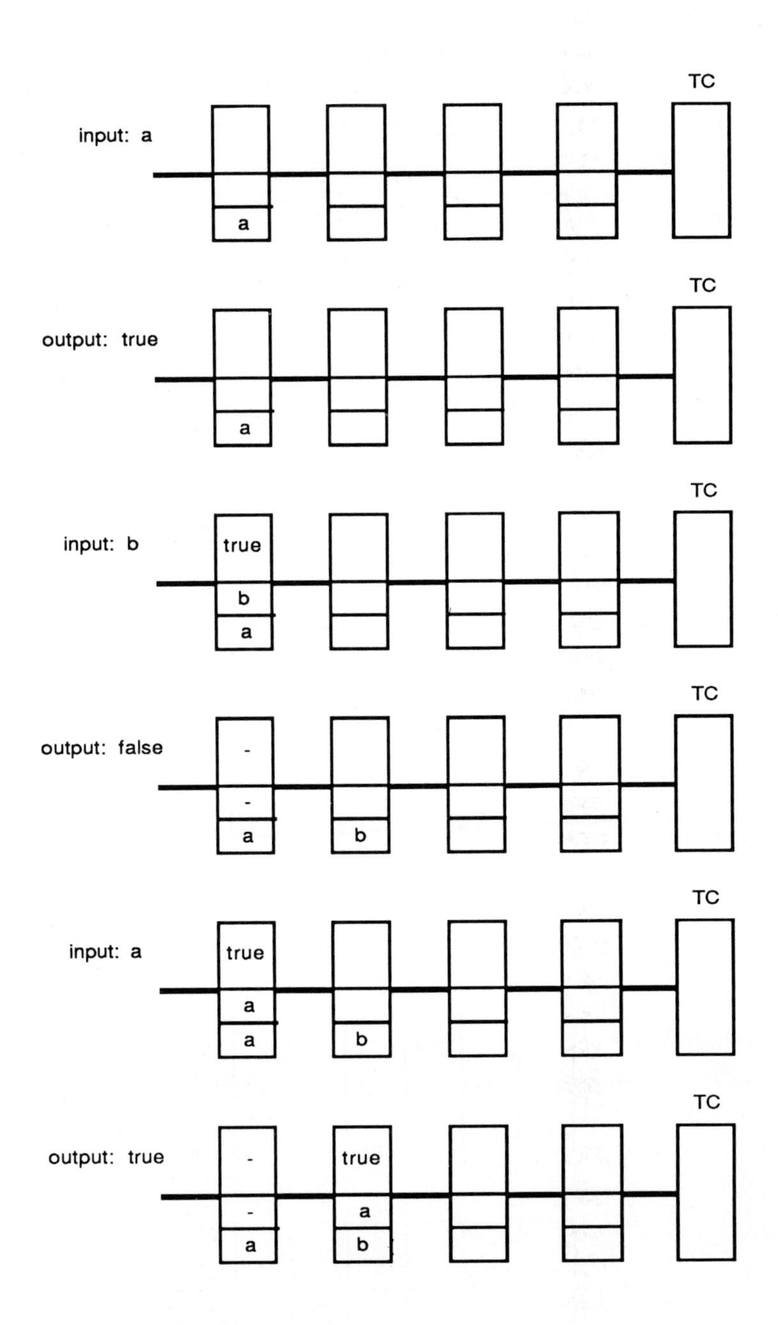

Figure 4.3. An example of the machine M(4).

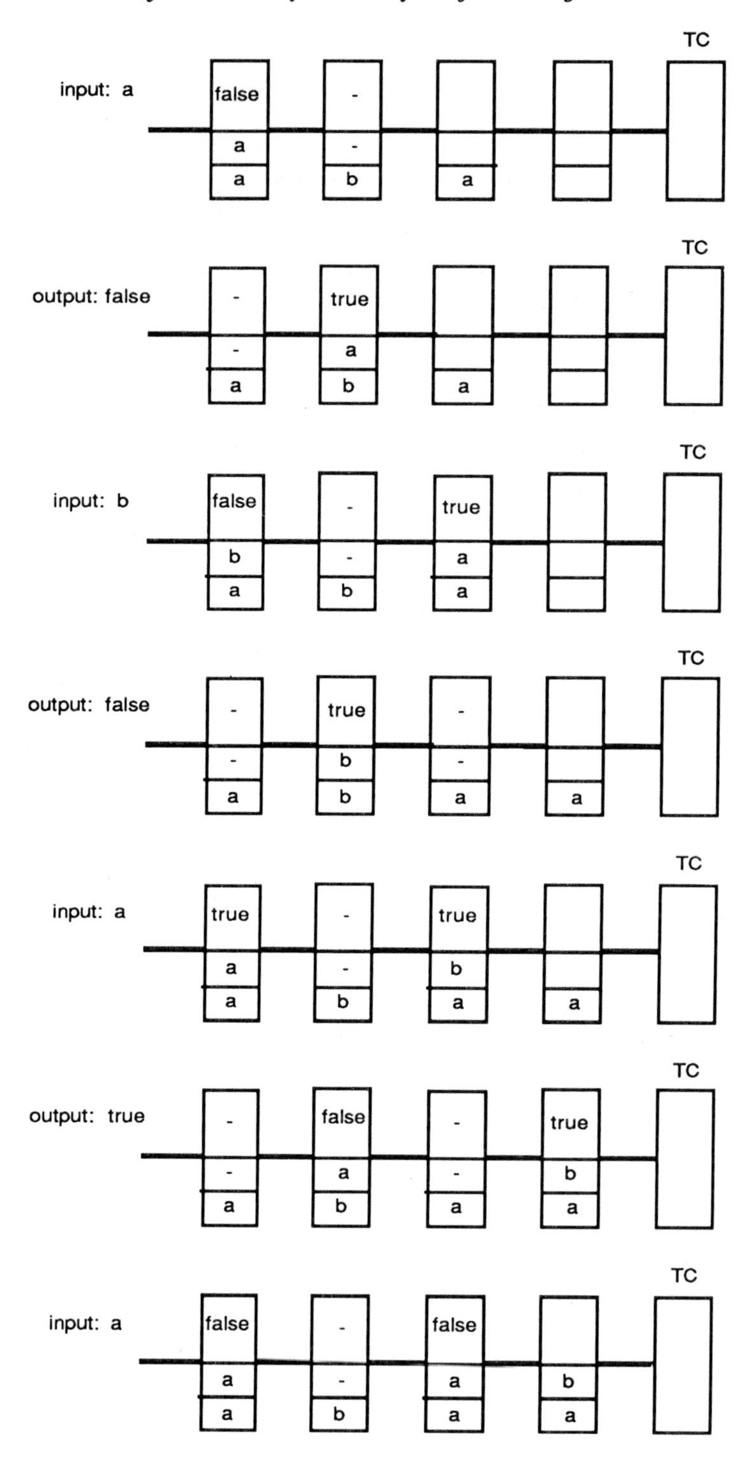

Figure 4.3. (continued).

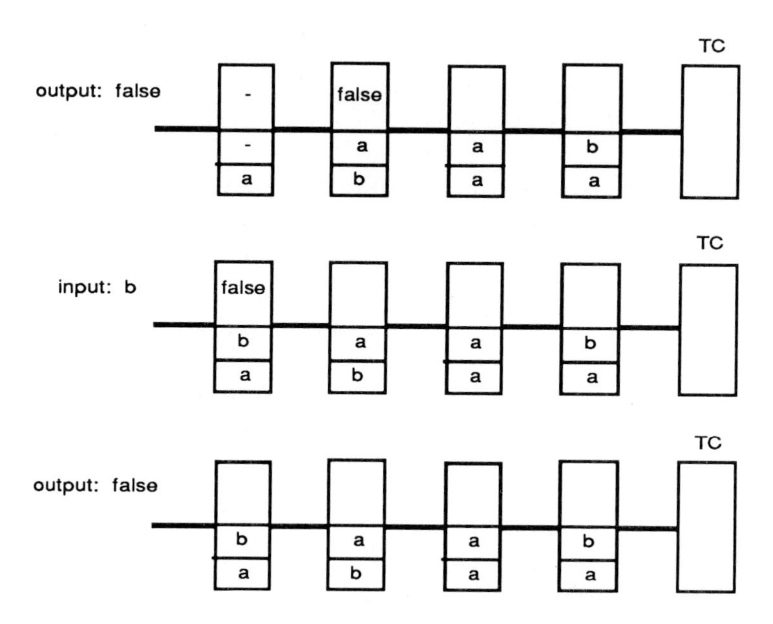

Figure 4.3. (continued).

Theorem $M(k) = \tau_I\partial_{H_k}(C_k \parallel \ldots \parallel C_1 \parallel \mathrm{TC}) = \mathrm{PAL}_k(\varepsilon)$.

Proof By induction on k.

k=0: $M(0) = \tau_I\partial_{H_0}(\mathrm{TC}) = \mathrm{TC}$, and using RSP we directly find $\mathrm{TC} = \mathrm{PAL}_0(\varepsilon)$.
k+1: We first prove $\tau_I\partial_{H_{k+1}}(C_{k+1} \parallel \mathrm{PAL}_k(\varepsilon)) = \mathrm{PAL}_{k+1}(\varepsilon)$. It is easily checked that the following two equations hold:

$$\begin{aligned}\tau_I\partial_{H_{k+1}}(C_{k+1} \parallel \mathrm{PAL}_k(\varepsilon)) = s_{k+2}(\mathrm{true}) \cdot \tau_I\partial_{H_{k+1}}(C_{k+1} \parallel \mathrm{PAL}_k(\varepsilon)) \\ + \Sigma_{x\in S} r_{k+2}(x) \cdot s_{k+2}(\mathrm{true}) \\ \cdot \tau_I\partial_{H_{k+1}}(C'_{k+1}(x) \parallel \mathrm{PAL}_k(\varepsilon));\end{aligned} \tag{1}$$

$$\begin{aligned}&\tau_I\partial_{H_{k+1}}(C'_{k+1}(x) \parallel \mathrm{PAL}_k(\varepsilon)) \\ &\quad = \tau \cdot \Sigma_{y\in S} r_{k+2}(y) \cdot \tau_I\partial_{H_{k+1}}(C''_{k+1}(x, y, \mathrm{true}) \parallel \mathrm{PAL}_k(\varepsilon)).\end{aligned} \tag{2}$$

We formulate as a lemma what is in fact the crucial induction hypothesis.

Lemma *For all symbols $x, y \in S$ and strings $v \in S^*$ with $|v| \leq 2k$, we have*

(i) $\tau_I \partial_{H_{k+1}}(C'_{k+1}(x) \parallel \{s_{k+1}(\text{ispal}(v)) \cdot \text{PAL}_k(v)\}) = \tau \cdot \text{PAL}_{k+1}(v \cdot x)$,

(ii) $\tau_I \partial_{H_{k+1}}(C''_{k+1}(x, y, \text{ispal}(v)) \parallel \text{PAL}_k(v)) = \tau \cdot s_{k+1}(\text{ispal}(y \cdot v \cdot x))) \cdot \text{PAL}_{k+1}(y \cdot v \cdot x)$.

Proof Define

$$Q(v,x) = \begin{cases} \tau_I \partial_{H_{k+1}}(C'_{k+1}(x) \parallel \{s_{k+1}(\text{ispal}(v)) \cdot \text{PAL}k(v)\}) & \text{if } |v| \leq 2k, \\ \delta & \text{if } |v| > 2k. \end{cases}$$

Now we prove that, for all $|v| \leq 2k$ (or equivalently: $|vx| < 2(k+1)$), we have

$$Q(v,x) = \tau \cdot \Sigma_{y \in S} r_{k+2}(y) \cdot s_{k+2}(\text{ispal}(y \cdot v \cdot x))) \cdot Q(y \cdot v, x)$$

and hence, by RSP and the specification of PAL_k in Table 4.8,

$$Q(v,x) = \tau \cdot \text{PAL}_{k+1}(v \cdot x).$$

For all $|v| \leq 2k$ we have:

$$\begin{aligned} Q(v,x) &= \tau_I \partial_{H_{k+1}}(\{[\Sigma_{y \in S} r_{k+2}(y) \parallel \Sigma_{b \in \{\text{true,false}\}} r_{k+1}(b)] \\ &\quad \cdot C''_{k+1}(x,y,b)\} \parallel \{s_{k+1}(\text{ispal}(v)) \cdot \text{PAL}_k(v)\})) \\ &= \Sigma_{y \in S} r_{k+2}(y) \cdot \tau_I \partial_{H_{k+1}}(\{\Sigma_{b \in \{\text{true,false}\}} r_{k+1}(b) \\ &\quad \cdot C''_{k+1}(x,y,b)\} \parallel \{s_{k+1}(\text{ispal}(v)) \cdot \text{PAL}_k(v)\}) \\ &\quad + \tau \cdot \tau_I \partial_{H_{k+1}}(\{\Sigma_{y \in S} r_{k+2}(y) \cdot C''_{k+1}(x,y,\text{ispal}(v))\} \parallel \text{PAL}_k(v)) \\ &= \tau \cdot \Sigma_{y \in S} r_{k+2}(y) \cdot \tau_I \partial_{H_{k+1}}(C''_{k+1}(x,y,\text{ispal}(v)) \parallel \text{PAL}_k(v)) \end{aligned}$$

using axiom T2.

(i) Suppose $|v| < 2k$, then we find

$$\begin{aligned} &\tau_I \partial_{H_{k+1}}(C''_{k+1}(x,y,\text{ispal}(v)) \parallel \text{PAL}_k(v)) \\ &\quad = s_{k+2}(x = y \text{ and } \text{ispal}(v)) \\ &\qquad \cdot \tau_I \partial_{H_{k+1}}(\{s_{k+1}(y) \cdot C'_{k+1}(x)\} \\ &\qquad \parallel \{\Sigma_{z \in S} r_{k+1}(z) \cdot s_{k+1}(\text{ispal}(z \cdot v))) \cdot \text{PAL}_k(z \cdot v)\}) \\ &\qquad + \tau \cdot \tau_I \partial_{H_{k+1}}(\{s_{k+1}(x = y \text{ and } \text{ispal}(v)) \cdot C'_{k+1}(x)\} \\ &\qquad \parallel \{s_{k+1}(\text{ispal}(y \cdot v))) \cdot \text{PAL}_k(y \cdot v)\}) \\ &\quad = \tau \cdot s_{k+2}(x = y \text{ and } \text{ispal}(v)) \cdot \tau_I \partial_{H_{k+1}}(C'_{k+1}(x) \\ &\qquad \parallel \{s_{k+1}(\text{ispal}(y \cdot v))) \cdot \text{PAL}_k(y \cdot v)\}) \\ &\quad = \tau \cdot s_{k+2}(x = y \text{ and } \text{ispal}(v)) \cdot Q(y \cdot v, x), \end{aligned}$$

since $|y \cdot v| \leq 2k$.

(ii) Suppose $|v| = 2k$, then we have

$$\begin{aligned}
&\tau_I \partial_{H_{k+1}}(C''_{k+1}(x, y, \text{ispal}(v)) \parallel \text{PAL}_k(v)) \\
&\qquad = \tau_I \partial_{H_{k+1}}(C''_{k+1}(x, y, \text{ispal}(v)) \parallel \delta) \\
&\qquad = \tau \cdot s_{k+2}(x = y \text{ and } \text{ispal}(v)) \cdot \delta \\
&\qquad = \tau \cdot s_{k+2}(x = y \text{ and } \text{ispal}(v)) \cdot Q(y \cdot v, x),
\end{aligned}$$

since $|y \cdot v| > 2k$.

Since $(x = y \text{ and ispal}(v)) \Leftrightarrow \text{ispal}(y \cdot v \cdot x))$ we have, for all $|v| \leq 2k$,

$$\begin{aligned}
&\tau_I \partial_{H_{k+1}}(C''_{k+1}(x, y, \text{ispal}(v)) \parallel \text{PAL}_k(v)) \\
&\qquad = \tau \cdot s_{k+2}(\text{ispal}(y \cdot v \cdot x))) \cdot Q(y \cdot v, x).
\end{aligned}$$

After substitution we find

$$Q(v, x) = \tau \cdot \Sigma_{y \in S} r_{k+2}(y) \cdot s_{k+2}(\text{ispal}(y \cdot v \cdot x))) \cdot Q(y \cdot v, x)$$

which is precisely what we wanted.

(iii) For $|v| > 2k$ we directly find

$$Q(v, x) = \delta = \text{PAL}_{k+1}(v \cdot x)$$

By RSP we have Lemma (i). Note that we implicitly proved (ii). □

Proof of correctness theorem (continued) Using the lemma, the proof of the theorem is easy.

With lemma (ii) and (2) we have

$$\begin{aligned}
&\tau_I \partial_{H_{k+1}}(C'_{k+1}(x) \parallel \text{PAL}_k(\varepsilon)) \\
&\qquad = \tau \cdot \Sigma_{y \in S} r_{k+2}(y) \cdot s_{k+2}(\text{ispal}(y \cdot x)) \cdot \text{PAL}_{k+1}(y \cdot x) \\
&\qquad = \tau \cdot \text{PAL}_{k+1}(x).
\end{aligned}$$

Finally with (1) we have

$$\begin{aligned}
&\tau_I \partial_{H_{k+1}}(C_{k+1} \parallel \text{PAL}_k(\varepsilon)) \\
&\qquad = s_{k+2}(\text{true}) \cdot \tau_I \partial_{H_{k+1}}(C_{k+1} \parallel \text{PAL}_k(\varepsilon)) \\
&\qquad\qquad + \Sigma_{x \in S} r_{k+2}(x) \cdot s_{k+2}(\text{true}) \cdot \tau \cdot \text{PAL}_{k+1}(x) \\
&\qquad = \text{PAL}_{k+1}(\varepsilon)
\end{aligned}$$

using RSP again.

So we have

$$\tau_I\partial_{H_{k+1}}(C_{k+1} \parallel \tau_I\partial_{H_k}(C_k \parallel \ldots \tau_I\partial_{H_1}(C_1 \parallel \mathrm{TC})\ldots)) = \mathrm{PAL}_{k+1}(\varepsilon).$$

It is easy to prove by induction, however, that

$$\alpha(C_{k+1}) \mid \{\alpha(C_k \parallel M(k-1)) \cap H_k\} = \emptyset,$$

and

$$\alpha(C_{k+1}) \mid \{\alpha(C_k \parallel M(k-1)) \cap I\} = \emptyset.$$

So, because $H_{k+1} \supseteq H_k$, using the conditional axioms CA1, CA2 and CA5, we find

$$\begin{aligned}&\tau_I\partial_{H_{k+1}}(C_{k+1} \parallel \tau_I\partial_{H_k}(C_k \parallel \ldots \tau_I\partial_{H_1}(C_1 \parallel \mathrm{TC})\ldots))\\ &\quad = \tau_I\partial_{H_{k+1}}(\tau_I\partial_{H_k}(\ldots \tau_I\partial_{H_1}(C_{k+1} \parallel C_k \parallel \ldots \parallel C_1 \parallel \mathrm{TC})\ldots)).\end{aligned}$$

Since $H_k \cap I = \emptyset$ for all k, we have

$$\begin{aligned}&\tau_I\partial_{H_{k+1}}(\tau_I\partial_{H_k}(\ldots \tau_I\partial_{H_1}(C_{k+1} \parallel C_k \parallel \ldots \parallel C_1 \parallel \mathrm{TC})\ldots))\\ &\quad = \tau_I \ldots \tau_I\partial_{H_{k+1}} \ldots \partial_{H_1}(C_{k+1} \parallel C_k \parallel \ldots \parallel C_1 \parallel \mathrm{TC})\end{aligned}$$

by axiom CA7 and finally with axioms CA5 and CA6 we find

$$\begin{aligned}&\tau_I \ldots \tau_I\partial_{H_{k+1}} \ldots \partial_{H_1}(C_{k+1} \parallel C_k \parallel \ldots \parallel C_1 \parallel \mathrm{TC})\\ &\quad = \tau_I\partial_{H_{k+1}}(C_{k+1} \parallel C_k \parallel \ldots \parallel C_1 \parallel \mathrm{TC})\end{aligned}$$

which is exactly $M(k+1)$. Therefore, we have $M(k) = \mathrm{PAL}_k(\varepsilon)$, for all k. □

Acknowledgements

Partial support received from the European communities under ESPRIT contract no. 432, An Integrated Formal Approach to Industrial Software Development (Meteor). I especially want to thank Jos Baeten who took the trouble to check this chapter several times before it was printed and who gave so much of his support in developing its contents.

REFERENCES

Baeten, Bergstra & Klop [1986]

J. C. M. Baeten, J. A. Bergstra and J. W. Klop(1986) *Conditional axioms and α/β-calculus in process algebra*, report FVI 86-17, Computer Science Department, University of Amsterdam 1986, to appear in Proceedings of the IFIP Conference on Formal Description of Programming Concepts, ed. M. Wirsing, North-Holland, Ebberup.

Baeten, Bergstra & Klop [1987]
J. C. M. Baeten, J. A. Bergstra and J. W. Klop (1987) *On the consistency of Koomen's Fair Abstraction Rule*, TCS 51 (1/2), pp. 129–76.

Bergstra & Klop [1985]
J. A. Bergstra and J. W. Klop (1985) *Algebra of communicating processes with abstraction*, TCS 37, pp. 77–121.

Bergstra & Klop [1986]
J. A. Bergstra and J. W. Klop (1986) 'Algebra of communicating processes', pp. 89–138 in *Mathematics & Computer Science II*, ed. J. W. de Bakker, M. Hazewinkel & J. K. Lenstra, CWI monograph 4, North-Holland, Amsterdam.

Hennessey [1986]
M. Hennessy (1986) *Proving Systolic Systems Correct*, TOPLAS 8 (3), pp. 344–87.

Hoare [1985]
C. A. R. Hoare (1985), *Communicating sequential processes*, Prentice-Hall.

Huang & Lengauer [1987]
C.-H. Huang and C. Lengauer (1987) 'An implemented method for incremental systolic design', *Proceedings of the PARLE Conference*, ed. J. W. de Bakker, A. J. Nijman and P. C. Treleaven, Eindhoven, Netherlands, LNCS 259, Springer-Verlag.

Kossen & Weijland [1987a]
L. Kossen and W. P. Weijland (1987) 'Correctness proofs for systolic algorithms: palindromes and sorting', report FVI 87-04, Department of Computer Science, University of Amsterdam, to appear in *Applications of process algebra*, ed. J. C. M. Baeten, CWI monographs.

Kossen & Weijland [1987b]
L. Kossen and W. P. Weijland (1987) 'Verification of a systolic algorithm for string comparison', to appear as CWI-report, Centre for Mathematics and Computer Science, Amsterdam.

Kung [1979]
K. T. Kung (1979) 'Let's design algorithms for VLSI systems', in *Proceedings of the Conference on VLSI: Architecture, Design and Fabrication*, California Institute of Technology, January, 1979.

Lipton & Lopresti [1985]
R. J. Lipton and D. Lopresti (1985) 'A systolic array for rapid string comparison', pp. 363–76 in *Proceedings of the Chapel Hill Conference on VLSI*, ed. H. Fuchs.

Mead & Conway [1980]
C. A. Mead and L. A. Conway (1980) *Introduction to VLSI-systems*, Addison-Wesley, Reading, Mass.

Milne [1983]
G. J. Milne (1983) 'CIRCAL: a calculus for circuit description', *Integration*, **1**, pp. 121–60.

Milner [1980]
R. Milner (1980) *A calculus of communicating systems*, LNCS 92, Springer-Verlag.

Mulder & Weijland [1987]
J. C. Mulder and W. P. Weijland (1987) 'Log-time sorting by square comparison', report CS-R8729, Centre for Mathematics and Computer Science, Amsterdam, to appear in *Applications of Process Algebra*, ed. J. C. M. Baeten, CWI monographs.

Rem [1983]
M. Rem (1983) 'Partially ordered computations with applications to VLSI-design', pp. 1–44 in *Proceedings on Foundations of Computer Science IV.2*, ed. J. de Bakker & J. van Leeuwen, MC tract 159, Amsterdam.

Weijland [1987]
W. P. Weijland (1987) 'A systolic algorithm for matrix-vector multiplication', pp. 143–60 in *Proceedings of the SION Conference on CSN*, CWI, Amsterdam.

Weijland [1988]
W. P. Weijland (1988) 'The algebra of synchronous processes', CWI-report CS-R8807, Amsterdam.

Part 2

Theory and methodology of design

In this part the design process itself is examined from three approaches.

In Chapter 5 design is modelled as transforming formal draft system designs, and the user specification process is examined in detail.

In Chapter 6 circuits are relations on signals, and design is achieved through the application of combining forms satisfying certain mathematical laws.

Chapter 7 treats the problem of the automatic synthesis of VLSI chips for signal processing, and the practical issues involved are discussed in greater depth.

5 The formal specification of a digital correlator

N. A. HARMAN AND J. V. TUCKER

Since our concern was speech, and speech impelled us
To purify the dialect of the tribe
And urge the mind to aftersight and foresight

T. S. Eliot
Little Gidding

ABSTRACT

We analyse theoretically the process of specifying the desired behaviour of a digital system and illustrate our theory with a case study of the specification of a digital correlator.

First, a general theoretical framework for specifications and their stepwise refinement is presented. A useful notion of the *consistency* of two general functional specifications is defined. The framework has three methodological divisions: an *exploration phase*, an *abstraction phase*, and an *implementation phase*.

Secondly, a mathematical theory for specifications based on *abstract data types, streams, clocks* and *retimings*, and *recursive functions* is developed. A specification is a function that transforms infinite streams of data. The mathematical theory supports formal methods and software tools.

Thirdly, a digital correlator is studied in considerable detail to demonstrate points of theoretical and practical interest.

1 INTRODUCTION

1.1 Overview

How can we precisely define the desired behaviour of a digital system? What rôle can such precise definitions have in the imprecise process of designing a digital system, and in its subsequent use?

We wish to formulate answers to these questions by theoretically analysing the first step of a design assignment, when it must be determined what is to be designed. This first step we model as a dialogue involving a *user*, or *client*, who commissions a digital device from a *designer*; an activity we term the *user specification process* of the design assignment.

We are interested in the specification of algorithms for digital systems that process an infinite sequence of inputs in time. We will propose a mathematical theory with which the behaviour of synchronous digital systems can be defined and analysed, and hence upon which a semi-formal model of the dialogue between user and designer can be founded. The theory is based on simple mathematical definitions of the ideas of *data type*; *clock*; *retiming*; *functional specification* of the *static behaviour* of a *processor*; *data stream*; *stream transformation*; *synchronisation schemes*; and *functional specification* of the *dynamic behaviour* of a processor. The mathematics supports formal methods and software tools that are helpful for the smooth completion of the user specification process.

With this mathematical theory we will conduct a detailed exploration of the user specification process of a design assignment to build a *digital correlator*. One aim of this case study is to explain and demonstrate the mathematical tools; and some of these will be introduced as and when they are required by the user specification process. Another aim of the exercise is to discuss and illustrate the *methodological structure* of the user specification process. This process will be divided into an *exploration phase*, an *abstraction phase*, and an *implementation phase*. Each phase will have its characteristic mathematical tools. We are interested in clarifying the relationship between the mathematical theory; the formal methods and software tools the theory suggests; and the semi-formal methodology that surrounds their use.

The digital correlator arose as a design assignment in work on spread spectrum radio by our colleague P. A. Matthews in the Department of Electrical and Electronic Engineering. A 32-bit correlator has been designed and successfully fabricated by one of us (Harman [1987]) with the specific aim of critically reviewing conventional design practices. These practices cannot be said to provide a systematic form, let alone constitute a methodological structure, for design; nor do they involve formal methods for algorithmic analysis; nor do they usefully separate the definition of the system from an implementation. A user specification process cannot be identified in current design practices, and seems to be neglected in the literature on VLSI design. This chapter reports on the first stage of what we hope will be a complete theoretical analysis of the design of this correlator. The second stage concerns the design of an architecture to implement the correlator. This involves further mathematical theory about synchronous concurrent systems and, in particular, for proving they meet specifications of the form provided in this chapter: see Harman, Hobley & Tucker [1990].

The user specification theory and case study of a correlator are part of a programme of research into a general theory of the design process for synchronous systems. The general theory includes three computationally equivalent mathematical models for verifying, simulating and laying-out synchronous systems; designs for software tools based on these theories; and methodological models to guide their use; see Section 3.10.

1.2 Results

We begin in Section 2 by describing two theoretical frameworks for investigating formally the two complementary processes of *digital algorithm design* and, in particular, *digital algorithm specification.*

In the framework for the user specification process, which underlies our work here, the main idea is to organise the specification of a component as the production of a series or *portfolio*

$$S_{initial} = S_0, S_1, \ldots, S_n = S_{final}$$

of *draft user specifications.* Each draft user specification is almost a formally defined object

$$S = (L, I, F, E)$$

wherein

> L is a formally defined level of computational abstraction which consists of a *hardware specification language* to define the component, equipped with a clock to measure the performance,
>
> I is an *informal description* of the component's behaviour and performance,
>
> F is a *formal specification* written in L, and
>
> E is a set of *experiments* or tests that F faithfully models I.

With this idea of a draft specification, the problem of making a theory or methodology for the user specification process reduces to the classification problem for transformations or *refinements*

$$S = (L, I, F, E) \rightarrow S' = (L', I', F', E')$$

in which one or more of the components of a draft are changed. The generation of a sequence of these transformations can be partially formalised by means of *formal specification transformations* that preserve the basic property of *consistency* (Section 2.2.6).

The user specification process can be divided into three phases: an *exploration phase* in which some initial formal specifications are obtained that meet with the user's approval; an *abstraction phase* when more suitable abstract refinements of the user's requirements are formulated; and an *implementation phase* when the abstract specifications are implemented in a systematic way to produce specifications of devices suitable for fabrications.

The concepts of drafts, refinements, and consistency, and the methodological division, constitutes a framework that is independent of any mathematical tools, formal methods, or specification languages, used in any particular design situation, including the new tools we introduce here.

In Section 3 we define the basic mathematical concepts we use to specify the behaviour and performance of a digital component. These include an *abstract data type* A and *clock* T, modelled as many-sorted algebras; a *data stream* a, modelled as a map $a : T \to A$; and a *specification* modelled as a stream transformer of the form

$$F : [T \to A] \to [T \to B \cup \{u\}]$$

where $[T \to A]$ and $[T \to B \cup \{u\}]$ are the sets of all data streams from data types A and $B \cup \{u\}$ respectively, and u indicates unspecified or invalid data. In particular, we define the mathematical structure of F in terms of schemes that apply to *any* class of functions on A, T and $[T \to A]$; hence, on applying them to certain recursive functions on A, T and $[T \to A]$ we obtain a specification language with which to define the level of abstraction of our draft specifications.

In Section 4 we give a careful informal description of the correlator. The remainder of the chapter is devoted to performing the three phases of its formal user specification. In Section 5 we accomplish the exploration phase in five consistent draft specifications using the tools of Section 3. In Section 6 we introduce further mathematical concepts, namely that of a *clock retiming* of T to a new clock R, being a function $\lambda : T \to R$ that is surjective and monotonic. This is a tool for expressing simple temporal abstractions and clock synchronisations. It is used in the abstraction phase in Section 7, and has *many* other applications. In Section 8 we describe the systematic implementation of clock retimings in specifications by means of boolean streams which are used in performing the implementation phase of the correlator specification process in Section 9.

The complete portfolio consists of 10 draft user specifications.

The prerequisites of this chapter are an acquaintance with abstract algebra and its application to the theory of abstract data types, and an interest in formal methods and methodologies for digital design.

We are pleased to thank P. A. Matthews and T. F. Buckley for introducing us to the correlator; and J. A. Bergstra, T. F. Buckley, K. M. Hobley, A. R. Martin, K. Meinke, and B. C. Thompson for useful discussions on the matters raised in this chapter.

2 DESIGN AND USER SPECIFICATION PROCESSES

Our analysis of the *user specification process* is intimately connected with an analysis of the *design process* for digital systems, which must be described first. The theory of user specifications given in Section 2.2 and Section 2.3 is intended to support this theory of design. We discuss the relationship between our design theory and design practice in Section 2.4. Finally, we attempt to survey certain literature relevant to our theoretical framework in Section 2.5.

2.1 Theoretical framework for the design of digital systems

2.1.1 What is an ideal digital algorithm? What are the characteristics of an ideal design for a digital algorithm A? We discern the following four types of property.

Logical abstraction The level of logical abstraction of the digital algorithm A should be clearly defined. This is determined by

(1) the level of abstraction of data, and the operations performed on data by the components of the algorithm, and
(2) the constructs used to store and communicate data, schedule and control basic operations, and hence to express the system architecture and the algorithm.

The formalisation of the data (including representation by words, bits and time cycles), and of the components or modules (such as counters, adders, multipliers and microprocessors) is possible by means of *abstract data types* (see Section 3.1). The formalisation of clock phases, registers, and controllers is possible by an appropriate algorithmic language for digital design i.e., an appropriate *hardware description language* HDL (incorporating the abstract data type facility). The algorithm A is formally represented in the language HDL.

Correctness by verification The user's task $TSpec$ of the digital algorithm A should by clearly defined and understood. We allow $TSpec$ to be informal, but require a formal specification $CSpec$ that *mathematically models* $TSpec$. Thus we suppose there is a *hardware specification language* HSL in which $CSpec$ is written; this language complements the hardware description language HDL so that $CSpec$ complements A. In particular HSL and HDL, and $CSpec$ and A, have the same level of abstraction. The introduction of HSL is intended to allow the designer to formulate the idea that *A is correct with respect to $CSpec$*, and to allow him or her to attempt to prove correctness.

We also expect the designer to be able to convince himself or herself that the formal object $CSpec$ faithfully models the informal description $TSpec$ and hence establish a sense in which the designer's algorithm A can accomplish the user's task $TSpec$. In this connection it is important that the specifications of HSL are formal and machine representable and so can be executed, or rather *animated*, by software, and measured or tested against informal understanding (Section 2.2.3).

The process of exploring the relationship between $TSpec$ and $CSpec$ is the subject of this chapter; the process of exploring the relationship between $CSpec$ and A is the subject of its sequel (Harman, Hobley & Tucker [1990]), see Figure 5.1.

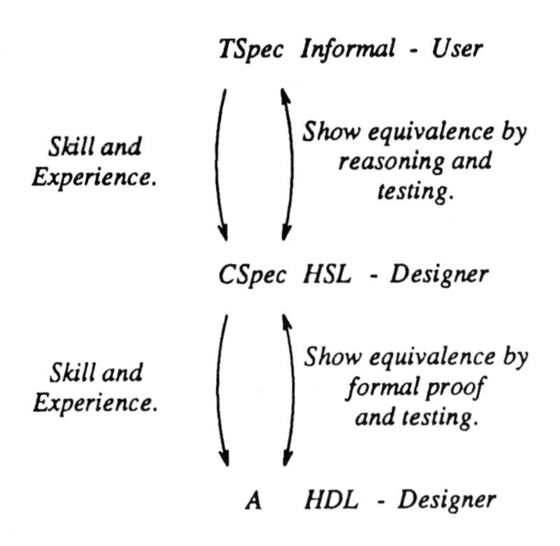

Figure 5.1. The relationship between $TSpec$, $CSpec$ and A

In thinking about an ideal design for a digital system, it is natural (as we have seen) to include ideas about the design process, and the designer and user. For example, for the designer to work on proving that A is correct with respect to $CSpec$, and on establishing that $CSpec$ correctly models the user's informal $TSpec$, certain software tools are necessary. Thus we must postulate a *design environment* based upon HDL and HSL containing verification tools and animation tools (see Figure 5.2). In practice, the existence and quality of such tools (and, by extension, of the environment) are determined by the formal definitions of HDL and HSL. Good tools are an essential criterion in the design of HDL and HSL, and place technical constraints on their structure. For example, the verification tools require sound (and possibly complete) formal correctness logics for reasoning about HDL and HSL.

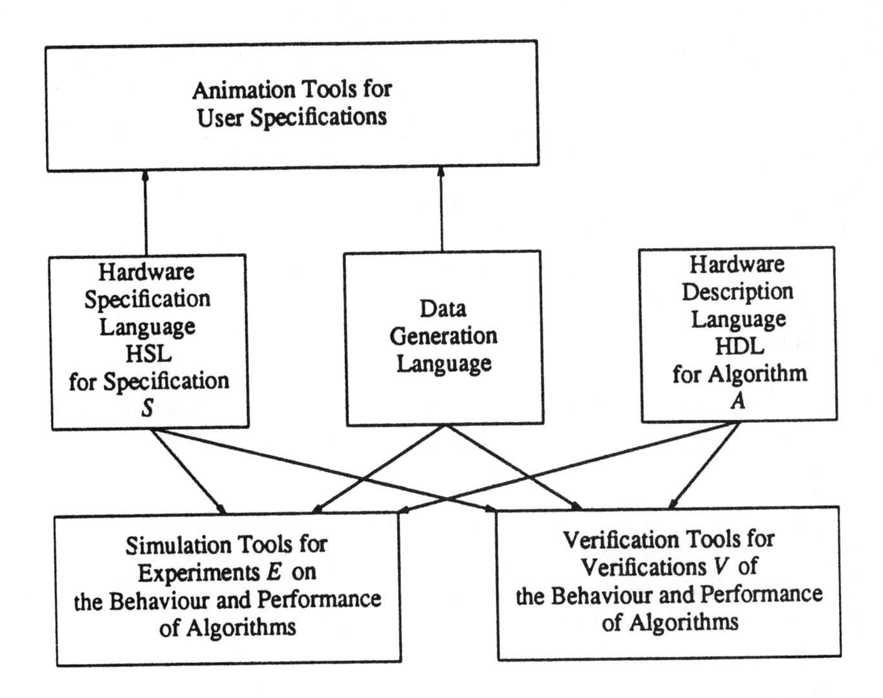

Figure 5.2. Components of digital design environment

Correctness by testing The correctness of A is a matter of both reasoning and testing. The introduction of $CSpec$ puts the process of testing A on a firm foundation. In testing A the aim is to attempt to *refute* $CSpec$. That is, the designer seeks input data upon which A fails to satisfy $CSpec$. If $CSpec$ is formalisable and machine representable, as is necessary for animation in the stage where the user validates it, then testing – the generation of data and the search for counter-examples – can in principle be automated.

We view testing by the designer as experimental computation, undertaken strictly within the level of abstraction of HDL. Thus to the design environment we add an *algorithm simulator* for HDL.

In the environment, the algorithm simulator and the specification animator should cooperate. The data used to test that $CSpec$ is a correct model of $TSpec$ can be used to test that A is a correct implementation of $CSpec$.

Performance The performance or complexity of the algorithm A should be analysed by the designer to assess the computation time or area required by A. This analysis should be made strictly in terms of the level of abstraction of HDL and HSL and should be focussed by a formal specification $PSpec$ of performance. We expect the designer to be able to prove that A meets (parts of) $PSpec$ and to test that A does not fail (parts of) $PSpec$. To do this a formal *performance* or *complexity model* must be added to the formal definition of the hardware description language HDL which contains clearly defined assumptions on the speed and size of the components and communication paths representable in the language. This formal complexity model can be incorporated in a *performance simulator* for HDL which complements the algorithm simulator in the design environment. The specification language HSL must also be enriched to accommodate the complexity model.

2.1.2 Formalising draft designs of digital algorithms The considerations above show that a digital algorithm, in isolation, is incomplete as a product of a scientifically based engineering design process. It is the formalised algorithm together with its formally defined level of abstraction, specifications, proofs and experiments that is to be designed. This package of formal objects we will term a *draft algorithm design* and was seen originally in Thompson & Tucker [1985].

A *draft algorithm design D* is made with five components.

(1) A *level of computational abstraction L* that is formally definable as a *hardware specification language HSL* and a *hardware description language HDL* with formally definable syntax, semantics and complexity model. The language HDL is the foundation upon which the rest of the draft design D is constructed. It formalises the data, components, communication and control employed in the algorithm; the clocks to measure L-time, and the scales to measure L-area, and any other quantities relevant to algorithm performance *as understood at the level of abstraction defined by L.*
(2) The *digital algorithm A* to be designed is the next component of D; A is a well-formed formula or program in the hardware description language HDL.
(3) The *specifications S* are of two kinds: a specification $CSpec$ for the logical correctness of algorithm A and a specification $PSpec$ for the performance correctness of A. The specifications are formal statements in the specification language HSL associated with L. Notice that we have left out the informal task specification $TSpec$.

(4) The *verifications* V are also of two kinds: proofs that A satisfies $CSpec$ and proofs that A satisfies $PSpec$. The proofs are mathematical arguments, capable of formalisation in appropriate *correctness* and *performance logics* tailored to the hardware specification and description languages. In practice, parts of $CSpec$ and $PSpec$ will be devoted to statements worthy of, or at least amenable to, formal proof, and parts will be devoted to statements for testing.

(5) The *experiments* E check the specifications $CSpec$ and $PSpec$ on data. The data sets generate computations that take place strictly within the level of abstraction L, producing results about L-time and L-space, for example. The formal definition of HDL is the foundation for the L-simulator which allows this essential practical work to be included in D.

Thus, in establishing a theoretical structure for the process of designing a digital algorithm we claim that the fundamental object of interest is the concept of a *draft algorithm design* D having the structure of a 5-tuple

$$D = (L, A, S, V, E)$$

as described above and summarised in Table 5.1. Note that D is a formal object based upon the formal definitions of HDL and HSL.

2.1.3 Formalising the design process Using the concept of the draft algorithm design D, we can model the design process as a process of generating a sequence

$$D_{initial} = D_0, D_1, \ldots, D_{n-1}, D_n = D_{final}$$

of draft algorithm designs. The sequence *together with explanations* of each *refinement* of D_i to D_{i+1} is called a *design history, design trajectory* or *design portfolio* and is written

$$D_{initial} = D_0 \to D_1 \to D_2 \to \ldots \to D_{n-1} \to D_n = D_{final}.$$

A *design theory* or *design methodology* for a particular class of system is based upon a classification of transformations or refinements

$$D = (L, A, S, V, E) \to D' = (L', A', S', V', E')$$

that generate design trajectories by changing one or more components in a draft.

Although each draft algorithm design D is a formal object, not all refinements $D \to D'$ will be formally definable (in terms of the languages HDL and HSL, for example). Each draft is required to be formal because it is intended to be machine processable.

Table 5.1. Summary of structure of draft algorithm design

Draft algorithm design

$D = (L, A, S, V, E)$

Level of computational abstraction

L = logical abstractions + performance abstractions
= (hardware description language HDL
+ hardware specification language HSL)
+ complexity model
= data + components + clocks + data streams + architectures
+ functional specifications + schedules + space-time models

Algorithms

A = formula or program in hardware description language HDL

Specifications

S = algorithm specifications $CSpec$+ performance specifications $PSpec$
= (proof specifications $CSpec_{proof}$+ test specifications $CSpec_{test}$)
+ (proof specifications $PSpec_{proof}$+ test specifications $PSpec_{test}$)

Verifications

V = proofs that A satisfies $CSpec_{proof}$ and $PSpec_{proof}$

Experiments

E = tests that A satisfies $CSpec_{test}$ and $PSpec_{test}$

In the practical work of analysing or organising a particular design discipline, the refinements may be rigorous, but very hard to formalise. This is unproblematic for our intended use of the framework.

Formally defined refinements are essential for design methodologies that emphasize correctness and formal verification. Furthermore, formally defined criteria expressing the relationship between D and D' are necessary. For example, in the simple optimisation of a circuit A, we may take $L = L'$ and $S = S'$, and must show that A' is equivalent to A in L. The *equivalence of algorithms* is a fundamental notion in a formal theory of refinements for draft designs.

2.2 A theoretical framework for the specification of digital systems

In this section we study a *user specification process* that is complementary to, but also *theoretically* independent of, the *design process* described in Section 2.1. This user specification process is concerned with the development of the formal specifications of $CSpec$ from the informal specification $TSpec$. We will first study the fine structure of formal user specifications (in the light of the above), and then, in the next section, consider the methodological structure of the specification process.

2.2.1 The origin of the initial draft design Given the design trajectory as presented in Section 2.1.3, how does the designer obtain $D_{initial}$? Clearly, any algorithm A in $D_{initial}$ must be developed from some specification S, and within some level of abstraction L, both in $D_{initial}$.

We expect that the design process will be preceded by a *user specification process*, when a series of formal user specifications within a number of levels of abstraction are developed, including the L and S in $D_{initial}$. We envisage that a series $S_0, \ldots, S_m$ of *draft user specifications* will be developed in a manner similar to the development of draft algorithm designs discussed in Section 2.1.3. The specification process can be considered to have ended, and the design process to have begun, when the designer has developed sufficiently accurate and formal specifications S_i and levels of abstraction L_i to successfully formalise algorithms A_i which are viable solutions to the user's problem.

2.2.2 Requirements on formal user specifications In considering the structure and development of formal user specifications, we must keep in mind that such specifications have four main purposes.

(1) They provide a target for *correctness proofs.* If designers are to formally prove an algorithm correct, they must show it meets a formal specification.
(2) They provide a target for *testing.* Algorithms may be automatically tested against formal specifications, so streamlining the testing procedure.
(3) They eliminate misunderstanding between the user and designer by forming the basis of a *formal contract*, stating precisely what a device is intended and should be able to do.
(4) They provide a basis for *product specifications* for other users.

Let us emphasise that we include testing. Providing a formal specification against which designs can be tested in a largely mechanised manner is useful because it will be more effective in finding errors. This is partly because many more tests may be run; the designer need only specify the test data, and let the results be checked automatically. But mainly it is because a formal specification ensures there is no possible ambiguity about what is, and is not, the correct result of a test.

Therefore we can see that user specifications must meet two widely separate requirements.

(1) They must be sufficiently formal to be used in proofs and to be machine-readable for automatic testing.
(2) They must also be understandable by the user.

The second requirement is necessary because we cannot reasonably expect users to agree to contracts on the basis of formal specifications they do not understand, and because *the involvement of the user is essential to the development of formal specifications* , since users will almost invariably supply only an *informal* specification. We cannot unambiguously formalise informal specifications because they are inherently imprecise.

2.2.3 The iterative nature of the user specification process We observed in Section 2.2.2 that the user must be involved in the development of formal specifications. Consequently, we propose a dialogue in which specifications are refined by the designer and criticised by the user.

From the user's *initial informal specification*, the designer will develop an *initial formal specification*. The user then *animates* the initial formal specification with appropriate software tools using data of his or her choosing. With the help of the process of animation, the designer obtains confirmation of the user's understanding of the formal specification (Section 2.2.2). However, we also expect that direct user understanding of formal specifications is possible. The process of animating and understanding the formal specification should also lead to test data for the final system.

It is probable that the animation process will reveal shortcomings in the formal specification. This may be caused by an inadequate informal specification; an incorrect interpretation of the informal specification by the designer; mathematical mistakes; or it may be deliberate (see Section 2.3.1). Whatever the cause, the user and designer should develop at least a new informal specification to correct any shortcomings or clarify any confusion revealed by the animation process. The designer can then revise the formal specification.

This process continues until the informal and formal specifications converge and meet with the user's satisfaction; that is, until the informal and formal specifications are *user consistent.* Notice that the specification process as outlined here produces a 'tighter', more precise, *informal* specification in addition to the formal specification at each refinement.

2.2.4 The structure of a formal user specification Given the above model of the specification process, how may we best consider the fine structure of a specification? The fine structure of a draft design D as developed in Section 2.1.2 is the 5-tuple (L, A, S, V, E) of which it is made. With reference to this model, we consider a *draft user specification* to be a 4-tuple, $S = (L, I, F, E)$, consisting of the following.

(1) *A level of computational abstraction L.* Clearly, a formal specification, just like a formal algorithm design, must be developed at a formally defined level of computational abstraction.
(2) *An informal specification I.* Recall from Section 2.2.3 that the proposed specification process involves successively developing *informal* as well as formal specifications. We therefore include an informal specification I as part of a specification S. An informal specification may contain text, pictures, and formulae.
(3) *A formal specification F.* This is a formalised version of I, expressed in a hardware specification language HSL based on the level of abstraction L.
(4) *Experiments E.* These include the results of animating F on selected data. They provide *support*, or *justification*, for the proposition: 'F faithfully models I'. Notice we cannot *prove* this proposition because of the imprecision of I.

In summary, a draft specification $S = (L, I, F, E)$ is summarised in Table 5.2.

2.2.5 Formalising the user specification process Using the concept of the draft specification S, we can model the specification process as a process of generating a sequence

$$S_{initial} = S_0, S_1, \ldots, S_{m-1}, S_m = S_{final}$$

of draft specifications. The sequence together with explanations of each refinement of S_i to S_{i+1} is called a *specification history, specification trajectory* or *specification portfolio* and is written

$$S_{initial} = S_0 \to S_1 \to \ldots \to S_{m-1} \to S_m = S_{final}.$$

A *specification theory* or *specification methodology* for a particular class of algorithmic tasks is based upon a classification of transformations or refinements

$$S = (L, I, F, E) \to S' = (L', I', F', E')$$

that generate specific trajectories by changing one or more components in the draft.

Table 5.2. Summary of structure of draft user specification

Draft user specification
$S = (L, I, F, E)$

Level of computational abstraction
L = logical abstractions + performance abstractions
= specification language + complexity model

Informal specification
I = definition of task

Formal specification
F = formal specification that models informal definition

Experiments
E = tests that F satisfies I

We will be concerned with four types of refinements $S \rightarrow S'$: *generalisations* in which the specification S' can be restricted in some way to simulate the behaviour of S; *specialisations* in which the behaviour of S' can be derived from S by restricting it in some way; and *implementations* and *abstractions* in which the specification S' is supposed to be equivalent to S in order to replace it in a design. In each of these cases we say that S' is a *consistent refinement* for S and, in the third and fourth cases, we will say that S' is a *correct refinement* of S.

The objectives and methods of refinements in digital design vary enormously. A draft specification S is not a formal object so only certain forms of refinements can be formalised (in terms of the language HSL, for example). However, basic to the classification of refinements is a classification of the relationships existing between the formal specifications in draft specifications.

2.2.6 Consistency of specifications Consider the situation when the formal specifications are functions (as will be the case in out theory of Section 3). Let $f_1 : X_1 \rightarrow Y_1$ and $f_2 : X_2 \rightarrow Y_2$ be *any* maps that we think of as formal specifications belonging to drafts S_1 and S_2. In terms of these functions, what does it mean to say that S_2 is consistent with S_1?

We define f_2 to be *consistent* with f_1 if there exist maps

$$i : X_1 \to X_2 \quad \text{and} \quad j : Y_2 \to Y_1,$$

which we term *representations* or *codings* of X_1 and Y_2 in X_2 and Y_1 respectively, such that f_1 can be simulated by f_2 in the sense that

$$f_1(x) = j(f_2(i(x))) \quad \text{for all } x \in X_1;$$

or, equivalently, such that Figure 5.3 commutes.

$$\begin{array}{ccc} X_1 & \xrightarrow{f_1} & Y_1 \\ i \downarrow & & \uparrow j \\ X_2 & \xrightarrow{f_2} & Y_2 \end{array}$$

Figure 5.3. Consistency of specifications

The two functions f_1 and f_2 are *consistent* if either f_1 is consistent with f_2 or f_2 is consistent with f_1.

This is a very general idea that is used to provide a general idea of a consistent refinement: we define f_2 to be the result of a *consistent refinement* of f_1 if the functions f_1 and f_2 are consistent. These ideas will receive further discussion and illustration through their application in Sections 3, 5, 7 and 9.

In each application of the definition of consistency the nature of the representation maps i and j is important, but little can be said in general. Suppose $i : X_1 \to X_2$ and $j : Y_1 \to Y_2$ are embeddings (injections). If f_1 is consistent with f_2 with respect to such i and j we will often say that f_2 is an *implementation* of f_1, or that f_1 is an *abstraction* of f_2. These representations i and j may also be called *implementations* of X_1 in X_2 and Y_2 in Y_1 respectively. If, in addition, i and j are isomorphisms (bijections) then we will say that f_2 is an *equivalent implementation* of f_1, or that f_1 is an *equivalent abstraction* of f_2.

Refinements that result in such implementations or abstractions are themselves called *implementation refinements*, or *abstraction refinements*, respectively. We will often refer to them as *correct implementations* rather than consistent refinements.

To illustrate the idea of consistency further, we develop the notions of *specialisations* and *generalisations*.

Consider first a simple case. Suppose that a specification $f : X \to Y$ is refined by imposing a precondition $P \subset X$. This results in the specification $g : P \to Y$ defined by

$$g(x) = f(x) \quad \text{for all } x \in P$$

which behaves as f on a subset of inputs and is a specialisation of f. Clearly, g is consistent with f: take $f_2 = g$ and $f_1 = f$, i to be set inclusion and j to be the identity in the definition.

More generally, a specification $f_1 : X_1 \times W \to Y_1$ can be specialised to make a specification $f_2 : X_2 \to Y_2$ by means of codings $i_0 : X_2 \to X_1$ and $j_0 : Y_1 \to Y_2$, fixed $w \in W$ and

$$f_2(x) = jf_1(i_0(x), w) \quad \text{for all } x \in X_2.$$

Clearly, f_1 is consistent with f_2: take $i : X_2 \to X_1 \times W$ to be defined by $i(x) = (i_0(x), w)$ and $j = j_0$ in the definition.

If f_2 is a specialisation of f_1 we say f_1 is a *generalisation* of f_2.

A complementary set of functions may be formulated for the situation when formal specifications are relations.

2.3 The three-phase model of user specification

Evidence from the correlator and other examples suggests that the specification process has three methodological phases.

2.3.1 Exploration phase During this first phase of the specification process, the designer is *exploring* the problem, attempting to understand and define precisely the user's requirements. The designer develops an initial formal specification from the user's informal specification, and proceeds by a process of refinement to produce a formal specification which is *user consistent.* We expect that the designer's initial attempts at formal specification do not meet the user's requirements. Indeed, it may be better to *deliberately* fail to meet the requirements during early attempts by excluding details that make it easier to capture the essential aspects of the specification. The resulting simpler specification may be more clearly seen to be correct, and the user may be better able to monitor the formal specification process in stages. Omitted details may then be added *incrementally.* At the end of this process we expect the designer to know precisely the user's requirements (through the formal specifications that are user consistent), and the user to be assured that the designer understands his or her needs (through the contemplation and animation of the formal specifications).

The exploration phase of the correlator consists of a portfolio of five specifications each of which is the result of a consistent refinement: see Section 5.

2.3.2 Abstraction phase At the end of the exploration phase, user and designer should be in agreement about what is required. However, the formal specification, while *accurately* reflecting the user's needs, may not be the *best* way of meeting them. We therefore propose that the exploration phase be followed by an *abstraction phase*, involving the replacement of low-level constructs by higher-level, *abstract*, ones. Two important forms of abstraction in digital design are *data abstraction* and *timing*, or *clock, abstraction.*

A typical example of a data abstraction concept is the set of integers

$$\mathbf{Z} = \{\ldots, -1, 0, 1, \ldots\}.$$

In hardware, we will normally represent $\mathbf{Z}$ (or more precisely, a subset of $\mathbf{Z}$) using some *binary encoding scheme* (for example, two's complement, one's complement, or sign and magnitude). While we must *ultimately* represent elements of $\mathbf{Z}$ in some way, there are good reasons for delaying this and hence *removing* a decision on representation that may have been taken by the user at the outset of the exploration phase.

First, by simplifying specifications we can make them easier to understand for both the designer and the user, and easier to prove the correctness of algorithms for the designer. To take full advantage of this we must be able to implement abstractions in behaviour preserving ways. We will return to this point in Section 2.3.3.

Secondly, we wish to eliminate *implementation assumptions.* By this we mean premature (sometimes erroneous and usually unconscious) assumptions that low-level implementations of abstract structures are essential parts of a specification. To continue the example of the integers $\mathbf{Z}$, a user may assume that these must be encoded by sign and magnitude. This could be because the user is unaware of any other scheme, or because a prototype has been built using standard devices which use this scheme. It may be that an alternative coding scheme would suit the user's need better. The abstraction phase should make such implementation assumptions clear and help eliminate them.

There will of course be many occasions when a device's environment will dictate the particular implementation to be used. However this should not prevent us from abstracting away from such details for the purposes of proof and understanding.

At the end of the abstraction phase we expect appropriate high-level specifications of the device to have been developed, that are devoid of implementation detail. We will study *timing abstraction* fully in Sections 6 and 7, and *data abstraction* in Section 9.

The abstraction phase consists of one specification of the correlator: see Section 7.

2.3.3 Implementation phase While abstract formal specifications are useful for the reasons discussed in Section 2.3.2, they do not describe the actual behaviour of the device to be manufactured. We therefore wish to develop *implementation specifications* from abstract specifications in a systematic way, by means of correct refinements. Thus the final stage of the user specification process is the *implementation phase.*

If the designer is to take advantage of the confirmation that the abstract specifications are user consistent, and later of correctness proofs performed using abstract specifications, these implementation specifications must be generated using techniques with proven behaviour-preserving properties. That is, abstract and implementation specifications should be equivalent in some sense. Otherwise, any proofs or experiments the designer has performed at higher levels of abstraction will be meaningless at the implementation level. Indeed, abstract specifications and implementation specifications will potentially have irreconcilable behaviour.

It would be tedious if the designer had to prove that implementation techniques are behaviour preserving for every specification, and it would negate much of the advantage of preforming proofs with abstract specifications. Fortunately however, there are some widely-used techniques for implementing timing and data abstraction (for example, one's complement, two's complement, sign and magnitude for implementing **Z**) which can be proved correct once only and then applied, systematically and straightforwardly, to generate implementation specifications from abstract specifications. Such techniques are the starting points for *specification compilers.* To properly accomplish this, a mathematical theory of specifications, and a formally defined hardware specification language are necessary.

We study techniques for implementing timing abstraction in Section 8 and for implementing data abstraction in Section 9.2.

The implementation phase of the correlator consists of four specifications, each of which is the result of a correct refinement: see Section 9.

2.4 The complete framework

What is the relationship between the specification process and the design process? It would be simplistic to assume that a *design process* linearly follows a *specification process.* The algorithm design process is likely to begin simultaneously with the specification process, even if early algorithm designs do not deserve a formal status.

2.4.1 Integration of user specification and design process Theoretically, it is possible for *each* specification to *induce*, or *initiate*, a design trajectory (Section 2.1.3). Since every design trajectory in the design process arises from a specification, the three phase classification of user specification induces a three phase classification of the design trajectories. We may therefore represent the specification and design processes by a matrix as shown in Figure 5.4. Consider how we have split the process into three phases to indicate the division of the specification process into exploration, abstraction and implementation phases. The key-words decorating the end of each stage are meant to indicate the essential aim of each stage. At the end of the exploration phase, the designer should have a precise understanding of the user's problem. At the end of the abstraction phase, new specifications which capture the essential crux of the user's problem should have been developed. And at the end of the implementation phase, a specification which represents the user's problem in concrete, implementable, terms should have been produced.

Of course, the situation in Figure 5.4 is idealised. First, in any model of practical activity we could hardly expect anything as clear or complete. For example, we would expect only a few specifications to induce useful design trajectories. In particular, many generated in the exploration phase are likely to be informally and incompletely expressed; they may well never get written down and exist only in the mind of the designer. Therefore there is unlikely to be a clearly defined $D_{initial}$. Secondly, many of the later designs (in the abstraction and implementation phases in particular) are likely to be the same algorithm represented at different levels of abstraction. That is, there are likely to be strong *vertical* links between design trajectories as well as the horizontal links between successive draft designs within a trajectory. More precisely, if S_i is correctly implemented by S_{i+1}, draft design D_i^j expresses algorithm A_i^j, and draft design D_{i+1}^l expresses algorithm A_{i+1}^l, then A_i^j and A_{i+1}^l should be provably equivalent. This will be true in the implementation phase, which largely consists of implementing more abstract specifications in standard, formally defined, ways.

2.4.2 Practical use of framework We propose that practising designers explicitly consider specifications and designs as tuples. We intend these models to be used for studying the design process for VLSI devices and, ultimately, by the designers and builders of CAD tools, see Section 3.10. The methodological model, complete with its specification and design tuples, is *a formalised descriptive framework for recording and reporting design activities.* The rôle of this type of design discipline may be likened to the role of *definitions, lemmas, theorems, proofs* and *remarks*

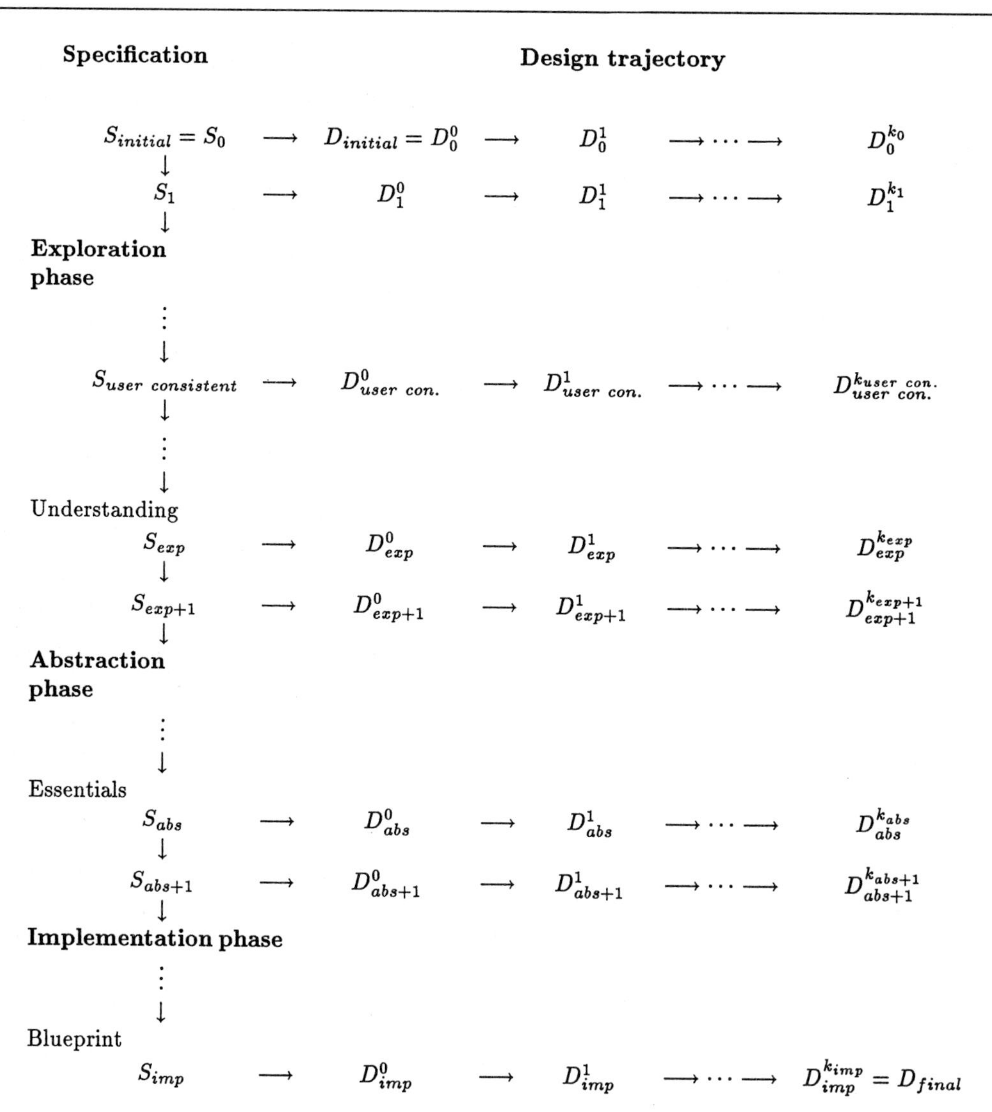

Figure 5.4. Specifications induce design trajectories

that dominate the recording and communication of mathematical knowledge amongst mathematicians. The professional mathematician's working practices are *shaped* by this logical discipline, but are certainly not modelled by it in psychological terms! A diagram such as Figure 5.4 is best considered an 'after the fact' representation of the specification and design processes, with dead ends, parallel paths, and errors omitted. Our presentations of instances of digital design, like the communicated proof of a theorem, will give few, if any, clues to the unprofitable avenues followed.

The framework is a yardstick for evaluating the scope and limits of apparently rigorous and well-understood practices in design. Such a framework is necessary to make digital design more scientifically satisfying and more *accountable*. In the case of designing safety critical systems this objective is essential and urgent.

2.4.3 Theoretical use of framework A unified framework for digital algorithm design focuses attention on neglected subjects such as the precise nature of refinements of both specifications and algorithms; and the verification of performance specifications. Typically, these theoretical topics arise at weak points in the practical application of the framework to some specific type of design problem, such as making systolic algorithms for signal processing. However, the framework also raises questions for the general theory of computation.

We have claimed (Section 2.1.2) that an algorithm is incomplete as a product of a scientifically based design process. Thus, replacing the central concept of an algorithm by that of a draft algorithm design, involving specifications and verifications, requires us to re-examine and, indeed, *replace* the *Church–Turing Thesis* on the scope and limits of computation. Consider the following question, for example:

What class of formal specifications of systems can be implemented by digital algorithms that can be completely specified and verified?

The answers to this and related questions are relevant to establishing the classes of specifications and algorithms that are suitable for safety critical computing.

2.5 Literature

The idea of a draft algorithm design as a 5-tuple first appeared in Thompson & Tucker [1985]. The ideas of a draft specification as a 4-tuple, consistency of refinements, and the three phase specification process are new. We will attempt to survey literature relevant to this theoretical framework.

Digital systems design The activity of digital systems design grew out of telephone systems engineering and the design of computers. A survey of popular textbooks since 1950 may begin with Keister *et al.* [1951], a fascinating text largely concerned with designing networks of *electro-mechanical* switches. This book uses C. E. Shannon's basic boolean algebra techniques seen first in Shannon [1938]. Noteworthy later texts include Flores [1963], a thorough treatment of computer arithmetic; Harrison [1965], an excellent account of the mathematical theory of switching circuits; Clare [1973], the first text to use T. E. Osbourne's *algorithmic state machine* (ASM) notation; Hill & Peterson [1973], an early application of a hardware description language (AHPL) for simulation and description of digital systems; and Mead & Conway [1980], the first text on structured digital systems design in VLSI. A good modern treatment of digital systems design is Ercegovac & Lang [1985], which stresses the current (mainly) informal but highly structured design methodologies. Other interesting works include Lewin [1968, 1977], Mano [1979], and Winkel & Prosser [1980].

Design theory The textbook tradition above reveals an increasing awareness of the process of design but is fundamentally naïve in its treatment of methodologies and formal models. By contrast, design theory for mechanical engineering is substantial: see Pahl & Beitz [1984], by which we have been influenced. Further related work on general design theory is contained in Yoshikawa [1986].

However, it is from work on programming methodology that the basic ideas that we have used to make the framework in this section have been imported. An independent methodological study with these origins is Cohen [1983]; see also Cohen [1982]. Cohen uses the idea of a dialogue structured by *contracts* and a Popperian interpretation of testing. Of course, there has emerged a text-book tradition in software engineering that is concerned exclusively with methodology.

Specification, derivation and verification for software At the heart of the framework is the concern for the formal specification and formal verification of algorithms, and this reflects the contemporary concerns of software engineering.

An early examination of program verification was undertaken by A. Turing in 1949 (see Morris & Jones [1984]), but the main sources for the subject are certain writings of J. McCarthy, P. Naur and especially R. W. Floyd, C. A. R. Hoare, E. W. Dijkstra, and N. Wirth in the period 1963–76. See the collection Gries [1978], and the monograph de Bakker [1981]; for introductory material see Gries [1981], Boyer & Moore [1981], and Backhouse [1986]

From this interest in the difficult activity of program verification there has, more recently, developed a large and diverse body of work on program specification. Significant sources are writings by D. Parnas, J. A. Goguen, R. Burstall, B. Liskov and C. A. R. Hoare. See the collections Gries [1978] and Gehani & McGettrick [1986]; for introductory material see Liskov & Guttag [1986] and the survey Cohen *et al.* [1986].

There are many formalisms for the formal specification of software based upon axiomatic definitions or model building. For example, the algebraic specification languages *OBJ* (for example, Goguen & Tardo [1979]), and *CLEAR* (for example Burstall & Goguen [1981]) are axiomatic formalisms, and the notation *Z* (Abrial *et al.* [1979]; see also Hayes [1987] for an introduction and case studies) is model-based. The semi-formal design methodology *VDM* (Vienna Development Method) is also model-based (see Bjorner & Jones [1982] for example).

Research on specific specification techniques has, more recently, led to research into the general ideas about specifications and operations on specifications. These attempts at general theories aim at more complex notations than those given in Section 2.2.6, and needed in this chapter, but they are relevant to the further development of the framework and its application to digital design. A particularly useful general study of refinements is Back [1981].

Much work is concerned with generalisations of logical approaches. For example, in Cohen *et al.* [1986] there are simple but general definitions concerning specifications modelled by sets of statements in formal systems. A simple study of operations on specifications is Sanella & Wirsing [1983]. A substantial *theory* of such specifications is J. A. Goguen and R. M. Burstall's *theory of institutions* (see Goguen & Burstall [1985, 1986]). Other approaches are: to work on themes of specifications modelled by relations as in, for example, Hoare & He [1985]; to analyse ideas about modularity such as in Back & Mannila [1982], or algebraically, as in Bergstra *et al.* [1986].

The transformation of programs is the basis of the technique of *program construction by stepwise refinement*, proposed by E. W. Dijkstra and N. Wirth, which is concerned with the informal development of programs from specifications. For the formal theories of correctness preserving refinements see: Back [1980], Back [1981], and the survey Mili *et al.* [1986].

The relationship between specifications and programs, and ideas about derivation, are particularly significant for functional programming (Backus [1978], Turner [1985]) and logic programming (Kowalski [1985]). A general study for procedural programming Hoare *et al.* [1987].

Specification, derivation and verification for digital systems The formal verification of digital hardware (and of programs) is most simply founded upon clearly defined logics such as *first-order logic*, *many-sorted first-order logic*, *higher-order logic*, *temporal logic*, *modal logic* and so on. Shannon's fundamental idea was, of course, to use *propositional logic*. Approaches based on first-order logics include Eveking [1985]; and Hunt [1986], who uses the Boyer–Moore theorem prover (Boyer & Moore [1979]) to verify a microprocessor based on the PDP-11.

The application of higher-order logics to formal specification and verification of digital hardware originated with the work of Hanna and Daeche (Hanna & Daeche [1984, 1985 and 1986]), who produced a theorem prover called *VERITAS*. The use of higher-order logics has been taken up by M. J. Gordon and his co-workers (Gordon [1986], Camilleri *et al.* [1986], and Gordon [1987]), and has been applied to the *VIPER* microprocessor. A partial correctness proof for *VIPER* appears in Cohn [1987]. Gordon's group have produced the interactive theorem prover *HOL*.

Temporal logics are logics for reasoning about events possibly separated in time and are an obvious choice for formally reasoning about hardware. Examples include Bochmann [1982], Malachi & Owicki [1981], Clarke & Mishra [1984], Moszkowski [1983] and Moszkowski [1986]. Temporal logic is usually used to write formal specifications only (though Moszkowski is an exception).

Another method is to build upon algebraic calculi, rather than formal logic. An early example is *CIRCAL* (see Milne [1985] for example). Further studies in which the theory of asynchronous communication are applied to hardware are Hennessy [1986] and Weijland [1987].

Another motivation for using formal methods is the ability to transform designs in a 'correctness preserving' way and hence perform refinements. One common aim is to convert simple but inefficient algorithms into complex but efficient ones; for example the *Karnaugh map* (Karnaugh [1953]) is used in digital logic design for transforming arbitrary boolean expression to canonical form. A recent general treatment is Sheeran [1983], which is based on Backus [1978]. Design transformation and derivations are emerging as a practical technique in research on special architectures such as systolic arrays: see the collections Moore *et al.* [1986], de Bakker *et al.* [1987a], de Bakker *et al.* [1987b].

Complexity theory Work germane to our ideas about performance specification includes complexity theory of circuits implementing boolean functions: see the surveys Harrison [1965], Savage [1976] and Wegener [1987]; and more recently, the complexity theory for VLSI of R. Brent, H. T. Kung and C. D. Thompson: see Thompson [1980], Ullman [1984], and the survey paper Baudet [1983]. These discrete models of computational performance contrast with physical models such as the *RC* model: see, for example, Mead & Conway [1980] and Dew, King, Tucker & Williams [1988]. For the semantic foundations of complexity theory see: Asveld & Tucker [1982], and Neilson [1984].

Most research and development in the areas of hardware design and software engineering are based on specific programming notations, specific verification techniques, and specific specification techniques. Our own work on the specification and verification of synchronous hardware is similarly specific. The point of the theoretical framework is to help us to formulate a general independent analysis of their scope and limits. We intend to apply the framework to our technical ideas as a criterion to measure their development.

3 BASIC TOOLS FOR FORMAL SPECIFICATIONS

We introduce mathematical concepts that formalise the ideas of *data types*, *clocks*, *streams*, *functional specifications* for *processors*, and *stream transformers*. A *user specification* for a system will be modelled as a stream transformer, derived from a functional description of a processor by surrounding it with a formal 'wrapper' that *schedules* input and output data with respect to a clock. The stream transformers are a model for a *specification language* which is the basis for the definition of the level of computational abstraction in the draft specifications. These theoretical tools will be sufficient for some initial attempts at the formal specification of the correlator in the exploration phase.

3.1 Abstract data types and many-sorted algebras

In a level of abstraction, data will be defined by an abstract data type. We need only the basic ideas of the algebraic theory of data types which we summarise here. A fuller treatment can be found in (amongst other places) ADJ [1978], Erhig & Mahr [1985] and Meseguer & Goguen [1985].

A *single-sorted algebra* consists of a set A called the *domain*, or *carrier*, of the algebra, together with some elements $c_1, \ldots, c_p$ from A called *constants*, and functions $f_1, \ldots, f_q$ defined on A called *operations*. We write

$$A = (A \mid c_1, \ldots, c_p, f_1, \ldots, f_q).$$

Notice we may denote both the algebra and its carrier by A.

Examples

(1) Consider the algebra **B** of booleans with carrier $\{tt, ff\}$, constants tt and ff, and operations $\neg$ (for *not*) and $\wedge$ (for *and*)

$$\neg : \{tt, ff\} \to \{tt, ff\} \quad \text{and} \quad \wedge : \{tt, ff\}^2 \to \{tt, ff\},$$

we write $\mathbf{B} = (\{tt, ff\} \mid tt, ff, \neg, \wedge)$.

(2) Consider the algebra **Bit** with carrier $\{0, 1\}$, constants 1 and 0, and functions

$$\begin{aligned} not &: \{0,1\} \to \{0,1\}; \\ and &: \{0,1\}^2 \to \{0,1\}, \\ nor &: \{0,1\}^2 \to \{0,1\}, \end{aligned}$$

we write **Bit** $= (\{0,1\} \mid 0, 1, \mathit{not}, \mathit{and}, \mathit{nor})$.

(3) Consider the algebra **N** of natural numbers with carrier $\{0,1,\ldots\}$, constant 0, and function $succ : \{0,1,\ldots\} \to \{0,1,\ldots\}$ defined by

$$succ(n) = n + 1;$$

we write $\mathbf{N} = (\{0,1,\ldots\} \mid 0, succ)$.

(4) Consider the ring **Z** of integers with carrier $\{\ldots,-1,0,1,\ldots\}$, constants $0,1$ and functions

$$\begin{aligned} + &: \{\ldots,-1,0,1,\ldots\}^2 \to \{\ldots,-1,0,1\ldots\}, \\ * &: \{\ldots,-1,0,1,\ldots\}^2 \to \{\ldots,-1,0,1\ldots\}; \\ - &: \{\ldots,-1,0,1,\ldots\} \to \{\ldots,-1,0,1\ldots\}, \end{aligned}$$

we write $\mathbf{Z} = (\{\ldots,-1,0,1,\ldots\} \mid 0, 1, +, *, -)$.

A *many-sorted algebra* extends the concept of a single-sorted algebra by allowing a number of carrier sets $A_1, \ldots, A_s$, constants from these sets, and functions of the form

$$f : A_{v_1} \times \cdots \times A_{v_n} \to A_u,$$

where $v_1, \ldots, v_n, u \in \{1, \ldots, s\}$. Again, we use A to denote both algebra and its carriers. For $v = v_1, \ldots, v_n$ we will write A^v to stand for $A_{v_1} \times \cdots \times A_{v_n}$.

Examples

(5) Consider the algebra made by joining algebras **Z** and **B** together with the relation of equality on integers. Such an algebra will have constants $0, 1 \in \mathbf{Z}$, $tt, f\!f \in \mathbf{B}$, and functions

$$\begin{array}{lll} + : \mathbf{Z}^2 \to \mathbf{Z}, & * : \mathbf{Z}^2 \to \mathbf{Z}, & - : \mathbf{Z} \to \mathbf{Z}, \\ \neg : \mathbf{B} \to \mathbf{B}, & \wedge : \mathbf{B}^2 \to \mathbf{B}, & = : \mathbf{Z}^2 \to \mathbf{B}. \end{array}$$

(6) Let us add to the algebra in Example (5), the algebra **N** of natural numbers, and the ring $\mathbf{Z}[X]$ of polynomials with integer coefficients. This results in a four-sorted algebra with carriers $\mathbf{B}, \mathbf{Z}, \mathbf{N}$ and $\mathbf{Z}[X]$. In addition to the operations we have already defined, we have

$$+ : \mathbf{Z}[X]^2 \to \mathbf{Z}[X], \quad * : \mathbf{Z}[X]^2 \to \mathbf{Z}[X]; \quad - : \mathbf{Z}[X] \to \mathbf{Z}[X];$$

the inclusion mapping

$$i : \mathbf{Z} \to \mathbf{Z}[X],$$

which allows integers to be considered as constant polynomials; the coefficient function

$$coef : \mathbf{Z}[X] \times \mathbf{N} \to \mathbf{Z},$$

such that $coef(p, i)$ is the ith coefficient of polynomial p; and the degree function

$$deg : \mathbf{Z}[X] \to \mathbf{N},$$

such that $deg(p)$ is the degree of polynomial p.

(7) This four sorted polynomial algebra of Example (6) can be generalised by replacing $\mathbf{Z}$ with any commutative ring R. Such structures arise naturally in the theory of signal processing, see Blahut [1985].

An *abstract data type* is a many-sorted algebra considered unique up to isomorphism.

3.2 Functional specifications

At the heart of each formal user specification will be a function, specifying the processor without reference to timing considerations. That is, we will define the device *statically* without reference to its *dynamic* behaviour.

Let A be a non-empty set. We specify a single-sorted processor that computes with data from A by means of a function

$$f : A^n \to A^m.$$

Observe that f is a *vector valued function.* We define the *coordinate functions* $f_1, \ldots, f_m$ of f to be those functions whose result is the ith element of f. Therefore

$$f(a_1, \ldots, a_n) = (f_1(a_1, \ldots, a_n), \ldots, f_m(a_1, \ldots, a_n)).$$

Example

(8) Consider the gcd function, which yields the greatest common divisor of two integers, $gcd : \mathbf{Z}^2 \to \mathbf{Z}$.

If we wish to express functional specifications which operate on data of different types (for example time), we must define functions on many-sorted algebras. Let $A_1, \ldots, A_s$ be any non-empty sets. We specify a many-sorted processor that computes with data from $A_1, \ldots, A_s$ by means of a function

$$f : A_{v_1}^{j_1} \times \cdots \times A_{v_n}^{j_n} \to A_{u_1}^{l_1} \times \cdots \times A_{u_m}^{l_m},$$

where $v_1, \ldots, v_n, u_1, \ldots, u_n \in \{1, \ldots, s\}$.

Examples

(9) Consider the *multiplexor* (*mux*) function, which selects one of two binary streams according to the value of a third boolean stream

$$mux : \{0,1\}^2 \times \mathbf{B} \to \{0,1\},$$
$$mux(x,y,b) = \begin{cases} x & \text{if } b = tt \\ y & \text{if } b = ff. \end{cases}$$

(10) Consider the many-sorted algebra in (6) augmented by polynomial evaluation

$$e : \mathbf{Z}[X] \times \mathbf{Z} \to \mathbf{Z},$$

which evaluates a polynomial $p(z)$ on an integer z.

(11) Let R be a commutative ring and consider the *convolution operation* $conv : R^n \times R^n \to R$, which forms the inner product of an *input word* $\underline{x} = x_1, \ldots, x_n$ and a set of *weights* $\underline{w} = w_1, \ldots, w_n$, defined by

$$conv(\underline{x}, \underline{w}) = x_1 w_1 + \cdots + x_n w_n.$$

Convolution is a signal processing operation commonly used in areas such as digital filtering and image processing.

The sets $A_1, \ldots, A_s$ together with the static specification f define an abstract data type.

3.3 Clocks and streams

To describe digital systems we need to be able to describe functional specifications acting in time. We now define the concept of a *clock* as the simplest abstract data type that can be used to formalise time.

A *clock* is an algebra $T = (T, 0, t+1)$ isomorphic with $\mathbf{N}$ in Example (3). That is, T is a set of natural numbers denoting discrete time, equipped with the successor function denoting the clock's counting of cycles. We can think of a *clock cycle* as being of an interval of time starting at t and ending at $t+1$. The constant 0 is the start of the first time cycle. All clocks are isomorphic. The relationship between clocks is studied in Section 6.

Let $T = \{0, 1, \ldots\}$ denote discrete timepoints or cycles of a clock T. Let A be any non-empty set. We define a *stream* over A to be an infinite sequence $a(0), a(1), \ldots$ of data from A, or a map $s : T \to A$. Let $[T \to A]$ be the *set of streams.*

Observe that $[T \to A]^n \equiv [T \to A^n]$: that is, a vector of streams $(a_1, \ldots, a_n) \in [T \to A]^n$ is equivalent to a stream of vectors $\underline{a} \in [T \to A^n]$. In other words, $\underline{a} = (a_1, \ldots, a_n)$ and $\underline{a}(t) = (a_1(t), \ldots, a_n(t))$.

3.4 Stream transformers

We conceptualise a user specification as a *stream transformer.* That is, a function which accepts one or more streams as input, and generates one or more streams as output. Since streams are themselves functions, our specifications will be higher order functions, or *functionals.*

Let A be a non-empty set. We specify a single-sorted processor that computes with data from A by means of a functional

$$F : [T \to A^n] \to [T \to A^m]$$

where

$$F(a_1, \ldots, a_n)(t) = \text{the output at time } t \text{ of a processor executing on input streams } a_1, \ldots, a_n.$$

The specification is *dynamic* in that it refers to computation in time.

Also of importance is the equivalent *applicative* form $\hat{F}$ of the stream transformer F. This is a function $\hat{F} : [T \to A^n] \times T \to A^m$ defined by

$$\hat{F}(a_1, \ldots, a_n, t) = F(a_1, \ldots, a_n)(t).$$

Given any function $f : A^n \to A^m$ that is a static specification for a processor and takes k clock cycles to compute a result (Section 3.2), we can specify a stream transformer of the form $F : [T \to A^n] \to [T \to A^m]$ that is a dynamic specification of the processor, by applying f repeatedly to the data constituting the input streams. That is $F(\underline{a})(t) = f(\underline{a}(t-k))$, where k is the *computation time* in clock cycles of the processor f. This is illustrated in Figure 5.5.

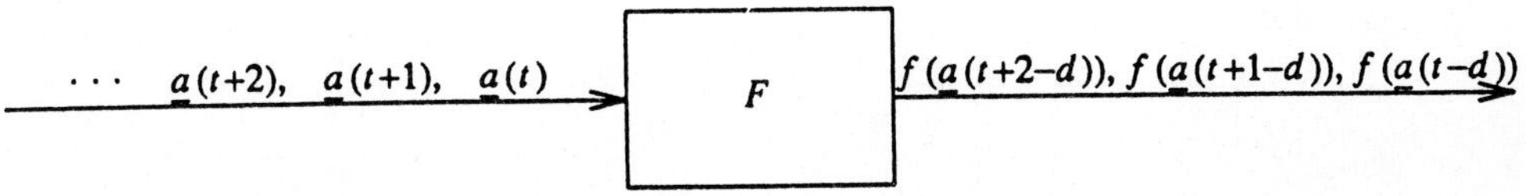

... $\underline{a}(t+2)$, $\underline{a}(t+1)$, $\underline{a}(t)$ → F → $f(\underline{a}(t+2-d))$, $f(\underline{a}(t+1-d))$, $f(\underline{a}(t-d))$

Figure 5.5. Creating a stream transformer from a functional specification

Notice that if the computation time k is non-zero, and if F is required to accept inputs every clock cycle, more than one computation will be in progress at any one time. This will require the computation of F to be *pipelined.* When the ith stage of the jth computation has been completed, the $(i+1)$th stage of the jth computation and the ith stage of the $(j+1)$th computation may proceed. Pipelining is required if the separation in time of the inputs to a processor is less than its computation time.

So for example, the implementation of a processor which requires 4 clock cycles to compute a result must be pipelined if inputs arrive every 3 clock cycles or less, but not if 4 or more clock cycles elapse between inputs, since this will allow a complete computation to carried out before the next input arrives.

In general, for computation time k and time l between inputs, $c = (k+1)/l$ computations must be in progress simultaneously. Clearly c will be fractional if k is not a multiple of l. This just means that at some times $t \in T$, $\lfloor c \rfloor$ computations will be in progress, and at others $\lceil c \rceil$ computations will be in progress.

The introduction of pipelining introduces another performance metric for systems in addition to the computation time*. This is the *output period*, which is the number of clock cycles separating valid outputs from a processor.

As a final point, note that it is perfectly meaningful for the computation time k to be zero. This just means that a processor can complete its computation within a single clock cycle.

Examples

(12) Consider the function $gcd : \mathbf{Z}^2 \to \mathbf{Z}$ of Section 3.2 (Example (8)). We can create a stream transformer $GCD : [T \to \mathbf{Z}]^2 \to [T \to \mathbf{Z}]$

$$GCD(a,b)(t) = gcd(a(t), b(t)).$$

Note here we have made use of the identity $[T \to A]^n \equiv [T \to A^n]$ (Section 3.3) so that we have expressed $GCD : [T \to \mathbf{Z}^2] \to [T \to \mathbf{Z}]$ as $GCD : [T \to \mathbf{Z}]^2 \to [T \to \mathbf{Z}]$.

(13) Consider the function $mux : \{0,1\}^2 \times \mathbf{B} \to \{0,1\}$ of Section 3.2 (Example (9)). We can create a stream transformer $MUX : [T \to \{0,1\}]^2 \times [T \to \mathbf{B}] \to [T \to \{0,1\}]$

$$MUX(x,y,b)(t) = mux(x(t), y(t), b(t)).$$

Let $A_1, \ldots, A_s$ be non-empty sets. We specify a many-sorted processor that computes with data from A^v by means of a functional

$$F : [T \to A_{v_1}] \times \cdots \times [T \to A_{v_n}] \to [T \to A_{u_1}] \times \cdots \times [T \to A_{u_m}]$$

analogously to the single-sorted case.

* In real systems the actual length of the clock cycle in real time units is also important.

Suppose that for given input streams $a_1, \ldots, a_n$ the device to be specified returns output streams $b_1, \ldots, b_m$ where $a_i \in [T \to A_{v_i}]$ and $b_i \in [T \to A_{u_i}]$. Each $b_i, i \in 1, \ldots, m$ is the value of the corresponding coordinate function F_i of F applied to $a_1, \ldots, a_n$

$$b_i(t) = F_i(a_1(t), \ldots, a_n(t)).$$

Stated differently,

$$F(a_1, \ldots, a_n) = (b_1, \ldots, b_m) \quad \text{and} \quad F(a_1, \ldots, a_n)(t) = (b_1(t), \ldots, b_m(t)).$$

In general, given a many-sorted function specification f of a processor, we can create a stream transformer F. Let $A_1, \ldots, A_s$ be any non-empty sets and let f be function specification as in Section 3.2. The stream transformer then has the form

$$F : [T \to A_{v_1}] \times \cdots \times [T \to A_{v_n}] \to [T \to A_{u_1}] \times \cdots \times [T \to A_{u_m}].$$

where $v_1, \ldots, v_n, u_1, \ldots, u_n \in \{1, \ldots, s\}$. In a simple case, F may be defined from f as

$$F(a_1, \ldots, a_n)(t) = f(a_1(t-k), \ldots, a_n(t-k)),$$

where k is computation time.

3.5 Stream transformation and synchronisation

We observe that devices generally have a common structure: a preliminary phase when the device performs some *initialisation*, followed by a main phase when the device *computes*. The computation phase can be described as the evaluation of some function f repeatedly over time on a stream of arguments. Since the inputs and outputs of f will often be distributed in time, we must surround it with enough machinery to enable us to correctly *schedule* data.

Let us consider more carefully the relationship between functional specification f and stream transformer F. The basic structure of F (in the single-sorted case) that we will use in user specifications is as follows.

Let A be any non-empty set, and let $f : A^n \to A^m$. Let T be a clock. Then

$$F : [T \to A^n] \to [T \to (A \cup \{u\})^m]$$

is defined from function f by

$$F(a_1, \ldots, a_n)(t) = \begin{cases} (u, \ldots, u) & \text{if } \neg V(t), \\ f(a_1(\delta_1(t)), \ldots, a_n(\delta_n(t))) & \text{if } V(t), \end{cases}$$

where $V : T \to \mathbf{B}$ and $\delta_1, \ldots, \delta_n : T \to T$. Here, u is a new object not in A, that stands for an *unspecified element* and is intended to mark output data arising from an implementation that is to be ignored, V is the *valid predicate*, and $\delta_1, \ldots, \delta_n$ are *selection functions.* The output of F is u while the valid predicate V is false. When V becomes true, the output of F at time t is $f(a_1(\delta_1(t)), \ldots, a_n(\delta_n(t)))$, where scheduling functions $\delta_1, \ldots, \delta_n$ control what data in streams $a_1, \ldots, a_n$ are read into the processor specified by f.

In addition, we may want to restrict the domain of streams to a subset $S \subset [T \to A]$. This is a much simplified scheme compared with the many-sorted case (see Section 3.6 below), but the salient points should be clear.

Example

(14) Suppose that any device we may build to compute the *gcd* function of Example (8) would have a computation time $k > 0$. The output of such a device at any time t would refer to data read at the previous time $t - k$. Additionally, for time $t < k$ the output would be unspecified. This is illustrated below

$$GCD : [T \to \mathbf{Z}]^2 \to [T \to \mathbf{Z} \cup \{u\}]$$

$$GCD(a, b)(t) = \begin{cases} u & \text{if } t < k \\ gcd(a(t-k), b(t-k)) & \text{if } t \geq k \end{cases}$$

Observe that $\delta_1(t) = \delta_2(t) = t - k$ and $V(t) = t \geq k$.

3.6 General specifications

We now introduce a general scheme for many-sorted stream transformers based on the approach outlined above. This scheme will also have a restriction namely, that only devices with *data independent initialisation and scheduling* may be specified, that is, devices in which scheduling is a function of time alone, and *not* of the input streams.

3.6.1 The type I synchronisation scheme Let $A_1, \ldots, A_s$ be non-empty sets representing the data types of elements of streams. Let T be a clock. Suppose

$$f : T \times A_{v_1}^{j_1} \times \cdots \times A_{v_n}^{j_n} \to A_{u_1}^{l_1} \times \cdots \times A_{u_m}^{l_m}$$

is a function statically describing the operation of the device to be specified in terms of elements of its input streams, where $v_1, \ldots, v_n, u_1, \ldots, u_m \in \{1, \ldots, s\}$.

Let $\delta_1^i, \ldots, \delta_{j_i}^i : T \to T$ for each $i = 1, \ldots, n$ be selection functions, one to control the distribution in time of elements of each input stream.

Let $V : T \to \mathbf{B}$ be the valid predicate. When this is *true*, the device is assumed to be capable of computing significant results. When it is *false*, the output of the device is *unspecified.*

We say that stream transformer

$$F : [T \to A_{v_1}] \times \cdots \times [T \to A_{v_n}] \to [T \to A_{u_1} \cup \{u\}] \times \cdots \times [T \to A_{u_m} \cup \{u\}]$$

is defined from function f by a *data independent synchronisation scheme of type I* iff

$$F(x_1, \ldots, x_n(t) = \begin{cases} (u, \ldots, u) & \text{if } \neg V(t), \\ f(t, x_1(\delta_1^1(t)), \ldots, x_1(\delta_{j_1}^1(t)), \ldots \\ \quad \ldots, x_n(\delta_1^n(t)), \ldots, x_n(\delta_{j_n}^n(t))) & \text{if } V(t). \end{cases}$$

The number j_i is the *degree of stream dependency* of the specification on stream x_i. By degree of stream dependency we mean the size of the 'window' or section of the stream x_i we may look at when computing f.

3.6.2 The type II synchronisation scheme The type I scheme is unsuitable for the specification of devices which must store data for indefinite periods. We therefore introduce the general scheme of type II, which relaxes the scheduling restrictions. For example, in a computation we may examine entire input streams, waiting until certain conditions occur before computation can begin.

Let $A_1, \ldots, A_s$ be non-empty sets. Let T be a clock. Suppose

$$f : T \times A_{v_1}^{j_1} \times \cdots \times A_{v_n}^{j_n} \to A_{u_1}^{l_1} \times \cdots \times A_{u_m}^{l_m},$$

where $v_1, \ldots, v_n, u_1, \ldots, u_m \in \{1, \ldots, s\}$, is a function statically describing the operation of the device to be specified as for the scheme of type I.

Let $\delta_1^i, \ldots, \delta_{j_i}^i : T \times [T \to A_{v_1}] \times \cdots \times [T \to A_{v_n}] \to T$ for each $i = 1, \ldots, n$ be selection functions controlling the distribution in time of elements of the input streams. These are as for the type I scheme *except they are now dependent on the input streams in addition to time t.*

Let $V : T \times [T \to A_{v_1}] \times \cdots \times [T \to A_{v_n}] \to \mathbf{B}$ be the valid predicate, which is *true* if the output of F is valid and *false* otherwise. We allow both our data and validity predicate to be data dependent.

We say that stream transformer F

$$F : [T \to A_{v_1}] \times \cdots \times [T \to A_{v_n}] \to [T \to A_{u_1} \cup \{u\}] \times \cdots \times [T \to A_{u_m} \cup \{u\}]$$

is defined from function f by a *data dependent synchronisation scheme of type II* iff

$$F(x_1,\ldots,x_n)(t) = \begin{cases} (u,\ldots,u) & \text{if } \neg V(t,x_1,\ldots,x_n) \\ f(t, x_1(\delta_1^1(t,x_1,\ldots,x_n)), \\ \quad \ldots, x_1(\delta_{j_1}^1(t,x_1,\ldots,x_n)), \\ \quad \ldots, x_n(\delta_1^n(t,x_1,\ldots,x_n)), \\ \quad \ldots, x_n(\delta_{j_n}^n(t,x_1,\ldots,x_n))) & \text{if } V(t,x_1,\ldots,x_n). \end{cases}$$

3.7 Role of abstract data types and functional languages in user specification

We wish to use the above synchronisation schemes to formally specify systems by means of stream transformations. Furthermore, we wish to precisely define classes of stream transformations in order to have mathematical models for the definition of a *specification description language SDL*. Recall that an *SDL* is required to formalise the level of computational abstraction in the draft user specifications for devices.

The data independent synchronisation schemes can be used with any class K of functions f, δ_i, predicates V, and sets S over A and T. Thus it is natural to define a class $\overline{K}$ of stream transformations by applying a scheme to a class K of functions. In particular, the schemes provide a theory of stream transformers, and hence of user specifications, from a theory of functions on data A and time T.

The case of data dependent scheduling is more complicated because we must have a class K of functions on data A, time T and streams $[T \to A]$.

Which classes of functions are of interest? A class K should be clearly definable; allow a mathematical theory; be straight-forward to apply; and allow machine animations of stream transformers.

For immediate practical results it is possible to take K to be the class of functions and predicates that are programmable on A in any given programming language L. For example, we could choose L to be Modula 2, CLU or Ada, since they support abstract data types and have functions. Alternatively, we could choose L to be a functional programming language such as ML, Hope or Miranda. Any such choice could serve as a starting point for a formal definition of an SDL, and hence for a level of computational abstraction.

It is possible to study and apply usefully the stream transformations in this chapter without a specific decision on K and to postulate the existence of an algorithmic language L that implements functions on abstract data types. However we are still left with certain problems over data dependent scheduling. We will now introduce a theory of functions on an abstract data type that is suitable for the theoretical analysis of specifications, and for preparing executable specifications, and allows us to treat data dependent schedules.

3.8 Computable functions on abstract data types
To define a class K of functions on A and T we will define (i) a set of basic functions on A and T; and (ii) a set of mechanisms for creating new functions from existing functions.

For (i) we take the operations of the data type A and clock T, together with some other functions which require us to adjoin the booleans $\mathbf{B}$ (as in Example (1)).

For (ii) we consider four mechanisms which allow us to define three classes of functions, namely: the *polynomial functions*; the *primitive recursive functions*; and the *inductively definable functions.*

Let $\mathbf{A}$ denote the $s+2$ sorted algebra consisting of the s sorted algebra A, the clock T and the booleans $\mathbf{B}$.

Let $A^v = A_{v_1} \times \cdots \times A_{v_n}$ and $A^u = A_{u_1} \times \cdots \times A_{u_m}$ be any Cartesian products of the domains A_i of $\mathbf{A}$, for $i \in \{1, \ldots, s\}$.

3.8.1 Basic functions The following are the basic functions

(1) for each constant c^A of $\mathbf{A}$ the function $f(a) = c^A$, where $a = (a_1, \ldots, a_n) \in A^v$;
(2) for each operation σ^A of $\mathbf{A}$ the function $f(a) = \sigma^A(a)$, where $a = (a_1, \ldots, a_n) \in A^v$;
(3) for each $n > 0$ and each i such that $n \geq i > 0$, the *projection function* $f(a) = a_i$, where $a = (a_1, \ldots, a_n) \in A^v$;
(4) for each domain A_i of $\mathbf{A}$ the *conditional function*

$$f(b, a_1, a_2) = \begin{cases} a_1 & \text{if } b = tt, \\ a_2 & \text{if } b = ff, \end{cases}$$

where $b \in \mathbf{B}$ and $a_1, a_2 \in A_i$.

3.8.2 Function building operations The following are the mechanisms of interest on (possibly partial) functions:

Parallel composition For each sequence of functions $f_1, \ldots, f_n$ with $f_i : A^v \to A_{u_i}$ the function $f : A^v \to A^u$ defined by

$$f(a) = (f_1(a), \ldots, f_n(a)),$$

where $a \in A^v$, is the result of *parallel composition* or *vectorisation.*

Sequential composition For functions $g : A^v \to A^u$ and $h : A^u \to A^w$ the function $f : A^v \to A^w$ defined by

$$f(a) = h(g(a)),$$

where $a \in A^v$, is the result of *sequential composition.*

Simultaneous primitive recursion on time For functions $g : A^v \to A^u$ and $h : T \times A^v \times A^u \to A^u$ the function $f : T \times A^v \to A^u$ defined by

$$f(0, a) = g(a),$$
$$f(t+1, a) = h(t, a, f(t, a)),$$

where $t \in T$ and $a \in A^v$, is the result of *simutaneous primitive recursion.*

Search on time For function $g : T \times A^v \to \mathbf{B}$ the function $f : A^v \to T$ defined by

$$f(a) = (least\ t)[g(t, a) = tt],$$

where $t \in T$, and $a \in A^v$, and $g(z, a) \downarrow$ and $g(z, a) = ff$ for all $z < t$, is the result of *search.*

3.8.3 Three classes of functions The basic functions closed under the operations of parallel and sequential composition constitute the class $P(\mathbf{A})$ of all *conditional polynomial functions* on $\mathbf{A}$.

The basic functions closed under the operations of parallel and sequential composition, and simultaneous primitive recursion, constitute the class $PR(\mathbf{A})$ of all *simultaneous primitive recursive functions* on $\mathbf{A}$.

The basic functions closed under all four operations constitute the class of $IND(\mathbf{A})$ of all *inductively definable functions* on $\mathbf{A}$.

The functions in $P(\mathbf{A})$ and $PR(\mathbf{A})$ are total whereas the functions in $IND(\mathbf{A})$ can be partial.

These classes of functions where introduced in Tucker & Zucker [1987] in order to identify the class of effectively calculable or programmable functions on an abstract data type. It was proved that $IND(\mathbf{A})$ is precisely the class of functions programmable by *while* programs on an abstract data type. Let us note that the functions programmable by *while* programs with arrays is the *larger* class $CIND(\mathbf{A})$, obtained by replacing simultaneous primitive recursion by *simultaneous course-of-values recursion.* This class constitutes the proper generalisation of the partial recursive functions from the natural numbers to an abstract data type. We do not define $CIND(\mathbf{A})$ since our specifications require $IND(\mathbf{A})$ only.

These classes of functions are useful for analysing the functions that arise in stream specifications; and they may be readily programmed or compiled for execution.

3.8.4 Theory of stream transformers Each class $P(\mathbf{A})$, $PR(\mathbf{A})$, $IND(\mathbf{A})$ of functions on A and T gives rise to a corresponding class of data independent stream transformations of type I, as described in Section 3.7. However, the classes of functions can be applied in a more significant way, that defines the data independent and dependent transformers uniformly.

Let us adjoin the unspecified element u to each domain A_i of $\mathbf{A}$ to make a new algebra $\mathbf{A}_u$. The operations of $\mathbf{A}$ extended to the value u are *strict*; that is

$$\sigma^A(a_1, \ldots, u, \ldots, a_n) = u$$

(see Tucker & Zucker [1987] for details of this algebraic construction).

Let the algebra $\mathbf{A}_u$, consisting of data type, clock, and booleans, be augmented with streams. That is, for each domain $A_i \cup \{u\}$ of $\mathbf{A}_u$ we add the set $[T \to A_i \cup \{u\}]$ of all streams over A_i with u, together with the operation

$$eval_i : [T \to A_i \cup \{u\}] \times T \to A_i \cup \{u\}$$

defined by

$$eval_i(a, t) = a(t)$$

to make the algebra $\overline{\mathbf{A}}_u$.

We may now apply the definitions of Sections 3.8.1 and 3.8.2 to $\overline{\mathbf{A}}_u$, in place of $\mathbf{A}$ to define the classes

$$P(\overline{\mathbf{A}}_u), \quad PR(\overline{\mathbf{A}}_u) \quad \text{and} \quad IND(\overline{\mathbf{A}}_u).$$

These classes formalise classes of stream transformers that include those defined from the data independent and dependent schemes of types I and II.

Recall that a stream transformer

$$F : [T \to A^v] \to [T \to A^u]$$

can be written in its equivalent applicative form

$$\hat{F} : [T \to A^v] \times T \to A^u.$$

This applicative form is suited to the defining mechanisms in the classes.

Theorem *Let F be a stream transformer defined from function f, validity predicate V and selection functions δ_i by schemes of type I or type II. Let $\hat{F}$ be its applicative form. Then if f, V and δ_i are in $P(\overline{\mathbf{A}}_u)$, $PR(\overline{\mathbf{A}}_u)$ or $IND(\overline{\mathbf{A}}_u)$, then $\hat{F}$ is in $P(\overline{\mathbf{A}}_u)$, $PR(\overline{\mathbf{A}}_u)$ or $IND(\overline{\mathbf{A}}_u)$ respectively.*

Proof Notice that the applicative form $\hat{F}$ is a polynomial function over $\mathbf{A}_u$ with the functions f, V, and δ_i adjoined as new operators. Thus if these functions are definable in a class then so is $\hat{F}$. □

Notice from the proof the complexity of a specification resides in f, V and δ_i.

Thus we may mathematically analyse stream transformations, and model constructs for a specification language, by means of the classes $P(\overline{\mathbf{A}}_u)$, $PR(\overline{\mathbf{A}}_u)$ and $IND(\overline{\mathbf{A}}_u)$ constructed for an abstract data type $\mathbf{A}$. In particular, a language allowing abstract data types and the function building mechanisms is suitable for expressing our specifications. This is easy to appreciate in the specifications of the correlator in Section 5, but is more complicated in the specifications of Sections 7 and 9.

3.9 Consistency and correctness of refinements

Finally, we comment on the refinement of stream transformations in the light of the concepts about specifications of Section 2.2.

Shortly we will consider refinements that drive the process of developing stream transformations which specify a digital correlator. From an initial stream transformation $CORR_1$, we will derive a portfolio of 10 stream transformations by means of refinements $CORR_i$ to $CORR_{i+1}$ for $i = 1, \ldots, 9$. These refinements vary in their objectives, depending upon their phase in the user specification process. In all but one case ($i = 5$), the stream transformations $CORR_i$ and $CORR_{i+1}$, and hence the refinements, will be seen to be *consistent* in the sense of Section 2.2.6. Furthermore, the theoretical work of Sections 6 and 8 concern timing abstractions and implementations that allow us to define refinements of stream transformations that are *correct* in the sense of Section 2.2.6.

As an example, consider the two stream transformations

$$F_1 : [T_1 \to A_1] \to [T_1 \to A_1] \quad \text{and} \quad F_2 : [T_2 \to A_2] \to [T_2 \to A_2]$$

and representation maps

$$i : [T_1 \to A_1] \to [T_2 \to A_2] \quad \text{and} \quad j : [T_2 \to A_2] \to [T_1 \to A_1]$$

which in the present context may be usefully termed *scheduling maps.* Then F_2 is *consistent* with F_1 means that

$$F_1(a)(t) = jF_2i(a)(t) \quad \text{for all } t \in T_1$$

or, equivalently, that the diagram of Figure 5.6 commutes.

$$\begin{array}{ccc} [T_1 \to A_1] & \xrightarrow{F_1} & [T_1 \to A_1] \\ i \downarrow & & j \uparrow \\ [T_2 \to A_2] & \xrightarrow{F_2} & [T_2 \to A_2] \end{array}$$

Figure 5.6. Consistency and correctness of refinements

3.10 Literature

These technical ideas, and those on retimings of clocks in Sections 6 and 8, are part of a general mathematical theory of synchronous concurrent algorithms and its application to hardware design. In particular, they complement ideas on the formal definition and verification of synchronous algorithms.

The three classes of computable functions on a many-sorted algebra are included in the study of computability theory on abstract data types and its application to the formal verification of programs in Tucker & Zucker [1987] (work of 1979). The functions were first applied to the verification of synchronous concurrent algorithms with respect to static specifications in Thompson & Tucker [1985]. A short account of Thompson and Tucker's methods, applicable to dynamic specifications involving streams, is contained in Eker & Tucker [1987], where it is applied to rasterising algorithms. The theory will be applied to architectures for the correlator in Harman, Hobley & Tucker [1990]. A comprehensive account of the functions and their use in defining and verifying synchronous concurrent algorithms is Thompson [1987].

An equivalent alternative account of synchronous algorithms is possible using von Neumann languages suited to simulation and testing. These languages are based on abstract data types, functional routines and concurrent assignments. In Thompson [1987] a language is formally defined and is proved to be computationally equivalent to the simultaneous primitive recursive functions (with respect to correctness and performance of implementing functions). An account of the practical development of this second typed language is Martin & Tucker [1987]. A third equivalent approach is to formally model the network architectures using graph theory. This account is suited to the formal study of architectures and layout: see Meinke [1988].

4 INFORMAL SPECIFICATION

We will specify the correlator informally. We will first describe the logical behaviour of the device, and then its physical characteristics and performance, as these are required by the user or are determined by the technology.

4.1 Reference word and data stream

The correlator is a signal processing device. It continually reads in data which arrives on an *input stream* $x = x(0), x(1), \ldots$ and applies a *correlation function* to this data. This function gives a measure of similarity between a segment of the stream x and a given *reference word* $y = y(0), \ldots, y(n-1)$. The results of this correlation function form the *output stream* $c = c(0), c(1), \ldots$ of the device. By a *stream*, we mean a sequence of values separated in time. For example, a sequence of bits sent along a wire, one after another, at distinct time instants, constitutes a stream. The streams we use are *synchronous* with respect to some *clock*; that is, elements arrive at well defined times controlled by this clock. We illustrate this simply in Figure 5.7.

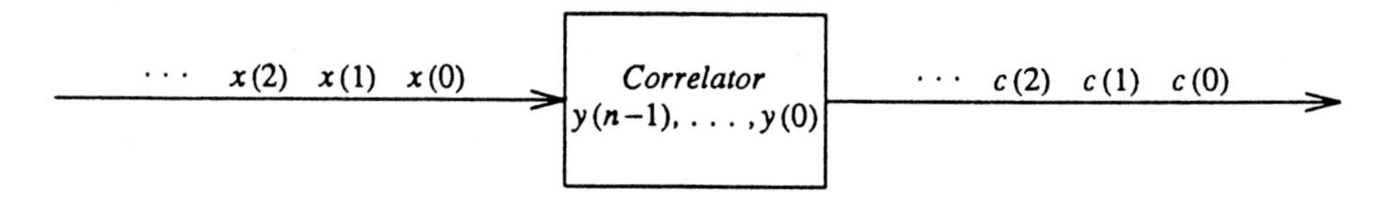

Figure 5.7. The correlator as a stream processor

The data of the input stream x are elements of a set X and the data of the output stream c are elements of a set C. Although items arrive on the stream x one at a time, the correlator operates on an *n-element segment* of x. By an n-element segment of a stream, we mean n values from the stream, which have arrived at times in the past, and have been stored. At any time t we are specifically interested in the *contiguous* segment consisting of the *last* n items which arrived on the stream x. We are effectively looking at the stream through a window of fixed size n, through which the stream moves one element every clock cycle as the stream is processed by the correlator. Since the value of n determines the length of the segment of the stream x wc will be correlating, n is a measure of the *size* of the correlator.

4.2 The correlation function

The set X is normally a finite subset of either the set $\mathbf{N} = \{0, 1, 2, \ldots\}$ of the natural numbers, or of the set $\mathbf{Z} = \{\ldots, -1, 0, 1, \ldots\}$ of the integers. These subsets are normally represented by the set $\{0,1\}$ of bits in one of the usual ways (for example, positive magnitude, or two's complement).

The stream c the correlation process generates is the result of applying a correlation function to an n-element segment of the stream x and a *reference word* $y(0),\ldots,$ $y(n-1)$. The elements of this reference word normally belong to a set Y which is also a finite subset of $\mathbf{N}$ or $\mathbf{Z}$. The sets X and Y need not be the same. Unlike the stream x, the reference word y is normally *fixed*; that is, it does not change with time. The output stream c computed by the correlator is defined to be

$$\begin{aligned} c(t+n+k-1) &= y(0).x(t) + y(1).x(t+1) + \cdots + y(n-1).x(t+n-1) \\ &= \sum_{i=0}^{n-1} y(i).x(t+i). \end{aligned} \qquad \text{(I)}$$

We add n to the time at which c is computed to show that the correlator will require n clock cycles to read in enough elements of x for computation to begin. That is, the correlator must ensure that all elements of the n-element segment (the 'sliding window') of x are defined. Therefore, n constitutes an *initialisation delay.*

We add k to show that the correlator must take a fixed number of clock cycles to compute a result: i.e., k is a (constant) *computation delay.* Like X, Y and n, the value for k, the computation time for the correlator, is part of the specification. If $k > 1$, the correlator must be *pipelined*, so a new result emerges every clock cycle (see Section 3.4). The relationship between data and reference words over time is shown in Figure 5.8.

4.3 Polarity correlation

Consider the special case when $X = Y = \{-1,1\}$, known as *polarity correlation* (see for example, Jordan [1986]). The multiplication function which is the basis of correlation function (I) in Section 4.2 is shown in Table 5.3. In this case, formula (I) for the correlator yields a pattern matcher: a high positive result ($\leq n$) indicates a good match, and a high negative result ($\geq -n$) a bad one.

The range of output values of correlation function (I) applied to $\{-1,1\}$ is not 'continuous'. The result will always be odd if n is odd, and even if n is even* (the more

* *Proof* Suppose, for a given correlation, there are p elements that match, and q that do not; $n = q + p$. The correlation result will be $p - q$ (see equation (I) and Table 5.3). (i) If n is even and q is even, p must be even ($n = p + q$) so $p - q$ is even. (ii) If n is even, and q is odd, p must be odd, so $p - q$ is even. (iii) If n is odd, and q is even, p must be odd, so $p - q$ is odd. (iv) If n is odd, and q is odd, p must be even, so $p - q$ is odd.

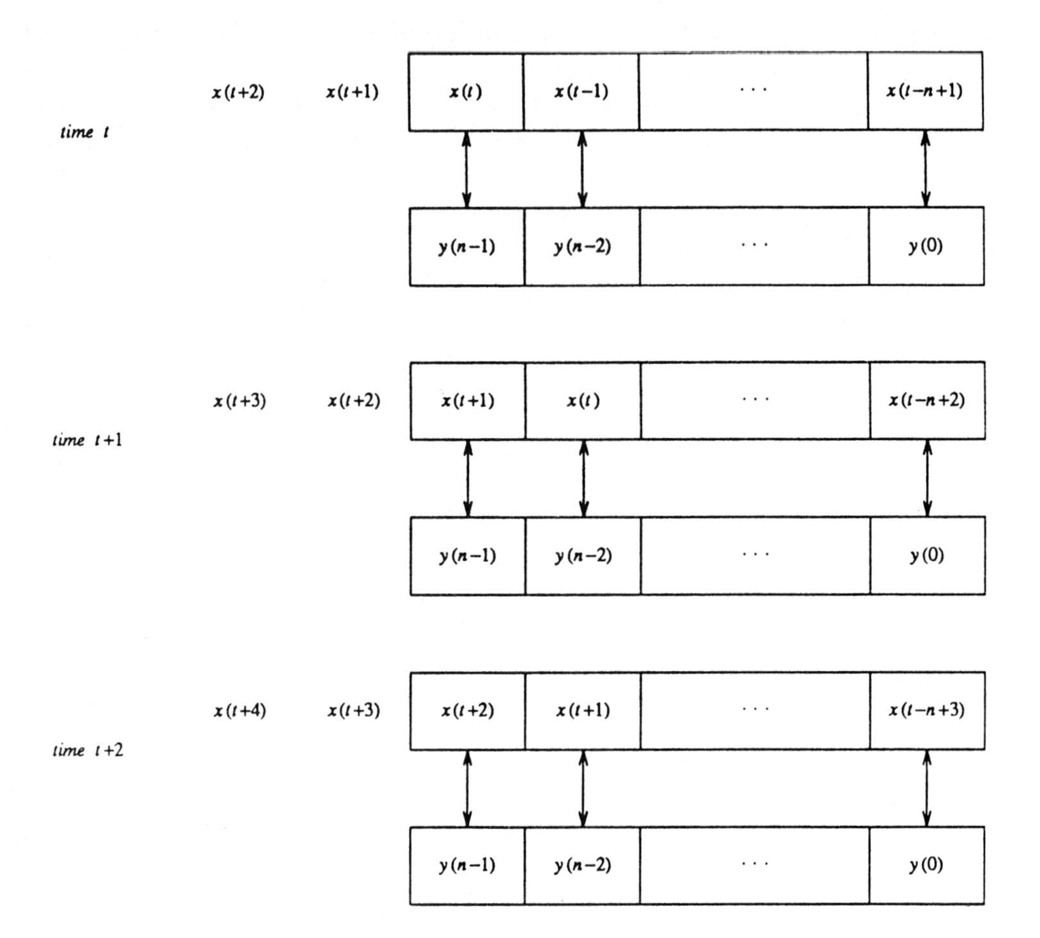

Figure 5.8. Relationship between reference and data over time

likely case since n will usually be a power of 2). Thus, if we adopt a common method of coding the correlator's output as bits, we must either accept that approximately half the possible output values will be unused, which is troublesome; or we must use a non-standard coding, which may lead to problems in interfacing with other devices.

We circumvent this problem by replacing multiplication on $\{-1, 1\}$ by an equality function on $\{0, 1\}$,

$$eq(x, y) = \begin{cases} 1 & \text{if } y = x, \\ 0 & \text{if } y \neq x, \end{cases}$$

and defining a new correlation function

$$c(t + n + k - 1) = \sum_{i=0}^{n-1} eq(y(i), x(t + i)). \tag{II}$$

With correlation function (II), a good match is indicated by a high positive result ($\leq n$), and a bad match by a low positive result (≥ 0). We have compressed the range of correlation function (I) over $\{-1, 1\}$ by half but we retain the same amount of information. In this way correlation over $\{-1, 1\}$ using correlation function (I) is normally considered to be equivalent to correlation over $\{0, 1\}$ using correlation function (II). It is in any case necessary to *represent* $\{-1, 1\}$ by $\{0, 1\}$ in a digital circuit, so this will not cause difficulties. It is a polarity correlator of the form (II) that we will formally specify.

Table 5.3. $\{-1, 1\} \times \{-1, 1\}$

$\times$	-1	1
-1	1	-1
1	-1	1

4.3.1 Example Consider correlation function (II) with $n = 4$. Suppose the reference word is 0 1 1 0 and consider the stream

$$x(t+2) = 0, \quad x(t+1) = 0, \quad x(t) = 1, \quad x(t-1) = 1, \quad x(t-2) = 0, \quad x(t-3) = 1.$$

The correlation results for times t to $t+2$ are shown in Figure 5.9. The correlation is simply the number of corresponding pairs in the reference and data words that match.

4.4 Reference word replacement

We have, until now, assumed that the reference word is fixed. In practice however, we will require the reference word to be replaced occasionally. A mechanism for this is to input *two* new streams y and s of elements of type Y and *booleans* $\mathbf{B} = \{tt, ff\}$ respectively. Stream y carries elements of the reference word and stream s is an *enable* signal, which controls reference word loading. When tt arrives on s, the reference word behaves like the stream x and the new item arriving on stream y causing the reference word to be shifted along with the oldest element being displaced. However when ff arrives on s, the item arriving on stream y is ignored and lost and hence the reference word is not changed. Thus we are able to *serially replace* the reference word. This is illustrated in Figure 5.10.

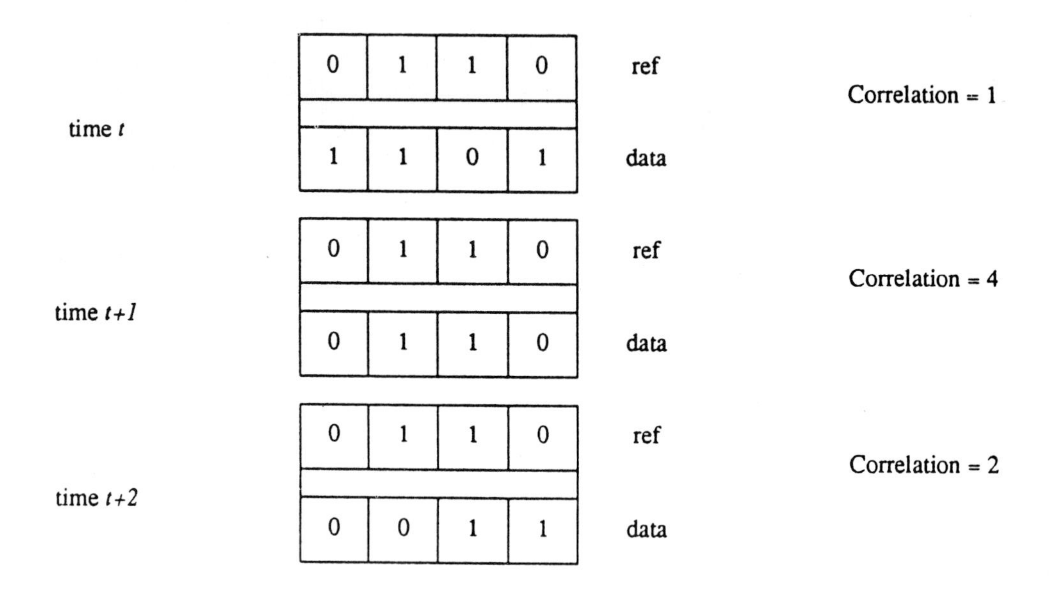

Figure 5.9. Example correlation

4.5 Terminology

We will now introduce some terminology commonly used in connection with correlators (see for example Corry & Patel [1983], Jordan [1986]).

A correlator which operates on an n-element segment of the stream x of data from X is termed an *n-point correlator*, or an *n-stage correlator.*

Suppose the elements of X and Y are represented by l and m bits respectively. That is, X is represented by $\{0,1\}^l$ and Y is represented by $\{0,1\}^m$. In this case, the device is often termed an *l-bit data, m-bit reference correlator.*

When $X = Y = \{-1,1\}$, a point or stage is sometimes termed a bit (somewhat confusingly). We must be particularly careful to distinguish between X and Y and their representations. In the example of interest $X = Y = \{-1,1\}$ but this data type will be *represented* by $\{0,1\}$. A 1-bit data, 1-bit reference, n-point correlator is then termed an *n-bit correlator.*

A description of a 1-bit reference, 4-bit data, 64-point correlator may be found in Corry & Patel [1983].

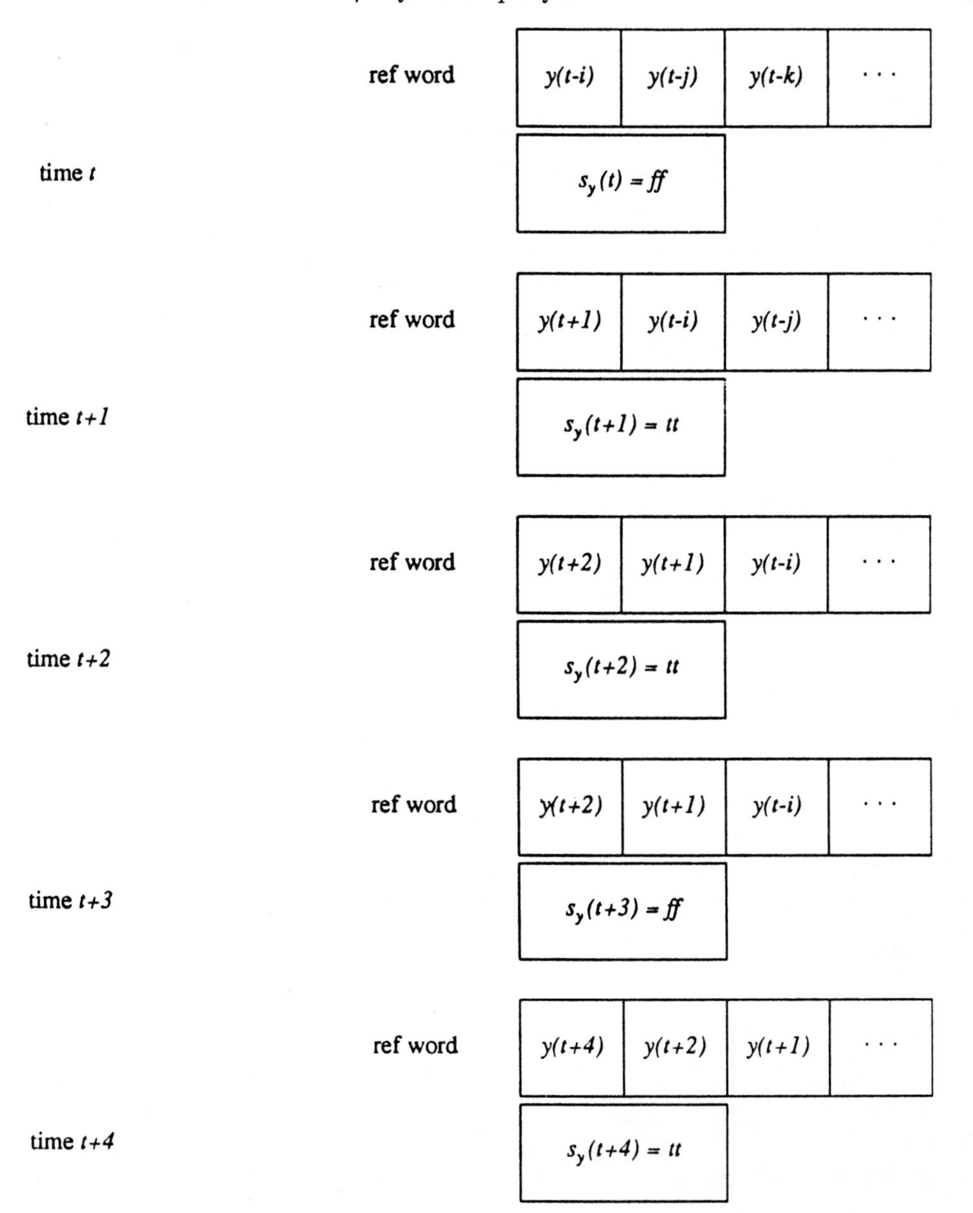

Figure 5.10. Reference word replacement

4.6 User specification

We propose to formally specify an n-bit correlator as defined above. That is, $X = Y = \{0, 1\}$, the correlation function is (II) above, $n = 32$ and computation time k is as small as possible. The correlator will have initialisation delay n and the reference word will be programmable by a boolean stream as described above. We illustrate this with Figure 5.11, and the user specification is summarised in Table 5.4.

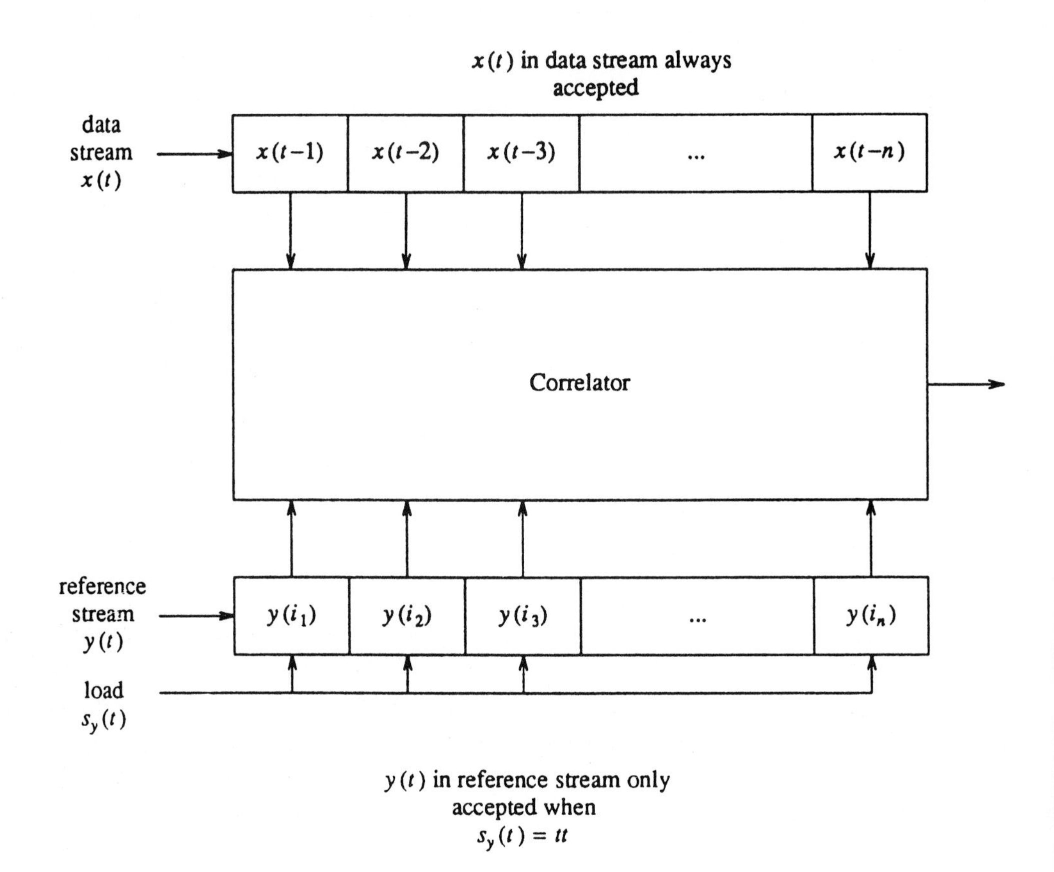

Figure 5.11. An abstract view of a general correlator

4.7 Physical specification

There are a number of *performance requirements* that must be met if the correlator is to be acceptable to the user and for the manufacture of the correlator to be feasible. These we call *user requirements* and *technology requirements* respectively. The latter are imposed by the choice of manufacturing technology and the physics of semiconductor devices.

The intended manufacturing technology for the correlator is nMOS (see Mead & Conway [1980] for an introduction).

Table 5.4. User specification

Characteristic	Specification
Bits, n	32
Input stream set, X	$\{0,1\}$
Reference word set, Y	$\{0,1\}$
Correlation function	(II) (Section 4.3)
Computation time, k	Small as possible
Initialisation delay	$n(=32)$

4.7.1 Speed of operation The user requires a *minimum clock rate* of 5 MHz. Therefore each *clock cycle* (see Section 3.3 below) can last no longer than 200 ns. A higher clock rate is acceptable and even desirable; however this may well be precluded by the *power dissipation* requirements (see Section 4.7.2 below).

The technology will impose an upper limit on the clock rate of the correlator. This will depend on several factors, including the *minimum feature size** and *fabrication line characteristics*†. However, this limit will certainly be greater than 5 MHz. In practice, the speed of the correlator will be limited by such things as signal propagation along long wires. However, we are only concerned with *specification* here, and cannot know or legitimately make assumptions about such considerations at this stage.

4.7.2 Power dissipation The 32-bit correlator is a pilot design for a larger correlator ($n = 512$; see Harman [1987]). Consequently, the user has no power dissipation requirements for the correlator.

The technology imposes a power dissipation limit of about 2 watts. This limitation is imposed by the need to radiate surplus heat while remaining below its operating temperature ceiling. A requirement for low power consumption directly conflicts with a requirement for high speed.

It is expected that the power consumption for the correlator will be about 0.25 watts.

* The size of the smallest manufacturable item on a chip.

† The precise details of the manufacturing process. This can vary between 'runs' of the process so the same circuit from different batches can have different limitations (Glasser & Dobberpuhl [1985] pp. 237–49)

We note that in this case, the technology requirement is stricter than the user requirement.

4.7.3 Area The user has essentially no area requirements.

The maximum permissible area of the chip is about 10 mm × 10 mm. This limit is largely due to the presence of flaws in the silicon substrate on which the chip is constructed. Larger chips are more likely to lie on a flawed area than small ones. The probability of a good chip, $P_0(N, A)$, is

$$P_0(N, A) = e^{-NA}$$

where N is the number of flaws per unit area, A is the chip area (see Mead & Conway [1980], p. 45).

It is expected that the correlator will be about 5 mm × 5 mm.

4.7.4 Summary In summary, we note that the speed requirements of 5 MHz is a user requirement and that there are effectively no user requirements on power and area (other than that the correlator should fit on a single chip). The power dissipation and area requirements are imposed on the designer by the technology. This is illustrated in Table 5.5.

4.7.5 The fabricated correlator As mentioned in Section 1, the correlator described here has been successfully fabricated. Details of the design and fabrication can be found in Harman [1987]. The correlator measured 5 mm × 5 mm, dissipated approximately 0.3 watts and operated at 2 MHz. Notice the speed requirement of the specification was not met.

5 EXPLORATORY FORMAL SPECIFICATIONS OF A DIGITAL CORRELATOR

We now begin the exploration phase of the specification process. In this phase, our intention is to develop the first formal user specifications of the correlator which accurately model the user's requirements, i.e., are *user-consistent*. The aim is to *understand* the user's problem by refining informal and formal specifications until the user is satisfied, by means of the incremental technique explained in Section 2.3.1 and 2.4. We have already developed an informal specification of the correlator to a relatively high level of precision in Section 4, and it should be born in mind that in practice an initial informal specification is unlikely to be as precise as that of the correlator given here.

Table 5.5. Requirements and constraints

Characteristic	Constraints	Type of constraint
Speed	The correlator should operate at a speed of at least 5 MHz	user
	The correlator can operate no faster than the switching time of its transistors (> 5 MHz).	technology
Size	The correlator should fit on a single chip.	user
	The correlator should not exceed the maximum die size (10 mm × 10 mm).	technology
Power	The correlator should dissipate no more power than is reasonable for a single chip.	user
	The correlator should dissipate no more power than can be comfortably radiated (about 2 watts).	technology

The incremental technique described in Section 2.3.1 suggests that we do not attempt to specify formally the device at the first attempt. Instead, we make simplifying assumptions which reduce the complexity of the specification and make the initial formal specification easier to develop. Then we add omitted aspects of the specification incrementally. Thus for the first draft we assume that entire words are loaded in each clock cycle, as opposed to single elements of words, and for the first and second drafts, we assume the reference word is *fixed* for any given data stream x (see Section 4.4 above), which means we are allowed to temporarily dispose of the control stream s.

Specifications will be presented in the form described in Section 2.2.4: namely, as a 4-tuple (L, I, F, E) which is refined from one specification to the next:

$$(L, I, F, E) \rightarrow (L', I', F', E').$$

The level of computational abstraction L will remain fixed throughout Section 5 (hence $L = L'$) and is described in Section 5.1. To avoid unnecessary repetition, the informal specification I will just state in what respects the current specification differs from that presented in detail in Section 4. The formal specification F is presented in full. To begin with experiments E are presented as a table showing the animation of the specification symbolically, but after the second formal specification, the component is too complicated and this approach becomes unsatisfactory: a software simulation is required and further experiments are omitted.

Each refinement will be seen easily to be consistent in the sense of Section 2.2.6

5.1 Level of abstraction of the specifications

The heart of the correlator is a function that compares pairs of data-elements from two n-element vectors and returns the number of elements that are the same.

Let A be any non-empty set. Consider the function $corr : A^n \times A^n \to \{0, \ldots, n\}$ defined by

$$corr(\underline{a}, \underline{b}) = |\{i : a_i = b_i\}|,$$

where $\underline{a} = (a_1, \ldots, a_n)$ and $\underline{b} = (b_1, \ldots, b_n)$. That is, $corr(\underline{a}, \underline{b})$ is the number of elements (cardinality) of the set of i's such that the ith elements a_i and b_i are equal. This is equivalent to correlation function (II) in Section 4.3, which is what the user requires (Section 4.6). We use A as the data type of elements of the input streams instead of $\{0, 1\}$ because should we require correlation function (I) instead, we would only need to replace *corr* in our specifications. In this way, we make our specifications largely independent of the actual form of the correlation function we are using. (Strictly speaking, this trivial data abstraction belongs to the next phase!)

We note that *corr* corresponds with f in the general schemes presented in Section 3.6 and is a static specification of the correlator.

As we have described them here, A and *corr* effectively define the level of computational abstraction L for Section 5. More precisely, let $\mathbf{A}$ be a 4-sorted algebra with carriers the data set A, numbers $\{0, \ldots, n\}$, time T and booleans $\{tt, ff\}$. Let the constants and operations of A be *corr*, the clock operations on T augmented by operations $t - 1$ and $t < t'$, and the usual operations on the booleans. Also, let n and k be constants of T. Then, in this section, our stream transformation specifications will be functions definable over the stream algebra $\overline{\mathbf{A}}$ as explained in Section 3.8

5.2 First correlator specification

We can now define a simple stream correlator. Let A be any non-empty set and let $\underline{w} : T \to A^n$ be a stream of vectors of length n. The idea is that at every clock cycle $t \in T$, new data-word $\underline{w}(t)$, composed of n data items, is generated and delivered to the processor *corr*. The reference word $r = (r_{n-1}, r_{n-2}, \ldots, r_0) \in A^n$ is fixed in this specification.

Thus we specify the correlator as a stream transformation

$$CORR_1 : [T \to A^n] \to [T \to \{0, \ldots, n\} \cup \{u\}],$$

$$CORR_1(\underline{w})(t) = \begin{cases} u & \text{if } t < k, \\ corr(\underline{w}(t-k), r_{n-1}, \ldots, r_0) & \text{otherwise,} \end{cases}$$

where $k \geq 0$ is the computation time for *corr*, and $(r_{n-1}, \ldots, r_0)$ is the reference word.

Notice the following.

(1) We assume that *corr* takes k clock cycles to compute a result and hence the start-up time for output takes k cycles.
(2) The *output period* of $CORR_1$ is 1; that is a new result emerges every clock cycle (see Sections 4.2 and Section 5.3.1 below). In the case $k > 0$, this would necessitate *corr* being pipelined (Section 3.4).
(3) The reference word is 'hardwired' to $(r_{n-1}, \ldots, r_0)$.

Synchronisation of input and output streams Table 5.6 shows how the input and output streams in $CORR_1$ are related. Input word $\underline{w}(t)$ arrives at time t.

Table 5.6. Input/output for first draft correlator specification

t	$CORR_1(\underline{w})(t)$
0	u
.	.
.	.
$k-1$	u
k	$corr(\underline{w}(0), r_{n-1}, \ldots, r_0)$
$k+1$	$corr(\underline{w}(1), r_{n-1}, \ldots, r_0)$
.	.
.	.
$k+j$	$corr(\underline{w}(j), r_{n-1}, \ldots, r_0)$

User requirements established in the specification The following user requirements are established in the specification $CORR_1$:

- the processor *corr* computes the correlation of two n-element words in constant k clock cycles;
- the output period of the system is 1 clock cycle.

User requirements ignored in the specification The following user requirements are not met by specification $CORR_1$:

- data words are read into the correlator serially, one data item per clock cycle;
- the reference word is not fixed but can be changed by the user.

$CORR_1$ does not define the stream correlator as we have informally described it because it formalises that a new data *word* arrives on the input stream every cycle as opposed to a new data *item*. Thus $CORR_1$ is not consistent with the user's requirements. We will now give a definition that refines the specification by describing the way a word of length n is read in by the clock T.

5.3 Second draft formal specification

Let A be any non-empty set and let $x \in [T \to A]$ be a stream. We replace the stream of words of the first draft, where a new n-tuple arrives each clock cycle, by a stream of data-items where a new data item arrives each clock cycle. This is done by *algebraically substituting* for $\underline{w}(t-k)$

$$\underline{w}(t-k) = (x(t-k), \ldots, x(t-k-n+1)).$$

5.3.1 Second draft correlator specification Now we give our new specification.

$$CORR_2 : [T \to A] \to [T \to \{0, \ldots, n\} \cup \{u\}],$$

$$CORR_2(x)(t) = \begin{cases} u & \text{if } t < k+n-1, \\ corr(x(t-k), \ldots, & \\ \quad x(t-k-n+1), r_{n-1}, \ldots, r_0) & \text{otherwise,} \end{cases}$$

where $k \geq 0$ is the computation time for *corr* and $(r_{n-1}, \ldots, r_0)$ is the reference word.

The idea is that we are correlating on a 'slice' of the stream x of length n; it takes n steps before we have acquired enough elements from the stream x to form a word of length n and we can begin correlating. So the initial delay before results emerge is made up of the *initialisation time* or *set-up time* n and the *computation time* k of the function *corr*. After this time, each result emerging is the result of a correlation that started k steps earlier.

Synchronisation of input and output streams Table 5.7 shows how the input and output streams of $CORR_2$ are related. From this point on, we will omit 'experiments' (in the form of tables showing input and output stream synchronisation) because they become too complex to represent in this form.

Table 5.7. Input/output for second draft correlator specification

t	$CORR_2(x)(t)$
0	u
.	.
.	.
$n-1$	u
n	u
.	.
.	.
$k+n-2$	u
$k+n-1$	$corr(x(n-1),\ldots,x(0),r_{n-1},\ldots,r_0)$
$k+n$	$corr(x(n),\ldots,x(1),r_{n-1},\ldots,r_0)$
.	.
.	.
$k+n+j$	$corr(x(n+j),\ldots,x(j+1),r_{n-1},\ldots,r_0)$

User requirements established in the specification The following user requirement is established in $CORR_2$ (in addition to those established in $CORR_1$):

- data words are loaded serially.

User requirements ignored in the specification The following user requirement is ignored by specification $CORR_2$:

- the reference word may be replaced by the user.

Consistency of the specification $CORR_2$ is consistent with $CORR_1$ in the sense that $CORR_2$ is an implementation of a *specialisation* of $CORR_1$. Consider the set

$$V = \{\underline{a} \in [T \to A^n] \mid (a_1(t+1), a_2(t+1), \ldots, a_n(t+1)) = (x, a_1(t), \ldots, a_{n-1}(t))\}$$

and specialise $CORR_1$ to V. We define a map $i : V \to [T \to A]$ by

$$i(\underline{a})(t) = a_1(t).$$

This i is an injection. To see this, consider a and $a' \in V$. If $i(a) = i(a')$ then, by definition of i, $\forall t, a_1(t) = a'_1(t)$ and, for i to be an injection, we require $a = a'$. Suppose for a contradiction that $a \neq a'$. Then $\exists t_0$ such that $a(t_0) \neq a'(t_0)$ and hence $a_j(t_0) \neq a'_j(t_0)$ for some $1 \leq j \leq n$. Therefore, $a_1(t_0 - j) \neq a'_1(t_0 - j)$, which is a contradiction, since $\forall t, a_1(t) = a'_1(t)$ by definition. Therefore, if $i(a) = i(a')$ then $a = a'$ and i is an injection. Then for all $a \in V$

$$CORR_2(i(\underline{a}))(t) = CORR_1(\underline{a})(t+n).$$

Structure of specifications In their applicative form $C\hat{O}RR_1$ and $C\hat{O}RR_2$ are conditional polynomial functions over $\overline{\mathbf{A}}$ with the constants $r_{n-1}, \ldots, r_0$ adjoined: recall Section 3.8.4. In addition, they are polynomials defined by data independent scheduling. By way of illustration, we rewrite $CORR_2$ in the strict form outlined for the general data independent scheme of type I (Section 3.6.1).

Processor:

$$f : T \times A^n \times A^n \rightarrow \{0, \ldots, n\},$$

$$f(t, \underline{x}, \underline{y}) = corr(\underline{x}, \underline{y}),$$

Scheduling:

$$\delta^1_1, \ldots, \delta^1_n : T \rightarrow T,$$

$$\delta^1_1(t) = t - k, \ldots, \delta^1_n(t) = t - k - n + 1,$$

Setup/initialisation:

$$V : T \rightarrow \mathbf{B},$$

$$V(t) = t < n + k - 1,$$

Stream transformer:

$$F : [T \rightarrow A] \rightarrow [T \rightarrow \{0, \ldots, n\} \cup \{u\}],$$

$$F(x)(t) = \begin{cases} u & \text{if } V(t) \\ corr(x(\delta^1_1(t)), \ldots, x(\delta^1_n(t)), r_{n-1}, \ldots, r_0) & \text{if } \neg V(t). \end{cases}$$

5.4 Third draft formal specification

We will now remove the simplifying assumption of a fixed reference word and define a correlator that meets our informal specification. We will allow the reference word $r_{n-1}, \ldots, r_0$ to be programmed by means of a stream $y \in [T \rightarrow A]$ and a control stream $s_y \in [T \rightarrow \mathbf{B}]$, and allow arbitrary acceptance or rejection of data items from stream y accordingly, as stream s_y carries values *true* or *false*.

The ability to select which elements of a stream will be accepted and 'correlated on' complicates the specification somewhat. Consider the situation depicted in Figure 5.12. Only those elements of stream y which arrive at the same time as a *true* element on stream s_y will be accepted (those not marked with a cross). All the others will be ignored. If we assume that $n = 4$, not until time $t = 6$ will the correlator have accepted enough elements to begin correlation.

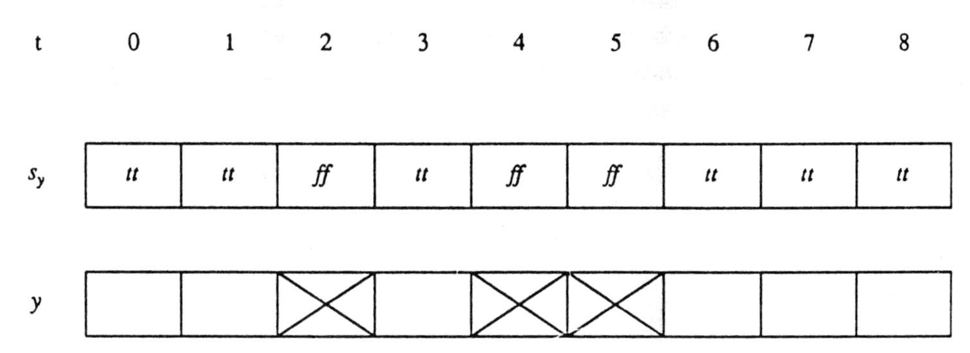

Figure 5.12. Streams of data may be accepted or ignored

This leads to two complications:

(1) We cannot assume that after n steps enough elements from the stream y will have been 'read in' for correlation to start.

(2) We may be correlating non-contiguous sections of the reference word input stream.

We examine these problems below.

5.4.1 Loading the reference word Observe that $y(0), y(1), \ldots$ are loaded or ignored according to the value of $s_y(0), s_y(1), \ldots$; that is, $y(i)$ is loaded iff $s_y(i) = tt$. The elements of the y stream which are loaded will go to make up an n-element word. Our problem is to identify which element of y will form the ith element of that word at time t, for $i = 1, \ldots, n$.

Let $\delta_i(t, s_y)$ denote the relevant time. That is, $y(\delta_i(t, s_y))$ is the element of the y stream that is the ith element of the reference word at time t. Therefore, we must define n partial functions $\delta_1, \ldots, \delta_n : T \times [T \to \mathbf{B}] \to T$ to denote these values. We define $\delta_i(t, s)$ by induction on i and t.

For the 1st element at time 0, we load the 0th element if $s(0) = tt$, otherwise, the 1st element is *undefined* ($\uparrow$)

$$\delta_1(0,s) = \begin{cases} 0 & \text{if } s(0) = tt \\ \uparrow & \text{if } s(0) = ff. \end{cases}$$

For the 1st element at time $t+1$, we load the $(t+1)$th element if $s(t+1) = tt$, otherwise, whatever was in the 1st element at the previous time t remains unchanged

$$\delta_1(t+1,s) = \begin{cases} t+1 & \text{if } s(t+1) = tt \\ \delta_1(t,s) & \text{if } s(t+1) = ff. \end{cases}$$

For $i = 2, \ldots, n$ the induction step is as follows. Only the 1st element can (possibly) become defined at time 0, all others are undefined

$$\delta_i(0,s) = \uparrow .$$

For $i = 2, \ldots, n$ if $s(t+1)$ is true, then we load the contents of the $(i-1)$th element at time t into the ith element. Otherwise, we retain the contents of the ith element at time t

$$\delta_i(t+1,s) = \begin{cases} \delta_{i-1}(t,s) & \text{if } s(t+1) = tt \\ \delta_i(t,s) & \text{if } s(t+1) = ff. \end{cases}$$

Lemma $\delta_i(t,s)$ *is defined iff* $t \geq d_i(s)$ *where* $d_i(s) = (least\ t)(|\{j : j \leq t \text{ and } s(j) = tt\}| = i)$. *In particular,* $\delta_1, \ldots, \delta_n$ *are defined iff* $t \geq d_n(s)$.

Proof Omitted.

(Intuitively, $d_i(s)$ is the least t such that i true (tt) items have arrived on stream s up to time t; that is, initialisation up to element i of the reference word is complete. Therefore, $d_n(s)$ is the first time t such that the whole reference word is initialised.)

5.4.2 Third correlator specification We now rewrite the specification to include $\delta_1, \ldots, \delta_n$. This will allow the user to replace the reference word and thus establish the only user requirement ignored in the previous specification $CORR_2$.

$$CORR_3 : [T \to A]^2 \times [T \to \mathbf{B}] \to [T \to \{0, \ldots, n\} \cup \{u\}],$$

$$CORR_3(x, y, s_y)(t) = \begin{cases} u & \text{if } t < d(s_y) + k \\ corr(x(t-k), \ldots, x(t-k-n+1), & \\ \quad y(\delta_1(t, s_y) - k), \ldots, y(\delta_n(t, s_y) - k)) & \text{otherwise} \end{cases}$$

where

$$\delta_1(0, s) = \begin{cases} 0 & \text{if } s(0) = tt \\ \uparrow & \text{if } s(0) = ff, \end{cases}$$

$$\delta_1(t+1, s) = \begin{cases} t+1 & \text{if } s(t+1) = tt \\ \delta_1(t, s) & \text{if } s(t+1) = ff, \end{cases}$$

$$1 < i \leq n,$$

$$\delta_i(0, s) = \uparrow,$$

$$\delta_i(t+1, s) = \begin{cases} \delta_{i-1}(t, s) & \text{if } s(t+1) = tt \\ \delta_i(t, s) & \text{if } s(t+1) = ff, \end{cases}$$

$$d : [T \to \mathbf{B}] \to T,$$

$$d(s) = (\mu t)(|\{i : i \leq t \text{ and } s(i) = tt\}| = n)$$

and k is the computation time for *corr*.

---

(1) The functions $\delta_1 \ldots, \delta_n : T \times [T \to \mathbf{B}] \to T$ are *partial*. However, by the lemma of Section 5.4.1, no attempt will be made to evaluate them on arguments for which they are undefined.
(2) The initialisation function d above is equivalent to d_n in the lemma of Section 5.4.1.

User requirements established in the specification The following user requirement is established in $CORR_3$ (over and above those established in $CORR_2$):

- the reference word may be replaced by the user.

Summary of user requirements established We summarise the user requirements established by specifications $CORR_1$, $CORR_2$ and $CORR_3$:

(1) the processor *corr* computes the correlation of two n-element words in constant k clock cycles;
(2) the output period of the system is 1;
(3) data-words are loaded into the correlator serially;
(4) the reference word may be replaced by the user.

Notice $CORR_3$ is consistent with the informal user specification. We therefore say that $CORR_3$ is a *user-consistent specification.* All specifications presented from this point will also be user-consistent.

User requirements ignored in the specification Specification $CORR_3$ ignores no *explicit* user requirements. However, at this point, we become aware of a number of *implicit* user requirements, which are not stated by the user but which are nevertheless present. For example, we may ask if the correlation results produced when the reference word is being changed have any validity (since they will be correlations of parts of two different reference words), and if it would not be better to explicitly state that correlation results are u during reference word reloading. We explore implicit user requirements in the next section.

Consistency of the specification $CORR_3$ is consistent with $CORR_2$ in the sense that $CORR_3$ is a *generalisation* of $CORR_2$. Given a stream $s \in [T \to \mathbf{B}]$ such that

$$s(0) = \cdots = s(n-1) = tt \text{ and } s(t) = ff, \forall t \geq n$$

and a stream $b \in [T \to A]$ such that

$$b(0) = r_0, \ldots, b(n-1) = r_{n-1},$$

then

$$CORR_3(a, b, s) = CORR_2(a).$$

Structure of specification In its applicative form $\hat{CORR}_3$ is inductively definable over $\overline{\mathbf{A}}$; and is defined by data dependent scheduling, of course. The partiality of the δ_i is *not* significant since they may easily be replaced by total functions. The function d involves an application of unbounded search.

5.5 The environment of the correlator

So far, we have considered the correlator in isolation, divorced from its intended *environment.* The operating environment of a device is important, and it may be the case that the device's intended environment is highly restricted, allowing us to greatly simplify what is a complex specification. This in turn would make it easier to understand and would simplify any correctness proofs.

The specification $CORR_3$ allows the user to partially reload the reference word y, and to wait potentially forever for the first reference word to be loaded. In practice, neither of these features is required; the user can always be expected to load a complete reference word, and can always be expected to load an initial reference word.

We model environment by *restricting the domain* of specifications. That is, by only allowing input streams from a subset S of the set of all possible streams for a given data type $A : S \subset [T \to A]$ Section 3.5.

5.5.1 Fourth draft correlator specification: initialisation condition Our first environment condition states that the first reference word will be loaded in the first n clock cycles. This allows us to eliminate the unbounded stream-search performed by the initialisation function d of $CORR_3$. We do this by taking the control stream s_y from

$$S = \{s \in [T \to \mathbf{B}] : s(t) = tt, 0 \le t < n\}.$$

We are now assured that a reference word will be loaded in the first n clock cycles.

Fourth draft specification We now present our fourth specification

---

Let $S = \{s \in [T \to \mathbf{B}] : s(t) = tt, 0 \le t < n\}$.

$$CORR_4 : [T \to A]^2 \times S \to [T \to \{0, \dots, n\} \cup \{u\}],$$

$$CORR_4(x, y, s_y)(t) = \begin{cases} u & \text{if } t < n + k - 1 \\ corr(x(t-k), \dots, x(t-k-n+1), \\ \quad y(\delta_1(t, s_y) - k), \dots, y(\delta_n(t, s_y) - k)) & \text{otherwise} \end{cases}$$

where $\delta_1(0, s) = 0$, since $s(0) = tt$ by definition of S,

$$\delta_1(t+1, s) = \begin{cases} t+1 & \text{if } s(t+1) = tt \\ \delta_1(t, s) & \text{if } s(t+1) = ff, \end{cases}$$

$$1 < i \le n, \quad \delta_i(0, s) = \uparrow,$$

$$\delta_i(t+1, s) = \begin{cases} \delta_{i-1}(t, s) & \text{if } s(t+1) = tt \\ \delta_i(t, s) & \text{if } s(t+1) = ff, \end{cases}$$

and $k \ge 0$ is the computation time for *corr*.

Notice that there is no stream dependency in the validity predicate of $CORR_4$, but $CORR_4$ remains stream dependent because of the stream dependency of the scheduling functions $\delta_1, \dots, \delta_n$.

User requirements established in the specification In $CORR_4$ we establish the following implicit user requirement:

- an initial reference word is loaded, starting at time $t = 0$.

Consistency of the specification $CORR_4$ is consistent with $CORR_3$ in the sense that $CORR_4$ is a *specialisation* of $CORR_3$. For all streams $a, b \in [T \to A]$ and $s \in S$

$$CORR_4(a, b, s) = CORR_3(a, b, s),$$

and for $s \notin S$

$$CORR_4(a, b, s) \uparrow \text{ but } CORR_3(a, b, s) \downarrow.$$

Structure of specification Recalling the structure of $CORR_3$ we notice that replacing the function d results in a function that is primitive recursive over $\overline{\mathbf{A}}$ augmented by a predicate for S.

5.5.2 Fifth draft formal specification: loading complete reference words We will now proceed to further restrict the domain of the correlator to *ensure only complete reference words are loaded.* To this end, we introduce the concept of *blocks.*

Blocks in boolean streams A *block* is a sequence of equal items arriving on a stream on successive clock cycles. The pair $(i, j) \in \mathbf{N}^2$ is called a block for stream s if

$$\begin{array}{ll} s(k) = s(k+1), k = i, \ldots, j-1 \cap s(j) \neq s(j+1) & \text{if } i = 0, \\ s(k) = s(k+1), k = i, \ldots, j-1 \cap s(i-1) \neq s(i) \cap s(j) \neq s(j+1) & \text{if } i \neq 0. \end{array}$$

That is, items $s(i), \ldots, s(j)$ are equal and distinct from items $s(i-1)$ (if $i \neq 0$) and $s(j+1)$. We denote the block (i, j) for stream s by $b_s(i, j)$.

We will only be concerned with *boolean* streams, $T \to \mathbf{B}$, which have blocks of two types.

(1) *Trueblocks* which consist of sequences of *true (tt)* values and can be defined

$b_s(i, j)$ is a *trueblock* for s iff

$$\begin{array}{ll} s(j+1) = ff \cap s(k) = tt, k = i, \ldots, j & \text{if } i = 0, \\ s(i-1) = ff \cap s(j+1) = ff \cap s(k) = tt, k = i, \ldots, j & \text{if } i \neq 0. \end{array}$$

(2) *Falseblocks* which consist of sequences of *false (ff)* values and can be defined

$b_s(i, j)$ is a *falseblock* for s iff

$$\begin{array}{ll} s(j+1) = tt \cap s(k) = ff, k = i, \ldots, j & \text{if } i = 0, \\ s(i-1) = tt \cap s(j+1) = tt \cap s(k) = ff, k = i, \ldots, j & \text{if } i \neq 0. \end{array}$$

Boolean streams may be considered to be lists of true and false blocks.

The *blocklength* $|b|$ of a block b is the number of clock cycles for which the block extends. It is defined by

$$|b_s(i,j)| = j - i + 1.$$

As an example, we will present the set $S = \{s \in [T \to \mathbf{B}] : s(t) = tt, 0 \le t < n\}$ introduced in Section 5.5.1 above in the block notation. S becomes

$$S = \{s \in [T \to \mathbf{B}] : b_s(0, n-1) \text{ a trueblock for } s\}.$$

That is, the set of boolean streams with an initial trueblock of length n.

Fifth draft formal specification We will now give a specification for a correlator which expects only those boolean streams with minimum trueblock length greater than or equal to n. In addition, we will require, as for $CORR_4$, that the first n elements of the control stream are *true*. That is, we are imposing the initialisation condition of Section 5.5.1 on $CORR_5$.

First we define a set Z of boolean streams with trueblock length a multiple of n, for all trueblocks

$$Z = \{s \in [T \to \mathbf{B}] : (\exists i > 0)|b_s| = i.n, \text{ for all } b \text{ a trueblock for } s\}$$

In addition, we require an initial trueblock of length n; that is, we require $S \cap Z$.

Let A be any non-empty set, let $S = \{s \in [T \to \mathbf{B}] : s(t) = tt, 0 \le t < n\}$ and let $Z = \{s \in [T \to \mathbf{B}] : (\exists i > 0)|b_s| = i.n, \text{ all } b_s \text{ a trueblock for } s\}$.

$$CORR_5 : [T \to A]^2 \times S \cap Z \to [T \to \{0, \ldots, n\} \cup \{u\}],$$

$$CORR_5(x, y, s_y)(t) = \begin{cases} u & \text{if } t < n + k - 1 \\ corr(x(t-k), \ldots, x(t-k-n+1), & \\ \quad y(\delta_1(t, s_y) - k), \ldots, y(\delta_n(t, s_y) - k)) & \text{otherwise} \end{cases}$$

where $\delta_1(0, s) = 0$ since $s(0) = tt$ by definition of $S \cap Z$,

$$\delta_1(t+1, s) = \begin{cases} t+1 & \text{if } s(t+1) = tt \\ \delta_1(t, s) & \text{if } s(t+1) = ff, \end{cases}$$

$$1 < i \le n, \qquad \delta_i(0, s) = \uparrow,$$

$$\delta_i(t+1, s) = \begin{cases} \delta_{i-1}(t, s) & \text{if } s(t+1) = tt \\ \delta_i(t, s) & \text{if } s(t+1) = ff, \end{cases}$$

and $k \ge 0$ is the computation time for *corr*.

User requirements established in the specification In $CORR_5$ the following implicit user requirement is established (in addition to that established in $CORR_4$):

- only complete reference words are loaded.

Consistency of the specification $CORR_5$ is consistent with $CORR_4$ in the sense that $CORR_5$ is a *specialisation* of $CORR_4$. For all streams $a, b \in [T \to A]$ and $s \in S \cap Z$

$$CORR_5(a, b, s) = CORR_4(a, b, s),$$

and for $s \in S$ but $s \notin Z$

$$CORR_5(a, b, s) \uparrow \text{ but } CORR_4(a, b, s) \downarrow .$$

Structure of specification As in the case of $CORR_4$, the specification $CORR_5$ is primitive recursive over $\overline{\mathbf{A}}$ augmented by a predicate for $S \cap Z$.

Summary of user requirements established We summarise the user requirements established by specifications $CORR_1$ to $CORR_5$:

- the processor *corr* computes the correlation of of two n-element words in constant k cycles;
- the output period of the system is 1 clock cycle;
- data words are loaded serially;
- the reference word may be replaced by the user;
- an initial reference word is loaded, starting at time $t = 0$;
- only complete reference words are loaded.

Exercise Adapt the appropriate correlator specification to explicitly state that the correlation output should be u while the reference word is being reloaded.

6 CLOCKS AND RETIMING

Before beginning the abstraction phase of the user specification process, we introduce some new theoretical tools. These tools allow us to express timing characteristics in specifications at different levels of abstraction and are based on the formal concept of a *clock retiming.*

6.1 Retimings

Recall that a clock is an algebra $T = (T, 0, t + 1)$. A clock $T = (T, 0, t + 1)$ is said to be at a *lower level of temporal abstraction* than a clock $R = (R, 0, r + 1)$ if each

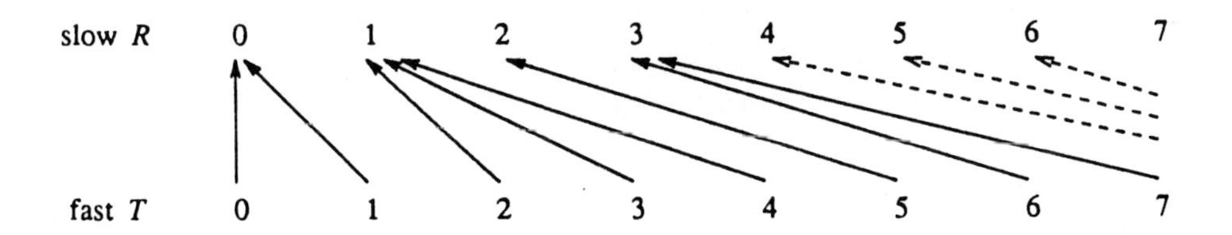

Figure 5.13. Map between clocks T and R

cycle of R corresponds with one or more cycles of T. We also say that T is a *faster* clock than R. Consider a function

$$\lambda : T \to R$$

that formalises this idea where $\lambda(t)$ is the time on clock R corresponding with time t on clock T. We define this λ to be a *retiming function that retimes T to R* if

(1) $\lambda(0) = 0$,
(2) λ is *surjective*, i.e., for all $r \in R$ there is a $t \in T$ such that $\lambda(t) = r$,
(3) λ is *monotonic*, i.e., for all $t, t' \in T$, $t \leq t'$ implies $\lambda(t) \leq \lambda(t')$.

Notice that (1) is implied by (2) and (3). Let $Ret(T, R)$ denote the set of all retimings of T to R.

Examples

(1) Consider the retiming depicted in Figure 5.13:

$$\lambda(0) = 0, \quad \lambda(1) = 0, \quad \lambda(2) = 1, \quad \ldots, \quad \lambda(7) = 3.$$

(2) Consider the case where each cycle of R corresponds with $k \in \mathbf{N}$ cycles of T. That is, clock T runs at exactly k times the rate of clock R. For example, consider the case where a single system clock is implemented by two separate, non-overlapping, clocks and $k = 2$ (Mead & Conway [1980], p. 65). We define $\lambda \in Ret(T, R)$ to be

$$\lambda(t) = \left\lfloor \frac{t}{k} \right\rfloor.$$

For any given $t \in T$, $\lambda(t)$ is just the integer division t *div* k.

(3) More generally, consider the function $f : T \to R$ defined by

$$f(t) = \lfloor \alpha t \rfloor$$

for $\alpha \in \mathbf{R}$ and $\alpha > 0$. Then f is a clock retiming if, and only if, $\alpha \leq 1$.

(4) As an informal example of a retiming, consider the Instruction-Fetch-Execute cycle of a simple computer. Instructions will typically take a variable number of system clock cycles to execute, depending on the instruction. We reconsider the retiming depicted in Figure 5.13 to model this situation with the system clock cycle timed by T, and the instruction cycle timed by the relatively slower, more irregular, R. In this example, the first instruction takes two system clock cycles, the second three, and so on. At any given time $t \in T$, the $\lambda(t)$th instruction is being executed. For more information on the architecture of computers, see for example Anceau [1986], Baer [1980] or Kuck [1978].

In using a retiming function we will need to have available certain other functions. Given a retiming $\lambda : T \to R$ of T to R, we define the corresponding *immersion* $\overline{\lambda} : R \to T$ by

$$\overline{\lambda}(r) = (least\ t \in T)(\lambda(t) = r).$$

Thus $\overline{\lambda}(r)$ is the first clock cycle of T corresponding with r in R. Clearly, $\overline{\lambda}$ is a *right inverse* of λ in the sense that

$$\lambda\overline{\lambda}(r) = r, \quad \text{for all } r \in R,$$

but is *not* a *left inverse*, i.e., for some t

$$\overline{\lambda}\lambda(t) \neq t,$$

see Figure 5.14. Notice that $\overline{\lambda}\lambda(t)$ yields the $t' \in T$ corresponding to the start of the current R clock cycle in which t lies. Hence we define $start(\lambda) : T \to T$ by

$$start(\lambda)(t) = \overline{\lambda}\lambda(t).$$

Next, we define the *length* of the retiming to be a function $l(\lambda) : R \to \mathbf{N}$ defined by

$$l(\lambda)(r) = \overline{\lambda}(r+1) - \overline{\lambda}(r).$$

That is, $l(\lambda)(r)$ is the number of clock cycles of T spanned by the clock cycle r of R.

These functions define four basic parameters of a retiming as illustrated in Figure 5.14.

Observe that the functionality of *start* is

$$start : Ret(T, R) \to [T \to T]$$

and the functionality of l is

$$l : Ret(T, R) \to [R \to \mathbf{N}].$$

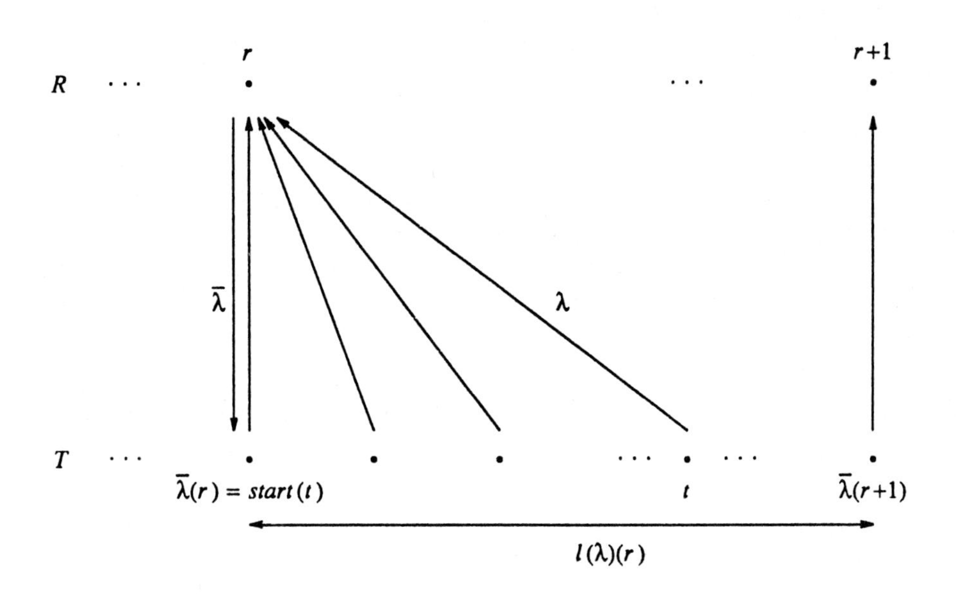

Figure 5.14. Basic structure of a retiming

Examples

(5) For example, the retiming in Figure 5.13 has

$$\overline{\lambda}(0) = 0, \overline{\lambda}(1) = 2, \overline{\lambda}(2) = 5, \overline{\lambda}(3) = 6$$

and

$$l(\lambda)(0) = 2, l(\lambda)(1) = 3, l(\lambda)(2) = 1.$$

(6) In the case where clock T runs k times as fast as clock R, $\overline{\lambda}$ is defined by

$$\overline{\lambda}(r) = kr$$

and the length is defined by $l(\lambda)(r) = k$.

(7) In the case of the retiming $\lambda(t) = \lfloor \alpha t \rfloor$ for $0 < \alpha \leq 1$ the immersion is defined by

$$\overline{\lambda}(r) = \left\lceil \frac{r}{\alpha} \right\rceil$$

and the length is defined by $l(\lambda)(r) = \lceil 1/\alpha \rceil$ (a constant function).

(8) Continuing our example of a simple computer, $\overline{\lambda}(r)$ is the system clock cycle marking the start time of the rth instruction to be executed. The number of system clock cycles the rth instruction takes to be executed is $l(\lambda)(r)$; and the system clock cycle marking the start of the instruction being executed at time t is $start(\lambda)(t)$.

6.2 Scheduling

Consider the relationship between streams over clocks $T = (T, 0, t+1)$ and $R = (R, 0, r+1)$ linked by a retiming $\lambda \in Ret(T, R)$. Given $\beta \in [R \to A]$ we define a stream $\alpha \in [T \to A]$ by

$$\alpha(t) = \beta\lambda(t).$$

The stream α is the result of *scheduling* β at times determined by λ. That is, α is the stream consisting of each element of β repeated or copied $l(\lambda)(r)$ times, for each $r \in R$. We define $sch(\lambda) : [R \to A] \to [T \to A]$ by

$$sch(\lambda)(\beta) = \alpha.$$

Conversely, given a stream $\alpha \in [T \to A]$ we define a stream $\beta \in [R \to A]$ by

$$\beta(r) = \alpha\overline{\lambda}(r).$$

The stream β is the result of *observing*, or *sampling* α at times $\overline{\lambda}(r)$ in clock T. That is, β is the stream consisting of the $\overline{\lambda}(r)$th elements of stream α, for all $r \in R$. Again, for λ we define $\overline{sch}(\lambda) : [T \to A] \to [R \to A]$ by

$$\overline{sch}(\lambda)(\alpha) = \beta.$$

Proposition $\overline{sch}$ *is a left inverse for sch, but not a right inverse. In particular,* $\overline{sch}$ *is an injection.*

Proof Omitted.

Observe that the functionality of sch is

$$sch : Ret(T, R) \to [[R \to A] \to [T \to A]]$$

and the functionality of $\overline{sch}$ is

$$\overline{sch} : Ret(T, R) \to [[T \to A] \to [R \to A]].$$

Example

(9) Consider the retiming depicted in Figure 5.13 (i.e., $\overline{\lambda}(0) = 0, \overline{\lambda}(1) = 2, \overline{\lambda}(2) = 5, \overline{\lambda}(3) = 6$) and the streams

$$\alpha \in [T \to \mathbf{N}], \quad \beta \in [R \to \mathbf{N}]$$

where

$$\alpha(0) = 1, \alpha(1) = 1, \alpha(2) = 2,$$
$$\alpha(3) = 3, \alpha(4) = 5, \alpha(5) = 8, \ldots \text{(the Fibonnaci numbers)}$$

and

$$\beta(0) = 0, \beta(1) = 1, \beta(2) = 4, \ldots \text{ (the squares of } \mathbf{N}\text{)}.$$

Then

$$\beta' = \overline{sch}(\lambda)(\alpha)$$

$$\beta'(0) = 1, \quad \beta'(1) = 2, \quad \beta'(2) = 8, \ldots;$$

and

$$\alpha' = sch(\lambda)(\beta)$$

$$\alpha'(0) = \alpha'(1) = 0, \quad \alpha'(2) = \ldots = \alpha'(4) = 1, \quad \alpha'(5) = 4, \ldots$$

6.3 Serialisation

We now present a larger example of scheduling to illustrate the concepts presented in Section 6.1 and Section 6.2. We discuss here *serialisation*, a technique we employ in Section 9 in the implementation phase of the specification process.

Consider how a stream $\beta \in [R \to A^n]$ of vectors of data from A may be represented as a stream $\alpha \in [T \to A]$ of single data items; for example as in the case when binary words are loaded bit-serially.

This process can be represented by a transformation called *serialisation*

$$ser(\lambda) : [R \to A^n] \to [T \to A]$$

where $ser(\lambda)(\beta)$ is determined by a retiming λ. Let $\lambda \in Ret(T, R)$ be a retiming and suppose that

$$l(\lambda)(r) \geq n, \qquad \text{for all } r \in R.$$

Then for *any* such λ, we can specify $ser(\lambda)(\beta)$ from β by defining

$$ser(\lambda)(\beta)(t) = \begin{cases} \beta_i\lambda(t) & \text{if } t - start(\lambda)(t) = i - 1 < n \\ * & \text{if } t - start(\lambda)(t) \geq n \end{cases}$$

where $\beta(r) = (\beta_1(r), \ldots, \beta_n(r)) \in A^n$, for all $r \in R$, and $*$ is *any element of* A, possibly dependent on $t \in T$. The stream $ser(\lambda)(\beta)$ consists of the components of the n-vectors of β in contiguous segments, possibly separated by padding, denoted by $*$. Our attention is drawn to the value of $l(\lambda)(r) - n$, being the length of padding allocated by λ after loading $\beta(r)$.

We note that $*$ differs from u in that $*$ specifies varying data from A, whereas u is *not* in A, but is employed in specifications to indicate the presence of data to be ignored. The construction is illustrated in Figure 5.15.

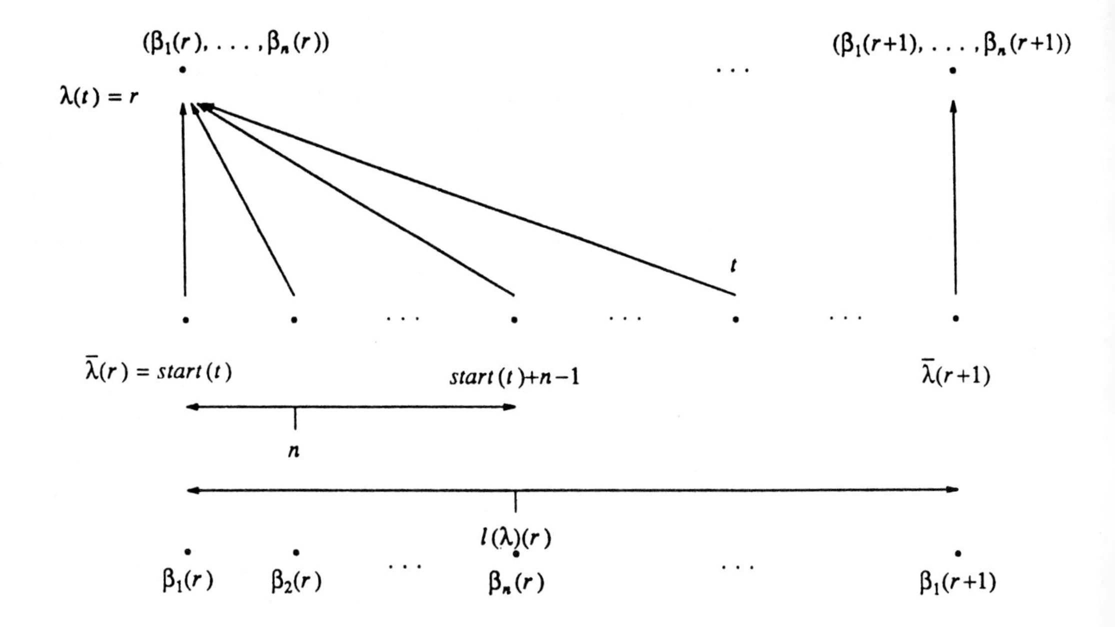

Figure 5.15. Serialisation

We may also define the reverse of serialisation $\overline{ser}(\lambda) : [T \to A] \to [R \to A^n]$ which maps a stream $\alpha \in [T \to A]$ of single data items into a stream $\beta \in [R \to A^n]$ of vectors

$$\overline{ser}(\lambda)(\alpha)(r) = \alpha(\overline{\lambda}(r) + n - 1), \ldots, \alpha(\overline{\lambda}(r)).$$

Proposition $\overline{ser}$ *is a left inverse of ser but is not a right inverse. In particular, ser is an injection.*

Proof Omitted.

Observe that the functionality of *ser* is

$$ser : Ret(T, R) \to [[R \to A^n] \to [T \to A]]$$

and the functionality of $\overline{ser}$ is

$$\overline{ser} : Ret(T, R) \to [[T \to A] \to [R \to A^n]].$$

Given such a stream, how do we know when one word ends and the next begins? Clearly, we must know the length of padding between words (and the word length of course). For example, we could consider regular cases when the length of padding is 0 or 1. In the case $l(\lambda)(r) - n = 0$, there is no padding and so the next word $\beta(r+1)$ is immediately loaded. In this case, the only way to distinguish between words is to know externally the wordlength n. In the case $l(\lambda)(r) - n = 1$, we may again distinguish words as above, by knowing their length. We may extend this case to include all those where the length of padding is some constant c.

We may also consider the case where the length of the padding depends on data contained in the serialised word itself.

Example

(10) Returning to our computer example, we note that instructions are of varying length and take a variable amount of time to execute. Consequently, the computer waits a variable amount of time between loading instructions. The length of the wait will depend on the instruction last loaded. In this case, since instructions are usually a variable number of computer-words in length, the first part of the serialised data to be loaded also controls how many further parts must be loaded to fetch the complete instruction. This example shows that serialisation as we have defined it need not necessarily mean bit-serial; but may apply to any size of sub-segment of data (in this case, the size of the computer word).

In general and in the case of the correlator, the length of padding may vary independently of β. Therefore, given the stream $ser(\lambda)(\beta)$ and knowing the length of the retiming, we must have a padding function $p : R \to \mathbf{N}$ where $p(r)$ is the amount of padding of $ser(\lambda)$ on loading $\beta(r)$. This information must be known externally to distinguish new words.

How can we code this information so it can be recovered from the stream? A first strategy is to let * be some distinguished element reserved to act as a marker. However, this disturbs the type-structure (by introducing an extra element in A) and is otherwise unattractive.

A standard method is to introduce a boolean control stream $pulse_n(\lambda)$ which codes the information by synchronising with $ser(\lambda)(\beta)$ as follows

$$pulse_n : Ret_n(T, R) \to B_n,$$

$$pulse_n(\lambda)(t) = \begin{cases} tt & \text{if } t - start(\lambda)(t) < n \\ ff & \text{otherwise,} \end{cases}$$

where $Ret_n(T, R) = \{\lambda \in Ret(T, R) : l(\lambda)(r) \geq n, \forall r \in R\}$, and

$$B_n^\infty = \{s \in [T \to \mathbf{B}] : \text{for any trueblock } b_s \text{ in } s \text{ there is an } i = i(b_s) \text{ such that } |b_s| = ni,\ b_s(0, nj-1) \text{ for some } j > 0 \text{ is a trueblock for } s, \text{ and there is no } t \in T \text{ such that for all } t' > t,\ s(t') = \mathit{ff}\}.$$

This definition is taken from Section 8.1.2 where we consider the generation of boolean control streams from retimings in a systematic manner.

6.4 Retimings and stream transformers

A retiming defines a relationship between computation at two levels of temporal abstraction. Let T and R be clocks related by retiming $\lambda \in Ret(T, R)$ and consider the two stream processors

$$\Phi_U : [R \to A] \to [R \to A] \quad \text{and} \quad \Phi_I : [T \to A] \to [T \to A].$$

Suppose that

$$\Phi_U(a)(r) = \overline{sch}(\lambda)(\Phi_I(sch(\lambda)(a)))(r),$$

or, equivalently, the diagram of Figure 5.16 *commutes*. This means stream transformer Φ_I is equivalent to Φ_U provided that input streams intended for Φ_U are *scheduled* before input to Φ_I and the result stream of Φ_I is *sampled*, by scheduling and sampling functions defined from λ in the manner of Section 6.2. (Recall from Section 6.2 that $\overline{sch}$ is a left inverse for sch.)

$$\begin{array}{ccc} [R \to A] & \xrightarrow{\Phi_U} & [R \to A] \\ sch \downarrow & & \overline{sch} \uparrow \\ [T \to A] & \xrightarrow{\Phi_I} & [T \to A] \end{array}$$

Figure 5.16. Relationship between user specification and implementation specification

Recalling Section 2.2.6 in such circumstances, we say Φ_I *implements* Φ_U or that Φ_U is an *abstraction* of Φ_I. The commutativity of Figure 5.16 is a statement about the correctness of the implementation of Φ_U by Φ_I.

7 ABSTRACT FORMAL SPECIFICATION OF A DIGITAL CORRELATOR

We will now use the tools developed in Section 6 to develop an *abstract* specification of the correlator. Recall from Section 2.3.2 that the abstraction phase aims to replace low-level constructs by higher-level ones. We do this in order

(1) to eliminate irrelevant and obstructive implementation details picked up from the informal specification, and
(2) to simplify the proof of any implementation.

Recall the (strictly speaking premature) decision to use A in place of $\{0, 1\}$ from the start; this assumption is an example of *data abstraction.* This abstraction is productive as it simplifies the specification, and increases flexibility by delaying a decision. Following Section 6, this section will concentrate on *timing abstraction* (data abstraction will be considered again in Section 9.3).

The level of computational abstraction L is based on $\overline{\mathbf{A}}$ in Section 5 enhanced with a new clock R and retimings of R to T. Let us now consider the informal specification I.

7.1 Informal specification of programming the reference word

Recall from Section 5.5, that in $CORR_4$ and $CORR_5$ we imposed conditions on the control stream s_y to the effect that (i) an initial reference word is always loaded, starting at time $t = 0$, and (ii) only complete reference words are loaded. Notice that one effect of these conditions is to illustrate that replacing the reference word is conceptually a *single* operation, rather than a series of n continuous operations. This is an example of an implementation assumption (Section 2.3.2) which complicates the specification and reduces its flexibility.

We would therefore like to replace the serial loading of the reference word with something more abstract, which explicitly shows reference word replacement to be a single operation. Of course, in developing an implementable specification later on (Section 9) we must replace this with a concrete mechanism. The precise details of this mechanism will depend on, for example, user preference, technological constraints, designer expertise, and so on.

How may we remove the serial loading assumption, and replace it with a more abstract structure? Every so often, a variable number of cycles of clock T, a new reference word is loaded. Therefore, we may replace the reference word-element stream y and the control stream s_y of $CORR_3$–$CORR_5$ by a stream $\underline{y} \in [R \to A^n]$ of reference words.

A second difficulty with $CORR_3$–$CORR_5$ is that, once initialised, the correlator will always produce 'valid' results, *even when the reference word is being changed* (Section 5.4.2). This was reasonable in terms of the specifications in the exploration phase, but from the point of view of the application, these results will be meaningless, being correlations of parts of two different reference words. Additionally, if we abstract away the implementation detail of serial loading, we cannot know that the correlation results produced during this period really *are* valid, since we do not know how reference word programming is implemented. We may correct this by adding a second term to the initialisation predicate.

7.2 Sixth draft formal specification

We will now present our sixth draft formal specification, incorporating the reference word programming abstraction mechanism discussed in Section 7.1.

Let R be the *reference word clock*, and let T be the *system clock* as before.
Let $\underline{y} \in [R \to A^n]$, with $\underline{y}(r) = (\underline{y}_1(r), \ldots, \underline{y}_n(r))$ for all $r \in R$, and
$Ret_i(T, R) = \{\lambda \in Ret(T, R) : l(\lambda)(r) \geq i, \forall r \in R\}$ (see Section 6.3).
$CORR_6 : [T \to A] \times [R \to A^n] \times Ret_i(T, R) \to [T \to \{0, \ldots, n\} \cup \{u\}]$,

$$CORR_6(x, \underline{y}, \lambda)(t) = \begin{cases} u & \text{if } t < \max(i, n) + k - 1 \text{ or} \\ & (t-k) - start(\lambda)(t-k) < i \\ corr(x(t-k), \ldots, & \\ \quad x(t-k-n+1), \underline{y}\lambda(t-k)) & \text{otherwise} \end{cases}$$

where max : $\mathbf{N}^2 \to \mathbf{N}$ returns the maximum of its arguments, and i is the *loading time* of the reference word.

Observe the following.

(1) We can determine the current $r \in R$ for any given $t \in T$ by $r = \lambda(t)$. Thus we can use $\underline{y}(\lambda(t))$ to determine the reference word at time t. This is equivalent to using $\lambda(t)$ to determine which instruction is being executed in the computer Example (4) in Section 6.1.

(2) Since we no longer assume the reference word is loaded serially, we can no longer assume it takes n cycles. We therefore use i to stand for the loading time for the reference word. This could be very short: for example, with a fully parallel loading scheme $i = 1$; or long: if reference words are loaded serially, one element every other system clock cycle $i = 2n$. In the first case ($i = 1$), the loading of the reference word takes much less time than reading in the first n elements of the data stream x, so it is reading in enough elements of x which determines the initialisation time. In the second case ($i = 2n$), loading the reference word takes twice as long as reading in enough elements of stream x, so it is reference word loading which determines the initialisation time. Hence the term $\max(i, n)$ which determines which of i and n is greater and hence

which determines the initialisation time. In all previous specifications where the reference word may be changed by serial loading $i = n$ so both loading the reference word and reading in n elements of stream x take the same time. The use of i for the reference word loading time allows $CORR_6$ to abstractly represent a wide range of methods for replacing the reference word.

(3) The output of $CORR_6$ is unspecified (u) when
(i) $t < \max(i, n) + k - 1$: that is, when either the initial reference word is not yet loaded or the first n data items have not yet been read in, whichever takes longest;
(ii) $(t - k) - start(\lambda)(t - k) < i$: that is, the reference word is being changed; $t - start(\lambda)(t)$ is the number of system clock cycles which have elapsed since a new clock cycle started on clock R. We use $t - k$ because correlation results take k steps to compute.

(4) We insist $l(\lambda)(r) \geq i$, for all $r \in R$ to ensure that there is always enough time to load a complete reference word. This may not be necessary, depending on how reference word programming is implemented. For example, if we adopt the serial loading method, it is safe to abandon the loading of one reference word part-way through, and commence loading another. However, it is not safe to assume this in general. For example, there may be mechanisms for reference word replacement where, once the loading of a new reference word has commenced, abandoning it before it has finished and restarting will result in a garbled reference word.

Implementation assumptions eliminated in the specification The following implementation assumption was eliminated in specification $CORR_6$:

- the reference word must be loaded serially.

Consistency of the specification Clearly $CORR_6$ is a user-consistent specification. We present no formal justification for the consistency of $CORR_6$ with earlier user-consistent specifications $CORR_3$–$CORR_5$. Rather, since it is based on a new intuition, we treat it as the result of a conceptual jump. We would expect that the acceptance-testing procedure, when the client-user animates a specification, would have to be repeated in this case. We use $CORR_6$ as the basis for a series of implementation specifications in Section 9. The development of this series of implementation specifications will be a formal process, and we clearly show that these implementation specifications are consistent with specifications $CORR_3$–$CORR_5$.

Structure of specification In its applicative form $CORR_6$ is a conditional polynomial over $\overline{\mathbf{A}}$ augmented by the clock R, streams $[R \to A]$ and retimings $Ret_i(T, R)$, together with certain obvious operations such as *start* etc. The structure of specifications in the case of multiple clocks and retimings is too complicated to discuss in any detail given the objectives of this chapter.

8 IMPLEMENTING RETIMINGS AS BOOLEAN STREAMS

We now introduce further theoretical tools which will allow us to proceed with the final stage of the specification process, the *implementation phase.* These tools will enable us to implement the reference word clock R and the schedule defined by the retiming $\lambda \in Ret(T, R)$, introduced in the abstraction phase, by means of boolean control streams. More generally, these tools can be applied to abstract specifications, based on similar timing abstractions, to generate implementable specifications in a systematic manner. This facility allows us to disregard details during much of the specification and design processes, and has implications for the proofs that implementations meet specifications.

8.1 Representing retimings by boolean streams

First we consider how we may define boolean control streams from retimings, in order to implement the retimings at the device level. We also consider the converse, i.e., how to abstract retimings from boolean streams.

Consider a retiming $\lambda \in Ret(T, R)$. We wish to define a boolean stream $b : T \rightarrow \mathbf{B}$ from λ to impart scheduling information contained in λ to a chip. That is, we wish to *implement* the retiming using a boolean stream. There are infinitely many ways we may do this, however only a few are of interest to us. We wish to use methods that (a) have practical utility, and (b) are *reversible.*

By a *reversible method* we mean a procedure κ for implementing a class of retimings for which there is an inverse procedure κ^{-1} that abstracts retimings from the boolean streams. Precisely, the method κ defines an *implementation map*

$$\kappa : \Delta \rightarrow \Gamma$$

from some class $\Delta \subset Ret(T, R)$ onto some class $\Gamma \subset [T \rightarrow \mathbf{B}]$ and possesses an inverse *abstraction map*

$$\kappa^{-1} : \Gamma \rightarrow \Delta$$

from Γ onto Δ so that for $\lambda \in \Delta$ and $b \in \Gamma$ we have

$$\kappa^{-1}\kappa(\lambda) = \lambda \quad \text{and} \quad \kappa\kappa^{-1}(b) = b$$

(recall Section 2.2.6). The method we consider first is the *single-pulse* method.

8.1.1 The single-pulse method In this method, the start of each new clock cycle $r \in R$ will be denoted by a single *tt* element on a boolean stream $b : T \rightarrow \mathbf{B}$. We define $pulse_1 : Ret(T, R) \rightarrow [T \rightarrow \mathbf{B}]$ by

$$pulse_1(\lambda)(t) = \begin{cases} tt & \text{if } start(\lambda)(t) = t \\ ff & \text{otherwise.} \end{cases}$$

Notice that the range of $pulse_1$ is *not* $[T \rightarrow \mathbf{B}]$. This is because the first element of any stream defined by $pulse_1$, will be *tt*, since time $t = 0$ will *always* mark the beginning of a new clock cycle on clock R (Section 6.1). Let $B_1 = \{b \in [T \rightarrow \mathbf{B}] : b(0) = tt\}$. Additionally, notice that $pulse_1$ will always produce boolean streams with an infinite number of *tt* elements. This is because if there were only a finite number of *tt* elements, there would be a $t \in T$ such that for all $t' > t$, $pulse_1(\lambda)(t') = ff$. This would in turn mean that for all $t' > t$, $\lambda(t') = \lambda(t)$ (by definition of $pulse_1$) and hence λ would *not* be surjective. We introduce

$$B_1^\infty = \{b \in [T \rightarrow \mathbf{B}] : b(0) = tt \text{ and there is no } t \in T \text{ such that for all } t' > t, b(t') = ff\}$$

to eliminate streams beginning with *ff* and with a finite number of *tt* elements. Now $pulse_1 : Ret(T, R) \rightarrow B_1^\infty$ is a bijection.

Example

(1) Consider the retiming λ depicted in Figure 5.17. The stream $pulse_1(\lambda)$ generated is also shown.

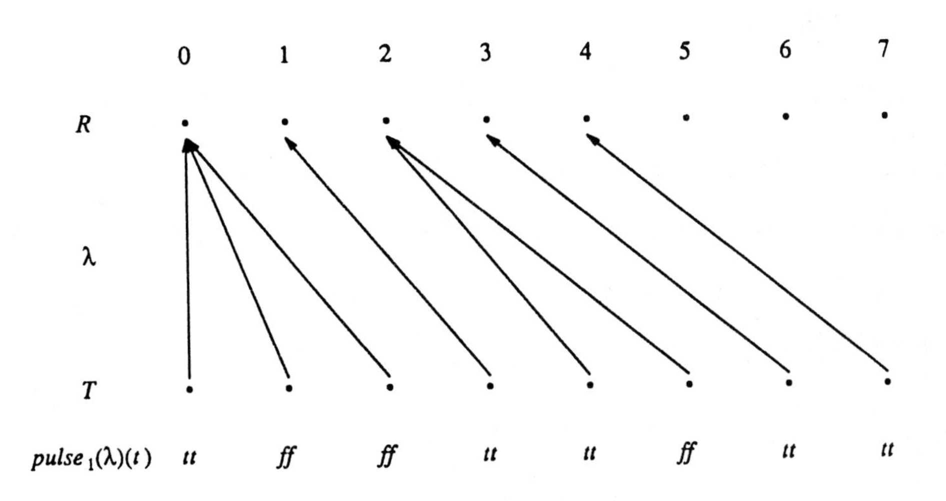

Figure 5.17. Single-pulse method

We define the inverse $pulse_1^{-1} : B_1^\infty \to Ret(T,R)$ as follows

$$pulse_1^{-1}(b)(t) = |\{i : b(i) = tt, \quad 0 < i \leq t\}|.$$

That is, the number of cycles of clock R which have elapsed up to time $t \in T$ is simply the number of tt elements that have arrived on stream $b \in B_1^\infty$ up to time t less 1 (because we measure time from 0).

Proposition *Given $\lambda \in Ret(T,R)$, and $b \in B_1^\infty$,*

$$\lambda = pulse_1^{-1} \circ pulse_1(\lambda) \quad \text{and} \quad b = pulse_1 \circ pulse_1^{-1}(b).$$

Proof Omitted.

8.1.2 The n-pulse method In this method, which generalises the single-pulse method, each new clock cycle $r \in R$ is represented by a block of n tt elements on a boolean stream s. For this, we require $l(\lambda)(r) \geq n$, for all $r \in R$; let

$$Ret_n(T,R) = \{\lambda \in Ret(T,R) : l(\lambda)(r) \geq n, \forall r \in R\}.$$

Consider the generalisation of B_1^∞,

$$\begin{aligned} B_n^\infty = \; & \{s \in [T \to \mathbf{B}] : \text{for any trueblock } b_s \text{ in } s \text{ there is an } i = i(b_s) \\ & \text{such that } |b_s| = in,\ b_s(0, jn-1) \text{ for some } j > 0 \\ & \text{is a trueblock for } s, \text{ and there is no } t \in T \\ & \text{such that for all } t' > t,\ s(t') = \mathit{ff}\}. \end{aligned}$$

We define $pulse_n : Ret_n(T,R) \to B_n^\infty$ as follows

$$pulse_n(\lambda)(t) = \begin{cases} tt & \text{if } t - start(\lambda)(t) < n \\ \mathit{ff} & \text{otherwise.} \end{cases}$$

That is, the start of each new clock cycle $r \in R$ is marked by n consecutive tt elements on stream $pulse_n(\lambda)$.

We define the inverse, $pulse_n^{-1} : B_n^\infty \to Ret_n(T,R)$, as follows

$$pulse_n^{-1}(b)(t) = \left\lceil \frac{|\{i : b(i) = tt, \quad n < i \leq t\}|}{n} \right\rceil.$$

Proposition *For any $\lambda \in Ret_n(T,R)$ and any $b \in B_n^\infty$,*

$$\lambda = pulse_n^{-1} \circ pulse_n(\lambda) \quad \text{and} \quad b = pulse_n \circ pulse_n^{-1}(b).$$

Proof Omitted.

Example

(2) Consider clock R defined from clock T by a retiming $\lambda \in Ret(T, R)$

$$\lambda(t) = \left\lfloor \frac{t}{10} \right\rfloor$$

and suppose $n = 5$. Then

$$\begin{aligned} pulse_n(\lambda)(0) = \cdots = pulse_n(\lambda)(4) = tt, \\ pulse_n(\lambda)(5) = \cdots = pulse_n(\lambda)(9) = ff, \\ pulse_n(\lambda)(10) = \cdots = pulse_n(\lambda)(14) = tt, \cdots . \end{aligned}$$

8.2 Linear retimings and periodic streams

We now consider other restricted classes of retimings and streams: *linear retimings* and *periodic streams.* A retiming $\lambda \in Ret(T, R)$ is *linear* if it has constant length

$$(\exists i > 0)(\forall r)(l(\lambda)(r) = i).$$

That is, each clock cycle $r \in R$ corresponds to the *same number* of clock cycles in T. Equivalently, a retiming is linear if and only if it is of the form $\lambda(t) = \lfloor t/\alpha \rfloor$, for some $\alpha \in \mathbf{N}$, $\alpha > 0$, given in Example (3) in Section 6.1, and for all r, $l(\lambda)(r) = \alpha$.

If λ is linear we write $l(\lambda)$ for its constant period.

Let A be any set. A stream $s : T \to A$ is *periodic of length* l if

$$(\forall i > 0)(s(0) = s(il), \ldots, s(l-1) = s(il + l - 1)).$$

That is, a stream consists of an initial *segment* $s(0), \ldots, s(l-1)$ of length l repeated infinitely many times.

The *period length* of a periodic stream is defined to be the least such l such that s is periodic of length l; let $l(s)$ be the period length of s.

8.2.1 Periodic boolean streams defined by linear retimings Let $LRet(T, R) = \{\lambda \in Ret(T, R) : \lambda \text{ linear}\}$ and $B_p^\infty = \{s \in [T \to \mathbf{B}] : s \text{ periodic}\}$. We define a new method $pulse_1 : LRet(T, R) \to B_p^\infty$, a map from linear retimings to periodic boolean streams as follows:

$$pulse_1(\lambda)(t) = \begin{cases} tt & \text{if } t \bmod l(\lambda) = 0, \\ ff & \text{otherwise.} \end{cases}$$

That is, a tt element is generated for every $t \in T$ such that t is a multiple of the period length of the retiming λ. We can also define a mapping $\overline{pulse_1} : B_p^\infty \to LRet(T, R)$ from periodic boolean streams to periodic retimings

$$\overline{pulse_1}(b)(t) = \left\lfloor \frac{t}{l(b)} \right\rfloor.$$

Notice we call this function $\overline{pulse_1}$ as opposed to $pulse_1^{-1}$ to make it clear that $\overline{pulse_1}$ is *not* a true inverse for $pulse_1$ but only a left inverse (an injection), i.e.,

$$\lambda = \overline{pulse_1} \circ pulse_1(\lambda), \forall \lambda \in LRet(T, R) \text{ but } b \neq pulse_1 \circ \overline{pulse_1}(b), \text{ for some } b \in B_p^\infty.$$

Proof Omitted.

Examples

(3) Suppose, as in Example (2), R is defined from T by retiming

$$\lambda(t) = \left\lfloor \frac{t}{10} \right\rfloor.$$

Then $l(\lambda) = 10$ and

$$\begin{aligned} pulse_1(\lambda)(0) &= tt, \\ pulse_1(\lambda)(1) &= \cdots = pulse_1(\lambda)(9) = ff, \\ pulse_1(\lambda)(10) &= tt, \cdots . \end{aligned}$$

(4) Consider the stream $b : T \to \mathbf{B}$

$$\begin{aligned} &b(0) = ff, \quad b(1) = tt, \quad b(2) = ff, \\ &b(3) = ff, \quad b(4) = tt, \quad b(5) = ff, \\ &b(6) = ff, \quad b(7) = tt, \quad b(8) = ff, \cdots \end{aligned}$$

consisting of the sequence ff, tt, ff repeated infinitely many times, with $l(b) = 3$. Then

$$\begin{aligned} \overline{pulse_1}(b)(0) &= \cdots = \overline{pulse_1}(b)(2) = 0, \\ \overline{pulse_1}(b)(3) &= \cdots = \overline{pulse_1}(b)(5) = 1, \\ \overline{pulse_1}(b)(6) &= \cdots = \overline{pulse_1}(b)(8) = 2, \ldots . \end{aligned}$$

with $l(\overline{pulse_1}(b)) = 3$, and

$$
\begin{array}{ll}
pulse_1 \circ \overline{pulse_1}(b)(0) = tt, & pulse_1 \circ \overline{pulse_1}(b)(1) = f\!f, \\
pulse_1 \circ \overline{pulse_1}(b)(2) = f\!f, & pulse_1 \circ \overline{pulse_1}(b)(3) = tt, \\
pulse_1 \circ \overline{pulse_1}(b)(4) = f\!f, & pulse_1 \circ \overline{pulse_1}(b)(5) = f\!f, \\
pulse_1 \circ \overline{pulse_1}(b)(6) = tt, & pulse_1 \circ \overline{pulse_1}(b)(7) = f\!f, \\
pulse_1 \circ \overline{pulse_1}(b)(8) = f\!f, & \cdots
\end{array}
$$

Observe that $pulse_1 \circ \overline{pulse_1}(b)$ consists of the sequence $tt, f\!f, f\!f$ repeated, and hence $b \neq pulse_1 \circ \overline{pulse_1}(b)$.

8.3 Retimings and levels of abstraction

The above methods allow us to formulate correctness criteria for implementations of specifications based on retimings by means of specifications based on boolean streams. For example, consider the stream transformer

$$\Phi_R : [R \to A] \times Ret(T, R) \to [T \to A]$$

which contains a retiming, and suppose we wish to generate an implementable version

$$\Phi_T : [T \to A] \times [T \to \mathbf{B}] \to [T \to A]$$

which only uses one clock. Suppose we implement the retiming $\lambda \in Ret(T, R)$ using method κ, and schedule the input stream $a \in [R \to A]$ to produce an input stream $a' \in [T \to A]$ where $a' = sch(\lambda)(a)$. Then, recalling Section 2.2.6, stream transformer Φ_T defined by

$$\Phi_T(x, b)(t) = \Phi_R(sch(\kappa^{-1}(b))(x), \kappa^{-1}(b))$$

correctly implements Φ_R, that is, the diagram of Figure 5.18 commutes.

$$
\begin{array}{ccccc}
[R \to A] & \times & Ret(T, R) & \xrightarrow{\Phi_R} & [T \to A] \\
 & sch \times Id \big\downarrow & & & \big\uparrow Id \\
[T \to A] & \times & Ret(T, R) & \xrightarrow{\Phi'_R} & [T \to A] \\
\big\downarrow Id & & \big\downarrow \kappa & & \big\uparrow Id \\
[T \to A] & \times & B_0^{\infty} & \xrightarrow{\Phi_T} & [T \to A]
\end{array}
$$

Figure 5.18. Relationship between retimed specification and implementation specification

As a technical point, observe in Figure 5.18 that the implied functionality of *sch* is

$$sch : [R \to A] \times Ret(T, R) \to [T \to A]$$

which is the applicative form of

$$sch : Ret(T, R) \to [[R \to A] \to [T \to A]]$$

given in Section 6.2. The two forms are equivalent.

9 IMPLEMENTABLE SPECIFICATION OF A DIGITAL CORRELATOR

We now begin the implementation phase of the specification process. We will apply the tools of Section 8 to develop an implementable specification from the abstract specification $CORR_6$ of Section 7.2.

9.1 Implementing timing abstraction

In Section 7 we eliminated the serial loading of the reference word by introducing a new clock R to control reference word programming. In the implementation phase of the specification process, we must decide how this abstract clock is to be realised on a chip.

Recall from Section 4.6, that the user of the correlator requires $n = 32$, where n is the size of the reference word and of the segment of stream x on which the correlation function operates. In other words, the reference word is a vector $(a_0, \ldots, a_{31}) \in A^{32}$ of 32 elements. The circuit packages available for the correlator were restricted to no more than 40 pins, of which many will be taken up by ancillary requirements (power, clock streams, etc.), and other input and output requirements (data stream x and the correlation result). Consequently, the reference word must be divided into sections and loaded over a number of clock cycles. In deciding on a suitable size for each section of the reference word, there is a wide choice, though we prefer to use powers of 2 for simplicity. For example, we could use sections of length 8 and hence load a complete reference word in 4 clock cycles, or sections of length 4, which would take 8 cycles to load. In practice however, the most straightforward way is to load each reference word element individually, taking 32 cycles. In other words, to adopt the technique of the informal specification and set $i = n = 32$ in $CORR_6$. Use of this technique will simplify both the correlator and the environment needed to support it.

In addition to deciding how the reference word should be loaded, we must decide how reference word loading is to be *controlled*. Currently this is achieved by the higher level clock R and retiming $\lambda \in Ret_i(T, R)$. In Section 8, we discussed methods for implementing clock retimings with boolean streams and we will adopt one of these methods. Of the methods presented in Section 8, those concerned with linear retimings and streams are inapplicable, since the retimings $Ret_i(T, R)$ are not necessarily linear. However, both the single-pulse method (Section 8.1.1) and the n-pulse method (Section 8.1.2) are suitable. The single-pulse method is applicable to any retiming λ. The n-pulse method requires $l(\lambda)(r) \geq n$, for all $r \in R$. If we set $i = n = 32$, this method can be chosen for the correlator, since a complete reference word will take 32 clock cycles to load serially. Of the two methods, we prefer the n-pulse method for simplicity, as we do not have to count clock cycles to know when a complete reference word has been loaded; we need only wait until a *ff* element arrives on the control stream. In the case of the correlator, the n-pulse method is also convenient for the user.

We will proceed as follows. First, having decided to load the reference word serially, we set the reference word loading time i in $CORR_6$ to be n and consequently the retiming of $CORR_6$ is now $\lambda \in Ret_n(T, R)$. In Section 9.1.1 we will substitute the reference word stream $\underline{y} : R \to A^n$ with a reference word element stream $y : T \to A$, using the $\overline{ser}$ function (Section 6.3). In Section 9.1.2 we will replace the retiming λ by a boolean control stream $b \in B_n^\infty$ such that $b = pulse_n(\lambda)$ and hence eliminate clock R. We will then show how the resulting specification $CORR_8$ can be rewritten in the same form as $CORR_5$.

The consistency and correctness of the process of implementing $CORR_6$ by $CORR_8$ is represented by the commutative diagram of Figure 5.19 (recall Section 2.2.6).

$$
\begin{array}{ccccccc}
[T \to A] & \times & [R \to A^n] & \times & Ret_n(T,R) & \stackrel{CORR_6}{\longrightarrow} & [T \to \{0,\ldots,n\} \cup \{u\}] \\
\downarrow Id & & & \overline{ser} \times Id \downarrow & & & \uparrow Id \\
[T \to A] & \times & [T \to A] & \times & Ret_n(T,R) & \stackrel{CORR_7}{\longrightarrow} & [T \to \{0,\ldots,n\} \cup \{u\}] \\
\downarrow Id & & \downarrow Id & & \downarrow pulse_n & & \uparrow Id \\
[T \to A] & \times & [T \to A] & \times & B_n^\infty & \stackrel{CORR_8}{\longrightarrow} & [T \to \{0,\ldots,n\} \cup \{u\}]
\end{array}
$$

Figure 5.19. From abstract specification to serial reference word loading

9.1.1 Seventh draft formal specification: serialising the reference word We wish to replace the reference word stream $\underline{y} \in [R \to A^n]$ by a reference word element stream $y \in [T \to A]$. That is, we require

$$CORR_7 : [T \to A] \times [T \to A] \times Ret_n(T, R) \to [T \to \{0, \ldots, n\} \cup \{u\}].$$

We achieve this by substituting $\overline{ser}(\lambda)(y)$ for $\underline{y}$ in $CORR_6$

$$CORR_7(x, y, \lambda) = CORR_6(x, \overline{ser}(\lambda)(y), \lambda).$$

Expanding this, we obtain the following

$$CORR_7 : [T \to A] \times [T \to A] \times Ret_n(T, R) \to [T \to \{0, \ldots, n\} \cup \{u\}],$$

$$CORR_7(x, y, \lambda)(t) = \begin{cases} u & \text{if } t < n + k - 1 \\ & \text{or } (t - k) - start(\lambda)(t - k) < n, \\ corr(x(t - k), \ldots, & \\ \quad x(t - k - n + 1), & \\ \quad \overline{ser}(\lambda)(y)(\lambda(t - k))) & \text{otherwise.} \end{cases}$$

We will now rewrite $CORR_7$ with $\overline{ser}$ expanded.

$$CORR_7 : [T \to A] \times [T \to A] \times Ret_n(T, R) \to [T \to \{0, \ldots, n\} \cup \{u\}],$$

$$CORR_7(x, y, \lambda)(t) = \begin{cases} u & \text{if } t < n + k - 1 \\ & \text{or } (t - k) - start(\lambda)(t - k) < n, \\ corr(x(t - k), & \\ \ldots, x(t - k - n + 1), & \\ y(\bar{\lambda}\lambda(t - k) + n - 1), & \\ \ldots, y(\bar{\lambda}\lambda(t - k))) & \text{otherwise.} \end{cases}$$

Observe that $\bar{\lambda}\lambda$ is $start(\lambda)$ (Section 6.1).

Implementation decisions taken in this specification In $CORR_7$ the following implementation decision is taken:

- the reference word is loaded serially.

Correctness of the specification The correctness of the transformation of $CORR_6$ to $CORR_7$ is illustrated by the top half of Figure 5.19 and expressed by the equation

$$CORR_6(x, y, \lambda) = CORR_7(x, ser(\lambda)(y), \lambda).$$

Structure of specification Clearly, $CORR_7$ has the same conditional polynomial structure as $CORR_6$.

9.1.2 Eighth draft formal specification: implementing the retiming We now wish to replace the retiming λ by a boolean control stream $b \in B_n^\infty$ such that $b = pulse_n(\lambda)$. That is, we require

$$CORR_8 : [T \to A] \times [T \to A] \times B_n^\infty \to [T \to \{0, \ldots, n\} \cup \{u\}].$$

We achieve this by substituting $pulse_n^{-1}(b)$ for λ in $CORR_7$, where $pulse_n^{-1}$ is the inverse of $pulse_n$ (Section 8.1.2). Set

$$CORR_8(x, y, b) = CORR_7(x, y, pulse_n^{-1}(b)).$$

and expanding this, we obtain

$$CORR_8 : [T \to A] \times [T \to A] \times B_n^\infty \to [T \to \{0, \ldots, n\} \cup \{u\}],$$

$$CORR_8(x, y, b)(t) = \begin{cases} u & \text{if } t < n + k - 1 \text{ or } (t-k) - start(pulse_n^{-1}(b))(t-k) < n, \\ corr(x(t-k), \ldots, x(t-k-n+1), y(start(pulse_n^{-1}(b))(t-k) + n - 1), \ldots, y(start(pulse_n^{-1}(b))(t-k))) & \text{otherwise.} \end{cases}$$

Notice we have used the fact that $\overline{\lambda}\lambda = start(\lambda)$ (Section 9.1.1).

While this specification implements the abstract specification $CORR_6$, it is possible to simplify it by eliminating the $pulse_n^{-1}$ terms. To do this, the following lemma is needed.

Lemma Given $b \in B_n^\infty$, and $\delta_1, \ldots, \delta_n$ from Section 5.4.1, then

$$start(pulse_n^{-1}(b))(t) + n - 1 = \delta_1(t, b), \ldots, start(pulse_n^{-1}(b))(t) = \delta_n(t, b), \qquad \text{(i)}$$

$$(t-k) - start(pulse_n^{-1}(b))(t-k) < n \text{ iff } b(t-k) = tt. \qquad \text{(ii)}$$

Proof Omitted.

We can now rewrite $CORR_8$ in the following form:

$$CORR_8 : [T \to A] \times [T \to A] \times B_n^\infty \to [T \to \{0, \ldots, n\} \cup \{u\}],$$

$$CORR_8(x, y, b)(t) = \begin{cases} u & \text{if } t < n + k - 1 \text{ or } \\ & b(t-k) = tt, \\ corr(x(t-k), \ldots, x(t-k-n+1), & \\ \quad y(\delta_1(t,b)-k), \ldots, y(\delta_n(t,b)-k)) & \text{otherwise.} \end{cases}$$

Implementation decisions taken in this specification In $CORR_8$ the following implementation decision is taken:

- the retiming λ is implemented by a $pulse_n$-type boolean stream.

Correctness of the specification The correctness of the transformation of $CORR_7$ to $CORR_8$ is illustrated by the bottom half of the commutative diagram of Figure 5.19 and expressed in the equation

$$CORR_7(x, y, \lambda) = CORR_8(x, y, pulse_n(\lambda)).$$

Let us note that if we had chosen to load the reference word in parallel, a different diagram would result, namely Figure 5.20.

$$\begin{array}{ccccccc}
[T \to A] & \times & [R \to A^n] & \times & Ret_n(T,R) & \xrightarrow{CORR_6} & [T \to \{0,\ldots,n\} \cup \{u\}] \\
\downarrow Id & & & sch \times Id \downarrow & & & \uparrow Id \\
[T \to A] & \times & [T \to A^n] & \times & Ret_n(T,R) & \xrightarrow{CORR_7} & [T \to \{0,\ldots,n\} \cup \{u\}] \\
\downarrow Id & & \downarrow Id & & \downarrow pulse_1 & & \uparrow Id \\
[T \to A] & \times & [T \to A] & \times & B_n^\infty & \xrightarrow{CORR_8} & [T \to \{0,\ldots,n\} \cup \{u\}]
\end{array}$$

Figure 5.20. From abstract specification to parallel reference word loading

Notice that in this case the reference word loading time i is 1 (so $\lambda \in Ret(T, R)$) and the $pulse_1$ method is used to implement λ. Notice also that the sch function (Section 6.2) is used to map $\underline{y} \in [R \to A^n]$ to $y \in [T \to A^n]$.

Structure of specification Notice that $CORR_8$ is exactly the same as $CORR_5$ except that an infinite number of true elements are required on the stream b for $CORR_8$ (Section 8.1.1). Thus, $CORR_8$ is primitive recursive over $\overline{\mathbf{A}}$ augmented by a predicate for B_n^∞.

9.2 Implementing data abstraction

Throughout we have been concerned with the correlation of elements of an unspecified abstract data type A. We must now select a particular data type on which to perform correlation, as well as choose some *representation* for this data type. In addition, we must choose representations of $\{0, \ldots, n\}$, for the results of the correlator, and of $\mathbf{B}$ for the control stream. This done, we can define a specification that is implementable. Actual digital devices can only communicate in terms of dual voltage levels which must be interpreted externally.

We proceed as follows. In Section 9.2.2 we choose the actual correlation data type A to be the set $\{0,1\}$ of bits (Section 4.6). Then, in Section 9.3, we select methods of representing streams of elements of A and $\{0, \ldots, n\}$, and elements of B_n^∞ using bit streams. We will represent $a \in \{0, \ldots, n\}$ as a positive magnitude number, where $L = \lceil \log n + 1 \rceil$ bits are required to represent elements in the set $\{0, \ldots, n\}$, and B_n^∞ as a single bit stream. Thus we can reduce or interpret specifications such as $CORR_8$ on $\overline{\mathbf{A}}$ to specifications over an algebra of bits $\{0,1\}$. Finally, we replace the data type of bits $\{0,1\}$ by a new data type to represent the high and low voltage levels which must be used by any real device.

9.2.1 Some tools for data abstraction We introduce **Bit** to represent elements of bit streams, where

$$\mathbf{Bit} \; = \; \{0,1\}.$$

It is convenient to also define $\mathbf{Bit}_u \; = \; \mathbf{Bit} \cup \{u\}$.

We define a function $pm : \{0, \ldots, n\} \cup \{u\} \to \mathbf{Bit}_u^L$ to map elements of $\{0, \ldots, n\} \cup \{u\}$ to their *positive magnitude bit vector representation*, where the coordinate functions

$$pm_0, \ldots, pm_{L-1} : \{0, \ldots, n\} \cup \{u\} \to \mathbf{Bit}_u$$

of pm are defined by

$$pm_i(a) = \begin{cases} u & \text{if } a = u, \\ 1 & \text{if } \lfloor a/2^i \rfloor \text{ odd}, \\ 0 & \text{otherwise}, \end{cases}$$

and consequently pm is defined by

$$pm(a) = (pm_0(a), \ldots, pm_{L-1}(a)).$$

We define the inverse $pm^{-1} : \mathbf{Bit}_u^L \to \{0, \ldots, n\} \cup \{u\}$ as

$$pm^{-1}(a_0, \ldots, a_{L-1}) = \begin{cases} u & \text{if } a_i = u \text{ for } i \in 0, \ldots, L-1 \\ a_0.2^0 + \cdots + a_{L-1}.2^{L-1} & \text{otherwise.} \end{cases}$$

This induces a function $corr' : \mathbf{Bit}^n \times \mathbf{Bit}^n \to \mathbf{Bit}^L$ defined by

$$corr'(\underline{a}, \underline{b}) = pm(corr(\underline{a}, \underline{b}))$$

that represents $corr$.

The functions pm and pm^{-1} are mappings between different data representations. Commonly we will require versions which map between *streams* of data with different representations. We generate stream transformer versions of pm and pm^{-1} in the manner of Section 3.4

$$PM : [T \to \{0, \ldots, n\} \cup \{u\}] \to [T \to \mathbf{Bit}_u^L],$$

$$PM(a)(t) = pm(a(t))$$

and

$$PM^{-1} : [T \to \mathbf{Bit}_u^L] \to [T \to \{0, \ldots, n\} \cup \{u\}],$$

$$PM^{-1}(a)(t) = pm^{-1}(a(t)).$$

In addition to pm and pm^{-1}, we need functions to map elements of $\mathbf{B} \cup \{u\}$ onto bits, where tt is represented by 1 and ff is represented by 0, namely

$$bbit : \mathbf{B} \cup \{u\} \to \mathbf{Bit}_u,$$

$$bbit(b) = \begin{cases} u & \text{if } b = u, \\ 1 & \text{if } b = tt, \\ 0 & \text{if } b = ff, \end{cases}$$

and the inverse

$$bbit^{-1} : \mathbf{Bit}_u \to \mathbf{B} \cup \{u\},$$

$$bbit^{-1}(b) = \begin{cases} u & \text{if } b = u, \\ tt & \text{if } b = 1, \\ ff & \text{if } b = 0. \end{cases}$$

Again we generate stream versions of $bbit$ and $bbit^{-1}$ in a straightforward manner:

$$BBIT : [T \to \mathbf{B} \cup \{u\}] \to [T \to \mathbf{Bit}_u],$$

$$BBIT(a)(t) = bbit(a(t))$$

and

$$BBIT^{-1} : [T \to \mathbf{Bit}_u] \to [T \to \mathbf{B} \cup \{u\}],$$

$$BBIT^{-1}(a)(t) = bbit^{-1}(a(t)).$$

We now have enough machinery to define a two-sorted algebra **Bit**, which consists of $\{0,1\}$ equipped with $corr'$, and the boolean operations, and clock T; and also the algebra's extension $\mathbf{Bit}_u$. Our remaining specifications will be defined over **Bit** and $\mathbf{Bit}_u$. The consistency and correctness of the implementations are represented in the top half of the commutative diagram in Figure 5.21.

$$\begin{array}{ccccccc}
[T \to \mathbf{Bit}] & \times & [T \to \mathbf{Bit}] & \times & B_n^\infty & \xrightarrow{CORR_8} & [T \to \{0,\ldots,n\} \cup \{u\}] \\
\downarrow Id & & \downarrow Id & & \downarrow BBIT & & \uparrow PM^{-1} \\
[T \to \mathbf{Bit}] & \times & [T \to \mathbf{Bit}] & \times & Bit_n^\infty & \xrightarrow{CORR_9} & [T \to \mathbf{Bit}_u^L] \\
\downarrow BVL & & \downarrow BVL & & \downarrow BVL & & \uparrow BVL_L^{-1} \\
[T \to \mathbf{VL}] & \times & [T \to \mathbf{VL}] & \times & VL_n^\infty & \xrightarrow{CORR_{10}} & [T \to \mathbf{VL}_u^L]
\end{array}$$

Figure 5.21. Representing data abstraction

9.2.2 Ninth draft formal specification: correlation data type We require $CORR_9$ to return a vector of bit streams. That is,

$$CORR_9 : [T \to \mathbf{Bit}] \times [T \to \mathbf{Bit}] \times Bit_n^\infty \to [T \to \mathbf{Bit}_u^L],$$

where $Bit_n^\infty = \{BBIT(b) : b \in B_n^\infty\}$. Recall from Section 8.1.2 that B_n^∞ is the set of streams $b \in [T \to \mathbf{B}]$ where all trueblocks are a multiple of n in length, there is an initial trueblock, and an infinite number of trueblocks. Therefore Bit_n^∞ is the image of the set B_n^∞ under the map $BBIT$.

Given $A = \mathbf{Bit}$, we may define $CORR_9$ by substitution in $CORR_8$

$$CORR_9(x,y,b) = PM(CORR_8(x,y,BBIT^{-1}(b))).$$

We expand this and define $CORR_9$

$$CORR_9 : [T \to \mathbf{Bit}] \times [T \to \mathbf{Bit}] \times Bit_n^\infty \to [T \to \mathbf{Bit}_u^L],$$

$$CORR_9(x,y,b)(t) = \begin{cases} pm(u) & \text{if } t < n+k-1 \text{ or } BBIT^{-1}(b)(t-k) = tt, \\ pm(corr(x(t-k), \ldots, x(t-k-n+1), y(\delta_1(t, BBIT^{-1}(b))), \ldots, y(\delta_n(t, BBIT^{-1}(b))))) & \text{otherwise.} \end{cases}$$

Notice we use pm rather than its stream equivalent PM in the expanded definition of $CORR_9$. This definition is unnecessarily complicated, and can be simplified in two ways

(1) by expanding and simplifying definitions, using

$$\begin{aligned} BBIT^{-1}(b)(t-k) &= tt \text{ iff } b(t-k)=1, \\ pm(corr(a,b)) &= corr'(a,b), \quad \text{and} \\ pm(u) &= u^L (= u, \ldots, u); \end{aligned}$$

(2) by modifying the definitions $\delta_1, \ldots, \delta_n$ so they accept streams of type **Bit**, using the following new versions

$$\delta'_1, \ldots, \delta'_n : T \times [T \to \mathbf{Bit}] \to T,$$

$$\delta'_1(0,s) = \begin{cases} 0 & \text{if } s(0)=1, \\ \uparrow & \text{if } s(0)=0, \end{cases}$$

$$\delta'_1(t+1,s) = \begin{cases} t+1 & \text{if } s(t+1)=1, \\ \delta'_1(t,s) & \text{if } s(t+1)=0, \end{cases}$$

$$1 < i \le n,$$

$$\delta'_i(0,s) = \uparrow,$$

$$\delta'_i(t+1,s) = \begin{cases} \delta'_{i-1}(t,s) & \text{if } s(t+1)=1, \\ \delta'_i(t,s) & \text{if } s(t+1)=0. \end{cases}$$

Lemma $\delta'_i(t,b) = \delta_i(t, BBIT^{-1}(b))$.

Proof The definition of δ'_i differs from the definition of δ_i only in that 1 has been substituted for tt, and 0 has been substituted for $f\!f$. By definition, $BBIT^{-1}(b)(t) = tt$ iff $b(t)=1$ and $BBIT^{-1}(b)(t) = f\!f$ iff $b(t)=0$.

We now rewrite $CORR_9$.

$$CORR_9 : [T \to \mathbf{Bit}] \times [T \to \mathbf{Bit}] \times Bit_n^\infty \to [T \to \mathbf{Bit}_u^L],$$

$$CORR_9(x,y,b)(t) = \begin{cases} u, \ldots, u & \text{if } t < n+k-1 \\ & \text{or } b(t-k)=1 \\ corr'(x(t-k), \ldots, x(t-k-n+1), & \\ \quad y(\delta'_1(t,b)), \ldots, y(\delta'_n(t,b))) & \text{otherwise.} \end{cases}$$

Implementation decisions taken in this specification In $CORR_9$ the following implementation decisions are taken:

- the data type A is $\mathbf{Bit} = \{0, 1\}$;
- the result set $\{0, \ldots, n\}$ of the correlator is represented by $\mathbf{Bit}_u^L$, where $L = \lceil \log n + 1 \rceil$, by means of positive magnitude numbers.

Correctness of the specification The correctness of the transformation of $CORR_8$ to $CORR_9$ is illustrated by the top half of the commutative diagram of Figure 5.21 and expressed in the equation

$$CORR_8(x, y, b) = PM^{-1}(CORR_9(x, y, BBIT(b))).$$

Structure of the specification The specification $CORR_9$ is primitive recursive over the algebra $\overline{\mathbf{Bit}}$ augmented by a predicate for Bit_n^∞.

9.3 Implementing voltage levels

We now introduce **VL** to represent high and low *voltage levels* where

$$\mathbf{VL} = \{low, high\}.$$

Again we also introduce a version which includes u

$$\mathbf{VL}_u = \mathbf{VL} \cup \{u\}.$$

We will represent bit streams in terms of **VL** in an obvious way

$$1 = high \text{ and } 0 = low.$$

It may seem unnecessary to introduce **VL** when 1 and 0 are commonly used to abstractly represent high and low voltage levels in discussing digital hardware. However, this is an abuse of notation because the reverse representation (*negative logic*) is also possible and is sometimes used, as is the combination of both representations (*mixed logic*, see Winkel & Prosser [1980]). Confusing *data* and its *representation* in this way is risky and a potential source of error.

The introduction of voltage levels results in a new two-sorted algebra **VL** that is isomorphic with **Bit**, and an extension $\mathbf{VL}_u$ that is isomorphic with $\mathbf{Bit}_u$. The isomorphism is, of course, the renaming map $bvl : \mathbf{Bit}_u \to \mathbf{VL}_u$ defined by

$$bvl(1) = high, \quad bvl(0) = low, \quad \text{and} \quad bvl(u) = u.$$

The inverse $bvl^{-1} : \mathbf{VL}_u \to \mathbf{Bit}_u$ and the stream transformations $BVL : [T \to \mathbf{Bit}_u] \to [T \to \mathbf{VL}_u]$ and $BVL^{-1} : [T \to \mathbf{VL}_u] \to [T \to \mathbf{Bit}_u]$ are defined in the obvious way.

We may now apply the machinery to $CORR_9$ to obtain

$$CORR_{10} : [T \to \mathbf{VL}] \times [T \to \mathbf{VL}] \times VL_n^\infty \to [T \to \mathbf{VL}_u^L],$$

where $VL_n^\infty = \{BVL(b) : b \in Bit_n^\infty\}$, defined by

$$CORR_{10}(x, y, b) = BVL_L(CORR_9(BVL^{-1}(x), BVL^{-1}(y), BVL^{-1}(b))).$$

This may be expanded following the expansion of Section 9.2.2.

Implementation decisions taken in this specification In $CORR_{10}$ the following implementation decision is taken:

- the type **Bit** is represented by voltage levels **VL** such that $1 = high$ and $0 = low$.

Correctness of the specification The correctness of the transformation of $CORR_9$ to $CORR_{10}$ is illustrated by the bottom half of the commutative diagram of Figure 5.21 and embodied in the statement

$$CORR_9(x, y, b) = BVL_L^{-1}(CORR_{10}(BVL(x), BVL(y), BVL(b))).$$

This concludes the derivation of the implementable specification of the correlator.

10 CONCLUDING REMARKS

The intention of this chapter has been to show how the process of specifying digital systems can be made more precise. To this end, the following concepts and tools have been introduced.

Specification driven design A *general purpose theoretical framework* for the *specification driven design* of digital systems. The main aspects of this are the formulation of the following concepts: a *draft algorithm design* D as a 5-tuple; a *draft specification* S as a 4-tuple

$$S = (L, I, F, E);$$

the idea of *specification trajectory*, each element of which may generate a *design trajectory*; and *refinements* $S \to S'$ that preserve *consistency*. The specification process has *three phases*, consisting of an initial *exploration phase*, where user and designer come to mutual understanding as to what is required, followed by an *abstraction phase*, where essential aspects of the user's requirements are isolated, uncluttered by extraneous detail, and a final *implementation phase*, where a specification appropriate for implementation is developed systematically from the abstract specification, subject to current technical constraints and the user's requirements.

Stream transformations over abstract data types as user specifications General language independent formats for writing formal specifications and mathematical models for a language for formal specifications. These schemes are sufficient for representing a large class of synchronous digital systems in a *standard* way, and are useful for *animating* formal specifications in most programming languages. The three classes of recursive functions on an abstract data type provide models for specification languages and for a mathematical theory for synchronous digital systems.

Refinement of user specifications and specification compilers A set of formal tools for the consistent and correct refinement of specifications by means of data and timing abstractions. Most of these tools revolve around the concept of a *clock retiming*, a simple method of representing *timing abstraction.* To allow retimings to be *implemented*, a class of methods for representing retimings, *boolean streams* were given. In addition, we introduced techniques for implementing data abstraction in terms of the *bit streams* employed in actual digital systems. The implementation techniques are designed to allow high-level specifications to be mapped into lower-level specifications in a largely mechanical way, such that precise *correctness criteria* are met at each step in the derivation. This systematic development suggests the possibility of a *specification compiler*, which would allow automatic or semi-automatic generation of the extremely complex specifications necessary for the implementation of digital systems.

The correlator case study Although a conceptually simple system, the correlator has many interesting and potentially difficult aspects when examined in a thorough, systematic, way. The correlator specifications presented in this chapter are summarised in Table 5.8. All three phases of the specification process described in Section 2 have been systematically followed, and the essential character of each stage revealed. The exploration phase of the correlator specification consists of a sequence of approximations to the required specification. The exploration phase is essentially one of gaining understanding, and specifications here may well not be correct. The abstraction phase involves a conceptual leap, where those aspects of the correlator not essential for its fundamental function are removed. Although the abstract specification $CORR_6$ is formal, the method by which it is reached is not. In contrast, each specification $CORR_7$ – $CORR_{10}$ of the exploration phase is a formalised progression from

Table 5.8. Summary of specification

Phase	**Specification**	**Characteristics**
Exploration	$CORR_1$	Period of 1 established. Reference word is fixed and structure of data stream ignored.
	$CORR_2$	Structure of the data stream is established. Reference word is fixed.
	$CORR_3$	Reference word made reloadable by serial loading. *User-consistent* specification. Environment of correlator ignored.
	$CORR_4$	Environmental condition that an initial reference word will always be loaded is included.
	$CORR_5$	Environmental condition that only complete reference words will be loaded is included.
Abstraction	$CORR_6$	The reference word replacement mechanism is abstracted away by means of a retiming. Not seen to be formally consistent with $CORR_5$.
Implementation	$CORR_7$	Serial loading is adopted for programming the reference word.
	$CORR_8$	The retiming of $CORR_6$ is replaced by a boolean stream and the second clock R in $CORR_6$ – $CORR_7$ is eliminated.
	$CORR_9$	The correlation data type is established to be $\{0, 1\}$. The output $\{0, \ldots, n\}$ of the correlator is represented as a binary *positive magnitude* number.
	$CORR_{10}$	The data type $VL = \{high, low\}$ is introduced to represent $\{0, 1\}$.

its predecessor: we replace the abstraction mechanisms in the abstract specification $CORR_6$ with implementable counterparts using the formal tools developed for the purpose. In this way, given the correctness of the abstract specification $CORR_6$, because the final implementation specification $CORR_{10}$ provably *implements* $CORR_6$ we are assured of the correctness of $CORR_{10}$.

REFERENCES

Abrial *et al.* [1979]
J. R. Abrial, S. A. Schuman and B. Meyer (1979) *Specification Language Z*, Massachusetts Computer Associates.

ADJ [1977]
J. A. Goguen, J. W. Thatcher, E. G. Wagner and J. B. Wright (1977) 'Initial algegra semantics and continuous algebras', *JACM,* **24**, 68–95.

ADJ [1978]
J. A. Goguen, J. W. Thatcher and E. G. Wagner (1978) 'An initial algebra approach to the specification, correctness and implementation of abstract data types', 80–149 in *Current Trends in Programming Methodology*, ed. R. T. Yeh, Prentice Hall.

Anceau [1986]
F. Anceau (1986) *The Architecture of Microprocessors*, Addison-Wesley.

Asveld & Tucker [1982]
P. R. J Asveld and J. V. Tucker (1982) 'Complexity theory and the operational structure of algebraic programming systems', *Acta Informatica,* **17**, 451–76.

Back [1980]
R. J. R. Back (1980) 'Correctness preserving program refinements: proof theory and applications', Mathematical Centre Tracts.

Back [1981]
R. J. R. Back (1981) 'On correct refinement of programs', *Journal of Computer and Systems Science,* **23**, 49–68.

Back & Mannila [1982]
R. J. R. Back and H. Mannila (1982) 'A semantic approach to program modularity', Department of Computer Science, University of Helsinki, Report C-1982-102.

Backhouse [1986]
R. C. Backhouse (1986) *Program Construction and Verification*, Prentice-Hall.

Backus [1978]
J. Backus (1978) 'Can programming be liberated from the von Neumann style? A functional style and Its algebra of programs.', *Communications of the ACM,* **21**, 613–41.

Baer [1980]
J. L. Baer (1980) *Computer Systems Architecture*, Pitman.

de Bakker [1981]
J. W. de Bakker (1981) *Mathematical Theory of Program Correctness*, Prentice-Hall.

de Bakker *et al.* [1987a]
J. W. de Bakker, A. J. Nijman and P. C. Treleaven (ed.) (1987) *Parallel Architectures and Languages Europe: Part I Parallel Architectures*, Lecture Notes in Computer Science 258, Springer-Verlag.

de Bakker *et al.* [1987b]
J. W. de Bakker, A. J. Nijman and P. C. Treleaven (ed.) (1987) *Parallel Architectures and Languages Europe: Part II Parallel Languages*, Lecture Notes in Computer Science 259, Springer-Verlag.

Baudet [1983]
G. M. Baudet (1983) 'Design and complexity of VLSI algorithms', in *Foundations of Computer Science IV: Distributed Systems Part 1, Algorithms and Complexity*, ed. J. W. de Bakker & J. van Leeuwen, Mathematisch Centrum, Amsterdam.

Bergstra *et al.* [1986]
J. A. Bergstra, J. Heering and P. Klint (1986) 'Module algebra', Centre for Mathematics and Computer Science, Report CS R8617.

Bjorner & Jones [1982]
D. Bjorner and C. B. Jones (1982) *Formal Specification and Software Development*, Prentice-Hall.

Blahut [1985]
R. E. Blahut (1985) *Fast Algorithms for Digital Signal Processing*, Addison-Wesley.

Bochmann [1982]
G. V. Bochmann (1982) 'Hardware specification with temporal logic: an example', *IEEE Transactions on Computers,* **31**, 223–31.

Boyer & Moore [1979]
R. S. Boyer and J. S. Moore (1979) *A Computational Logic*, Academic Press.

Boyer & Moore [1981]
R. S. Boyer and J. S. Moore (ed.) (1981) *The Correctness Problem in Computer Science*, Academic Press.

Burstall & Goguen [1981]
R. M. Burstall and J. A. Goguen (1981) 'An informal introduction to specifications using CLEAR', 185–213 in *The Correctness Problem in Computer Science*, ed. R. S. Boyer & J. S. Moore, Academic Press.

Camilleri *et al.* [1986]
A. Camilleri, M. Gordon and T. Melham (1986) 'Hardware verification using higher-order logic', in *Proceedings IFIP International Working Conference on: From HDL Descriptions to Guaranteed Correct Circuit Designs*, Springer-Verlag, Grenoble, September 1986.

Clare [1973]
C. R. Clare (1973) *Designing Logic Systems Using State Machines*, McGraw-Hill.

Clarke & Mishra [1984]
E. Clarke and B. Mishra (1984) 'Automatic verification of asynchronous circuits', in *Proceedings, CMU Workshop on Logics of Programs*, ed. E. Clarke & D. Kozen, Lecture Notes in Computer Science 164, Springer-Verlag.

Cohen [1982]
B. Cohen (1982) 'Justification of formal methods for system specifications', *IEE Software and Microsystems,* **5**.

Cohen [1983]
B. Cohen (1983) 'On the impact of formal methods on the VLSI community', 469–79 in *VLSI 83*, ed. F. Anceau & E. J. Aas, North-Holland

Cohen *et al.* [1986]
B. Cohen, W. T. Harwood and M. I. Jackson (1986) *Specification of Complex Systems*, Addison-Wesley.

Cohn [1987]
A. Cohn (1987) 'A proof of correctness of the VIPER microprocessor: the first levels', 27–72 in *VLSI Specification, Verification and Synthesis*, ed. G. Birtwistle & P. A. Subrahmanyam, Kluwer Academic Publishers.

Corry & Patel [1983]
A. Corry and K. Patel (1983) 'Architecture of a CMOS correlator', *GEC Journal of Research*, **1**.

Dew, King, Tucker & Williams [1990]
P. M. Dew, E. S. King, J. V. Tucker and A. Williams (1990) 'The prioritizer experiment: estimation and measurement of computation time in VLSI', this volume.

Eker & Tucker [1987]
S. M. Eker and J. V. Tucker (1987) 'Specification, derivation and verification of concurrent line drawing algorithms and architectures', University of Leeds, Centre for Theoretical Computer Science Report 10.87.

Ercegovac & Lang [1985]
M. D. Ercegovac and T. Lang (1985) *Digital Systems and Hardware: Firmware Algorithms*, Wiley.

Erhig & Mahr [1985]
H. Erhig and B. Mahr (1985) *Fundamentals of Algebraic Specification I: Equations and Initial Semantics*, EATCS Monograph Vol. 6, Springer Verlag.

Eveking [1985]
H. Eveking (1985) 'The application of CHDLs to the abstract specification of hardware', 167–78 in *Proc. 7th Int. Conf. Computer Hardware Description Languages and Their Applications*, ed. C. J. Koomen & T. Moto-oka, North-Holland, Amsterdam.

Flores [1963]
I. Flores (1963) *The Logic of Computer Arithemetic*, Prentice-Hall.

Gehani & McGettrick [1986]
N. Gehani and A. D. McGettrick (ed.) (1986) *Software Specification Techniques*, Addison-Wesley.

Glasser & Dobberpuhl [1985]
L. Glasser and D. Dobberpuhl (1985) *The Design and Analysis of VLSI Circuits*, Addison-Wesley.

Goguen & Burstall [1985]
J. A. Goguen and R. M. Burstall (1985) 'Introducing institutions', 221–56 in *Logics of Programs*, ed. E. Clarke & D. Kozen, Lecture Notes in Computer Science 164, Springer-Verlag.

Goguen & Burstall [1986]
J. A. Goguen and R. M. Burstall (1986) 'A study in the foundations of programming methodology: specifications, institutions, charters and parchments', 313–33 in *Category Theory and Computer Programming,*, ed. D. Pitt, S. Abramsky, A. Poigne & D. Rydeheard, Lecture Notes in Computer Science, Springer-Verlag.

Goguen & Meseguer [1983]
J. A. Goguen and J. Meseguer (1983) 'An initiality primer', draft paper.

Goguen & Tardo [1979]
J. A. Goguen and J. J. Tardo (1979) 'An introduction to OBJ: a language for writing and testing formal algebraic program specifications', 170–89 in *Proceedings of the Conference on Reliable Software.*

Gordon [1986]
M. Gordon (1986) 'Why higher-order logic is a good formalism for specifying and verifying hardware', 153–77 in *Formal Aspects of VLSI*, ed. G. J. Milne & P. A. Subrahmanyam, North-Holland, Amsterdam.

Gordon [1987]
M. Gordon (1987) 'HOL: a proof generating system for higher-order logic', 73–128 in *VLSI Specification, Verification and Synthesis*, ed. G. Birtwistle & P. A. Subrahmanyam, Kluwer Academic Publishers.

Gries [1978]
D. Gries (ed.) (1978) *Programming Methodology - A Collection of Articles by Members of IFIP WG2.3*, Springer-Verlag.

Gries [1981]
D. Gries (1981) *The Science of Programming*, Springer-Verlag.

Hanna & Daeche [1984]
F. K. Hanna and N. Daeche (1984) 'The VERITAS theorem-prover and its formal specification', Technical Report, University of Kent.

Hanna & Daeche [1985]
F. K. Hanna and N. Daeche (1985) *Specification and verification using higher-order logic*, North-Holland, Tokyo.

Hanna & Daeche [1986]
F. K. Hanna and N. Daeche (1986) 'Specification and verification using higher-order logic: a case study', 179–213 in *Formal Aspects of VLSI*, ed. G. J. Milne & P. A. Subrahmanyam, North-Holland, Amsterdam.

Harman [1985]
N. A. Harmam (1985) 'Formal methodologies in VLSI design', First Year Report, Dept. of Computer Studies, Univ. of Leeds.

Harman [1987]
N. A. Harman (1987) 'The design of a digital correlator', Department of Computer Studies Report No. 219, University of Leeds.

Harman, Hobley & Tucker [1990]
N. A. Harman, K. M. Hobley and J. V. Tucker (1990) 'The formal specification of a digital correlator II: algorithm and architecture verification', Centre for Theoretical Computer Science Report, University of Leeds, to appear.

Harrison [1965]
M. A. Harrison (1965) *Introduction to switching and automata theory*, McGraw-Hill.

Hayes [1987]
I. Hayes (ed.) (1987) *Specification Case Studies*, Prentice-Hall.

Hennessy [1986]
M. Hennessy (1986) 'Proving systolic systems correct', *ACM Transactions on Programming Languages and Systems,* **8**, 344–87.

Hill & Peterson [1973]
F. J. Hill and G. R. Peterson (1973) *Digital Systems: Hardware Organisation and Design*, Wiley.

Hoare & He [1985]
C. A. R. Hoare and He Jifeng (1985) 'Weakest prespecifications', Technical Monograph PRG-44, Programming Research Group, Oxford University.

Hoare *et al.* [1987]
C. A. R. Hoare, I. J. Hayes, He Jifeng, C. C. Morgan, A. W. Roscoe, J. W. Saunders, I. H. Sorenson, J. M. Spivey and B. A. Sufrin (1987) 'Laws of programming', *Communications of the ACM,* **30**, 672–86.

Hunt [1986]
W. A. Hunt (1986) 'FM8501: A verified microprocessor', The University of Texas at Austin, Institute for Computing Science Technical Report 47.

Jordan [1986]
J. R. Jordan (1986) 'Correlation algorithms, circuits and measurement applications', *IEE Proceedings G (Electronic Circuits and Systems),* **133**, 58–74.

Karnaugh [1953]
G. Karnaugh (1953) 'The map method for systems of combinatorial logic circuits', *AIEE Transactions on Communications and Electronics Part I,* **72**, 593–9.

Keister *et al.* [1951]
W. Keister, A. E. Ritchie and S. H. Washburn (1951) *The Design of Switching Circuits*, D. Van Nostrand.

Kowalski [1985]
K. Kowalski (1985) 'Logic programming and specification', 11–24 in *Mathematical Logic and Programming Languages*, ed. C. A. R. Hoare & J. C. Sherperdson, Prentice-Hall.

Kuck [1978]
D. J. Kuck (1978) *The Structure of Computers and Computations, vol. I*, Wiley.

Lewin [1968]
D. Lewin (1968) *Logical Design of Switching Circuits*, Nelson.

Lewin [1977]
D. Lewin (1977) *Computer Aided Design of Digital Systems*, Edward Arnold.

Liskov & Guttag [1986]
B. Liskov and J. Guttag (1986) *Abstraction and Specification in Programming Development*, MIT Press.

Malachi & Owicki [1981]
Y. Malachi and S. S. Owicki (1981) 'Temporal specification of self-timed systems', 203–12 in *VLSI Systems and Computations*, ed. H. T. Kung, B. Sproull & G. Steele, Computer Science Press.

Mano [1979]
M. M. Mano (1979) *Digital Logic and Computer Design*, Prentice-Hall.

Martin & Tucker [1987]
A. R. Martin and J. V. Tucker (1987) 'The concurrent assignment representation of synchronous systems', in *Parallel Architectures and Languages Europe Vol. II: Parallel Languages*, ed. J. W. de Bakker, A. J. Nijman & P. C. Treleaven, Lecture Notes in Computer Science 259, Springer-Verlag.

Mead & Conway [1980]
C. Mead and L. Conway (1980) *Introduction to VLSI Systems*, Addison-Wesley.

Meinke [1988]
K. Meinke (1988) 'A graph-theoretic model of synchronous parallel computation', Ph.D. Thesis, Department of Computer Studies, University of Leeds.

Meseguer & Goguen [1985]
J. Meseguer and J. A. Goguen (1985) 'Initiality, induction and computation', 459–541 in *Algebraic Methods in Semantics*, ed. M. Nivat & J. Reynolds, Cambridge University Press.

Mili *et al.* [1986]
A. Mili, J. Desharnais and J. R. Gagne (1986) 'Formal models of stepwise refinements of programs', *ACM Computing Surveys*, **18**, 231–76.

Milne [1985]
G. J. Milne (Apr 1985) 'CIRCAL and the representation of communication, concurrency and time', *ACM Transactions on Programming Languages and Systems*, **7**, 270–98.

Moore *et al.* [1986]
ed. W. Moore, A. McCabe and R. Urquhart (1986) *Systolic Arrays*, Adam Hilger.

Morris & Jones [1984]
F. L. Morris and C. B. Jones (1984) 'An early program proof by Alan Turing', *Annals of the History of Computing*, **6**, 139–43.

Moszkowski [1983]
B. Moszkowski (1983) *A temporal logic for multi-level reasoning about hardware*, North-Holland.

Moszkowski [1986]
B. C. Moszkowski (1986) *Executing Temporal Logic Programs*, Cambridge University Press.

Neilson [1984]
H. R. Neilson (1984) 'Hoare logic's for run-time analysis of programs', Ph.D. Thesis, University of Edinburgh.

Pahl & Beitz [1984]
G. Pahl and W. Beitz (1984) *Engineering Design*, The Design Council/Springer-Verlag.

Sanella & Wirsing [1983]
D. T. Sanella and M. Wirsing (1983) 'A kernel language for algebraic specification and implementations', Department of Computer Science, University of Edinburgh, Report CSR 131-83.

Savage [1976]
J. Savage (1976) *The Complexity of Computers*, Wiley.

Shannon [1938]
C. E. Shannon (1938) 'A symbolic analysis of relay and switching circuits', *AIEE Transactions*, **57**, 713–23.

Sheeran [1983]
M. Sheeran (1983) 'μFP, an algebraic design language', Technical Monograph PRG-39, Programming Research Group, Oxford University.

Thompson & Tucker [1985]
B. C. Thompson and J. V. Tucker (1985) 'Theoretical considerations in algorithm design', in *Fundamental Algorithms for Computer Graphics*, ed. R. A. Earnshaw, Proceedings of NATO ASI, Springer-Verlag.

Thompson [1987]
B. C. Thompson (1987) 'A mathematical theory of synchronous concurrent algorithms', Ph.D. Thesis, Department of Computer Studies, University of Leeds.

Thompson [1980]
C. D. Thompson (1980) 'A complexity theory for VLSI', Ph.D. Dissertation, Carnegie Mellon University, Computer Science Department.

Tucker & Zucker [1987]
J. V. Tucker and J. I. Zucker (1987) *Program correctness over abstract data types with error state semantics*, North-Holland.

Turner [1985]
D. A. Turner (1985) 'Programs as executable specifications', 29–50 in *Mathematical Logic and Programming Languages*, ed. C. A. R. Hoare & J. C. Sherperdson, Prentice-Hall.

Ullman [1984]
J. Ullman (1984) *Computational Aspects of VLSI*, Computer Science Press.

Wegener [1987]
I. Wegener (1987) *The Complexity of Boolean Functions*, Wiley-Teubner.

Weijland [1987]
W. P. Weijland (1987) 'A systolic algorithm for matrix-vector multiplication', University of Amsterdam Computer Science Department, Report FVI 87-08.

Winkel & Prosser [1980]
D. Winkel and F. Prosser (1980) *The Art of Digital Design*, Prentice-Hall.

Yoshikawa [1986]
ed. H. Yoshikawa (1986) *Design Theory for CAD*, North-Holland.

6 Describing and reasoning about circuits using relations

MARY SHEERAN

1 INTRODUCTION

One of the natural ways to model circuit behaviour is to describe a circuit as a function from signals to signals. A signal is a stream of data values over time, that is, a function from integers to values. One can choose to name signals and to reason about their values. We have taken an alternative approach in our work on the design language μFP (Sheeran [1984]). We reason about circuits, that is functions from signals to signals, rather than about the signals themselves. We build circuit descriptions by 'plugging together' smaller circuit descriptions using a carefully chosen set of combining forms. So, signals are first order functions, circuits are second order, and combining forms are third order.

Each combining form maps one or more circuits to a single circuit. The combining forms were chosen to reflect the fact that circuits are essentially two-dimensional. So, they correspond to ways of laying down and wiring together circuit blocks. Each combining form has both a behavioural and a pictorial interpretation. Because they obey useful mathematical laws, we can use program transformation in the development of circuits. An initial obviously correct circuit can be transformed into one with the same behaviour, but a more acceptable layout. It has been shown that this functional approach is particularly useful in the design of regular array architectures (Sheeran [1985, 1986], Luk & Jones [1988a]).

However, sometimes a relational description of a circuit is more appropriate than a functional one. (Note: we will use the word *relation* in the mathematical sense of *binary relation.*) Attempting to describe a circuit with complicated data flow using function composition as the main connective can lead to a rather contorted description. This is because function composition corresponds to a **unidirectional** flow of data between the composed functions. The output of one function becomes the input of the next. If we want to describe cells which communicate with each other in both directions, then there are two possibilities. We can remain in the functional setting and introduce special combining forms to describe the various possible arrangements for communication between cells. Our work on μFP has been an exploration of this approach. The second possibility is to generalise from functions to relations and to remove the distinction between input and output. The same circuit may have a more natural and readable description if we use relations to describe circuits and relational composition (and combining forms based on it) to plug them together. We have observed that many of our circuit transformations depend more on the connectivity of the circuit than on the direction in which data flows. This suggests that using relational descriptions will simplify both the initial circuit description and the choice of transformations.

This chapter presents a generalisation of our earlier work, in which circuits are described as binary relations (rather than functions) between signals. As in the functional case, we use combining forms to build hierarchical circuit descriptions. These combining forms have structural as well as behavioural interpretations. Our aim is to capture information about the (abstract) floorplan of a circuit, as well as about its behaviour.

Interestingly, the set of combining forms that arises naturally is not quite the same as for the functional case. The relational combining forms are more symmetric, because they are not biased towards a particular direction of data flow.

The relational combining forms also obey various algebraic laws. We will give some examples of these laws and of how they are proved and used. A single relational law often subsumes several μFP laws. This is encouraging. Whether design is done manually or using an automatic assistant, it is important that the number of transformation rules be kept small.

We present this combination of relations, combining forms and laws using some simple examples. Finally, we discuss the advantages and disadvantages of this approach in the context of related work in the field.

2 DESCRIBING CIRCUITS USING STREAMS

In both μFP (Sheeran [1983]) and Ruby, the language presented here, circuits operate on signals, which are streams of data values over time. In μFP, a circuit is a function from an input signal to an output signal.

$$\text{Signal} : \text{Int} \rightarrow \text{Val}$$

$$\mu\text{FPcct} : \text{Signal} \rightarrow \text{Signal}.$$

In Ruby, a circuit is a binary relation between signals, and inputs and outputs are not distinguished. This is the essential difference between the two languages.

$$\text{Rubycct} : \text{Signal} \leftrightarrow \text{Signal}.$$

We need to use signals because we want to describe and reason about sequential circuits. Streams provide a simple and elegant way of describing and reasoning about networks of sequential processes, without having to describe explicitly the state of the whole system (Kahn [1974]). It is important to note that we do not reason about the signals themselves, but about the circuits which operate on them. The signals are implicit; it is not necessary to reason about the individual elements of signals or about time indices. So, we avoid both explicit time and explicit state in our circuit descriptions. These are important simplifications, which ease the problems of reasoning about composite circuits.

However, when proving the algebraic laws themselves correct, it is often necessary to refer to and to decompose individual signals. The set of data values consists of atomic data values and tuples of data values (atomic or non-atomic). For example, $(a, b, (c, d))$ represents the tuple whose first element is a, second is b, and third is the tuple (c, d). A signal is a homogeneous stream of such data values. The nth element of a signal s is written $s(n)$. We will represent the construction of signals using angle brackets. ($\langle\rangle$ is the empty signal.) Thus,

$$\begin{aligned} \langle a, b, \langle c, d\rangle\rangle(t) &= (a(t), b(t), (c(t), d(t))) \\ \langle\rangle(t) &= () \quad \text{(the empty or null value).} \end{aligned}$$

We introduce a 'signal append' operator $\hat{}$, similar to normal sequence concatenation. For example,

$$\langle a, b\rangle \hat{} \langle c, d\rangle = \langle a, b, c, d\rangle \qquad \langle\rangle \hat{}\, a = a \qquad \langle a, b\rangle \hat{} \langle\langle c, d\rangle\rangle = \langle a, b, \langle c, d\rangle\rangle.$$

In the semantic equations for μFP given in Sheeran [1984], we frequently have to use the matrix transpose function (*zip*) to transform streams of tuples into tuples of streams and vice versa. This is because the input to a composite circuit (a stream of tuples) must be transposed to give the individual streams which must be passed to the circuit's components. Similarly, the outputs of the components must be recombined to give the stream of tuples which is the output. Here, we have avoided a proliferation of zips by careful choice of notation. As a result, the semantic equations are much easier to read.

For convenience, we introduce an abbreviation.

Definition 1 $\qquad a : b \;=_{def}\; \langle a\rangle \hat{}\, b$

Functional programmers may wish to read this as 'cons'. In a later section, we will show how these operations on signals are used in inductive proofs of algebraic laws. Circuit descriptions and algebraic laws do not themselves mention signals.

3 COMBINING FORMS FOR BUILDING CIRCUIT DESCRIPTIONS

Having decided to represent circuits as stream transformers, the next question is 'How should we construct the description of a circuit from the descriptions of its components?' In most formal hardware description languages, the channels or wires joining sub-circuits are named and a parallel composition operator which 'automatically' joins wires of the same name is used (Gordon [1986], Milne [1983]). This can lead to loss of information about circuit structure, and also fails to distinguish between local and long-distance communication in the circuit. We wish to capture information about circuit structure as well as behaviour and to reason about circuits rather than about data, so we choose not to name wires. Instead, we introduce combinators or combining forms which correspond to the ways in which circuits are actually plugged together in two dimensions. These combining forms have both structural and behavioural interpretations. In the structural interpretation, we can think of a circuit description as corresponding to a rectangular tile. The simplest tiles have connections only on two parallel sides. The connections corresponding to the domain of the relation are on one side, and those corresponding to the range are on the other. So, the relation captures information not only about the behaviour of the circuit, but about how the signals are grouped in space. Signals are not named. The structural interpretation of a combining form describes how to plug together the tiles for its components (and possibly some wiring tiles) to build the final circuit. The resulting abstract floorplan tells us how our primitive circuits are wired together and how the final circuit is laid out on the plane. There is no notion of distance involved.

We introduce combining forms both for the two-sided tiles and for tiles with connections on all four sides. Once we have chosen a convention for the orientation of these tiles in space, we can produce abstract floorplans from our circuit descriptions. A similar approach to the capturing of layout information is used in μFP, and a floorplan generation program has been implemented by Jones and Luk at Oxford. A Ruby floorplan generator would use exactly the same techniques.

For simplicity, we have chosen to represent both two-sided and four-sided tiles by binary relations. Our convention for producing floorplans will be that two-sided tiles will have their domain and range signals either on the left and right or on the top and bottom edges. This means that they match well with four sided tiles, which have their domain signals on the left and top edges, and their range on the bottom and right edges. So, the domain will always be above the bottom-left to top-right diagonal through the circuit, and the range below it. Switching between the different kinds of tile is simply done by bending some of the wires. The properties of binary relations and of operators on them are very well known and we wanted to take advantage of this. We have experimented with four-way relations, but found the notation unwieldy.

The behavioural interpretation of a circuit description tells us what the behaviour of the final circuit is, in terms of the behaviours of its components. The final circuit is again a stream transformer. So, our circuit descriptions encode information both about the structure and the behaviour of the circuit. This duality is central to our approach, and distinguishes it from most other work in the field.

The combining forms are chosen to obey simple mathematical laws. These laws are of the form 'Circuit A has the same behaviour as circuit B'. Circuits A and B may, of course, have different floorplans. The presence of these laws encourages a style of design in which an initial (but possibly inefficient) circuit is progressively transformed by the application of algebraic laws into a more satisfactory circuit. The laws guarantee that the final circuit has the same behaviour as the original one. Both μFP and Ruby are designed to be used during the design process as a means of checking design decisions. We are not interested in the post hoc verification of existing circuits.

In circuit descriptions using named wires, it is usual to identify individual elements of buses or wavefronts using indices (e.g., Weiser & Davis [1981]). Thus, the behaviour of each process in a rectangular grid might be given as a function of the coordinates of the process in the grid. We find such descriptions cumbersome, and wish to avoid explicit mention of the position on the plane of circuit elements. We therefore use generic combining forms each of which describes a class of circuits of the same shape, rather than a single circuit of a particular size. So, for example, we have a combining form for describing linear arrays. It is not necessary to indicate how many elements must appear in the array. The algebraic laws about arrays apply to arrays of all sizes. We design and reason about the circuit without making any assumptions about its size. The actual size of the circuit (number of inputs etc.) is decided as late as possible in the design process. Thus, most or all of the proof of correctness can be reused when a similar circuit of a different size is required.

So, we regard circuits as stream transformers and build circuit descriptions using 'structural' combining forms. Explicit mention of time, state and position is avoided, making it easier to reason about our circuit descriptions.

4 RELATIONAL VS FUNCTIONAL CIRCUIT DESCRIPTIONS

In this chapter, we take the view that any useful circuit computes a total function from its inputs to its outputs. This rules out circuits that behave nondeterministically, giving more than one possible output for a given input. Why then, do we wish to describe circuits using relations? It was the observation that many of our circuit transformations depend more on connectivity than on the direction in which data flows which led us to consider using relational descriptions. If direction of data flow doesn't matter, why not abstract away from it? In a functional circuit description, inputs and outputs are always clearly distinguished, and function composition (which is unidirectional) is the main connective. This can lead to problems when describing circuits with complicated data flow. To solve these problems, we will represent a circuit not as a function from input signals to output signals but as a binary relation between signals. The choice of which signals should combine to form the domain and which should form the range will depend on the grouping of the signals in space, not on the distinction between input and output. This means that the inputs and outputs will be spread throughout the domain and the range and a circuit description must, in general, be a relation, even if that circuit computes a function of its input signals.

When one ignores the distinction between inputs and outputs and uses relational composition as the main connective, it becomes easier both to describe the initial circuit and to choose appropriate transformations. The number of applicable laws is reduced because one Ruby law subsumes several μFP laws. Each allowed assignment of direction to the signals in a Ruby law will result in a separate μFP law. The reduction in the number of laws eases the burden on the designer. The combining forms in Ruby are more symmetric than those in μFP, because directionality is no longer imposed. As a result, some of the new combining forms have nicer algebraic properties than the old. In addition, circuit symmetries which were obscured by the directionality of μFP become evident. This can reduce the complexity of circuit transformations, particularly for regular circuits (Sheeran & Jones [1987]).

Of course, there is a price to pay. Not every circuit that we can describe in Ruby is 'reasonable'. Because we are ignoring the distinction between inputs and outputs, it is possible to plug circuits together incorrectly. In our final circuit, every wire must have a unique direction, and we must not attempt to drive our primitives (which are functional) in the wrong direction. It doesn't make sense to place a 1 (representing true) on the output of an *And* gate and expect 1s to appear on the inputs. Unfortunately, it is also possible, using relational composition, to plug together two circuits which compute total functions of their inputs, and to get a circuit that does not even compute a function of its inputs. So, we have to check that our final circuit is actually implementable. We will return to this point.

It should be noted that we are not attempting to describe circuits with true bidirectionality, in which wires can 'reverse' their directions. In HOL, the description of such circuits is one of the main motivations for the use of relations (Gordon [1986]). We argue here that relations are useful, even in the absence of genuine bidirectionality.

5 PRIMITIVE CIRCUITS IN RUBY

All Ruby circuits, including primitives, are relations between signals. An *And* gate is represented as a binary relation between streams of pairs of booleans and streams of booleans. In this case, the intended input of the circuit corresponds to the domain, and the output to the range. This will not always be the case. Combinational primitives operate **pointwise** over the elements of the signals in the domain and range. So, if **and** is the standard boolean function (and hence relation), then an *And* gate **spreads** that relation over time.

$$\begin{aligned}\langle a, b\rangle \; And \; c &\Leftrightarrow \forall t. \langle a, b\rangle(t) \textbf{ and } c(t)\\ &\Leftrightarrow \forall t. (a(t), b(t)) \textbf{ and } c(t)\end{aligned}$$

($a \; R \; b$ can be read as 'a is in the relation R to b'.) All of our combinational circuits spread a combinational function or relation (on data values) over signals in this way. Conventionally, F will denote the circuit corresponding to the combinational relation **f**.

Definition 2 $\quad F$ spreads $\mathbf{f} \; =_{def} \; (aFb \Leftrightarrow \forall t. a(t) \; \mathbf{f} \; b(t))$

In this way, we can construct a set of primitive combinational circuits. The exact choice of primitives depends on the level at which the circuit is being designed.

We also need constant circuits. A constant circuit acts as a source of a particular data value. It maps **any** signal to the signal that is a stream of the required value. For example, the circuit *Zero* is a source of 0s. Note the symmetry of the definition.

$$a \; Zero \; b \;\Leftrightarrow\; \forall t. a(t) = b(t) = 0$$

Finally, we need a sequential primitive if we are to describe sequential circuits. We introduce an abstract delay element, D. D delays any signal by one (abstract) time unit.

Definition 3 $\quad a \; D \; b \; =_{def} \; \forall t. a(t-1) = b(t)$

Delaying a tuple of signals corresponds to delaying each individual signal. For example,

$$\begin{aligned}\langle a, b, c\rangle \; D \; \langle d, e, f\rangle &\Leftrightarrow \forall t. \langle a, b, c\rangle(t-1) = \langle d, e, f\rangle(t) && \text{(definition of D)}\\ &\Leftrightarrow \forall t. (a(t-1), b(t-1), c(t-1)) = (d(t), e(t), f(t))\\ &\Leftrightarrow (a \; D \; d) \; \& \; (b \; D \; e) \; \& \; (c \; D \; f) && \text{(definition of D)}\end{aligned}$$

D is our only sequential primitive. There is no other way of introducing state into circuit descriptions.

6 COMBINING FORMS IN RUBY

6.1 Composition

Having decided to describe circuits as relations, the most obvious combining form is relational composition (represented by ;). The standard definition of ; is

Definition 4 $\quad a\ F;G\ c\ =_{def}\ \exists b.(aFb)\ \&\ (bGc)$

and it is this definition that we use. Intuitively, the circuit F ; G is the circuit constructed by connecting the range of F to the domain of G. The internal signal (b in the definition) is hidden, and the resulting circuit is again a relation between signals. This corresponds exactly to the use of & (and) for parallel composition and ∃ (existential quantification) for hiding in HOL.

By convention, we will draw the diagrams of our circuits with the domain on the left and the range on the right. This means, for example, that standard gates will be drawn as shown in Figure 6.1. The composition F ; G will be drawn with F on the left and G on the right, with the intermediate wires joined.

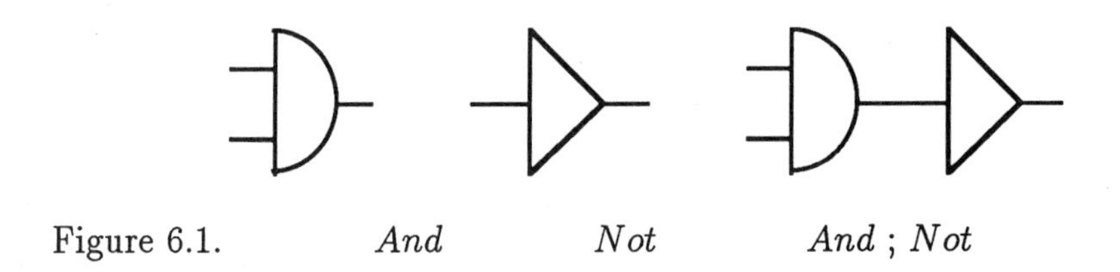

Figure 6.1. *And* *Not* *And ; Not*

We introduce a standard abbreviation for repeated composition. (*id* is the identity relation on signals.)

Definition 5

$$F^0 =_{def} id$$
$$F^n =_{def} F;F^{n-1} \quad (n > 0)$$

6.2 Inverse

Our convention (domain on left, range on right) means that relational inverse corresponds to flipping over the circuit; the domain and range are swapped.

Definition 6 $\quad a\ F^{-1}\ b\ =_{def}\ b\ F\ a$

From this definition, our first algebraic law follows immediately.

Law 1 $(F^{-1})^{-1} = F$

Our second algebraic law relates relational composition and relational inverse in the standard way.

Law 2 $(F;G)^{-1} = (G^{-1};F^{-1})$

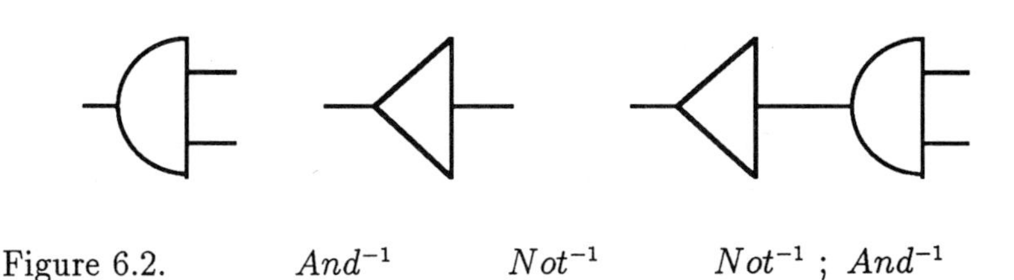

Figure 6.2. And^{-1} Not^{-1} Not^{-1} ; And^{-1}

Each circuit shown in Figure 6.2 is the inverse of the corresponding circuit in Figure 6.1.

Note that in general it is not the case that $R;R^{-1}$ is the identity. However, for a circuit F that is a 1-1 relation, $F;F^{-1}$ is the identity on the signals in the domain of F. This is true of our sequential primitive, D. Note that all signals are in the domain of D.

$$D;D^{-1} = D^{-1};D = id.$$

D^{-1} is a circuit in which the signal in the range is a 'predicted' version of the signal in the domain. Remember that because we abstract away from inputs and outputs, we have no notion of causality. We do not imply that the range is caused by the domain. So, D^{-1}, the anti-delay, makes as much sense as D, the delay. Both are useful for describing and reasoning about the behaviour of synchronous circuits. In any final circuit, we will have to make sure to drive all latches forwards, and all anti-latches backwards; we know how to implement delay, but not prediction.

Our circuit primitives and combining forms have been chosen so that for any circuit, F

Law 3 $D;F;D^{-1} = F.$

This law states that the behaviour of F does not depend on the exact time at which any particular calculation is done. Any circuit can be **retimed** (Leiserson [1983]) without upsetting its behaviour. It is useful that this property holds even when F is a relation with its inputs and outputs spread throughout the domain and range. In fact, $D;F;D^{-1}$ always corresponds to delaying all inputs and 'anti-delaying' all outputs, no matter which signals are designated as inputs and outputs. We prove this retiming property by structural induction. We first show that it holds for any combinational circuit.

$D; F; D^{-1} = F$ for any combinational F.

Proof Let f be the combinational relation which F spreads over time

$$F \text{ spreads } f \Leftrightarrow (a\ F\ b \Leftrightarrow \forall t.a(t)\ f\ b(t))$$

$$\begin{aligned}
a(D; F; D^{-1})e &\Leftrightarrow \exists b, c.(a\ D\ b)\ \&\ (b\ F\ c)\ \&\ (c\ D^{-1}\ e) && \text{(definition of ;)} \\
&\Leftrightarrow \exists b, c.\forall t.a(t-1) = b(t) \\
&\quad\ \&\ \ b(t)\ f\ c(t)\ \&\ c(t) = e(t-1) && \text{(definition of } D,\ ^{-1}, \text{ spread)} \\
&\Leftrightarrow \forall t.a(t-1)\ f\ e(t-1) && (\exists \text{ elim.}) \\
&\Leftrightarrow \forall t.a(t)\ f\ e(t) \\
&\Leftrightarrow a\ F\ e && (F \text{ spreads } f)
\end{aligned}$$

Next, we show that the retiming property holds for the sequential primitive D.

Proof $D; D; D^{-1} = D$ since $D; D^{-1} = id$

Finally, we must show that each combining form introduced preserves the retiming property. We will take composition (;) as an example. We must show that if F and G obey the retiming property, then so does $F; G$. We write this as

$$\frac{D; F; D^{-1} = F \qquad D; G; D^{-1} = G}{D; (F; G); D^{-1} = F; G.}$$

Proof

$$\begin{array}{lcccccccccccl}
D; F; G; D^{-1} & = & D & ; & F & ; & D^{-1} & ; & D & ; & G & ; & D^{-1} \quad (D; D^{-1} = id) \\
& = & & & F & & & ; & & & G & & \quad (\text{assumptions})
\end{array}$$

A similar proof must be done each time a combining form is introduced. All of the combining forms used here preserve the retiming property. The property will be useful when we come to transform circuits by moving latches about.

6.3 Par

Our next combining form, par, combines n circuits into a single circuit, whose domain and range contain signals carrying n-tuples, as illustrated in Figure 6.3. The definition for pairs extends naturally to tuples.

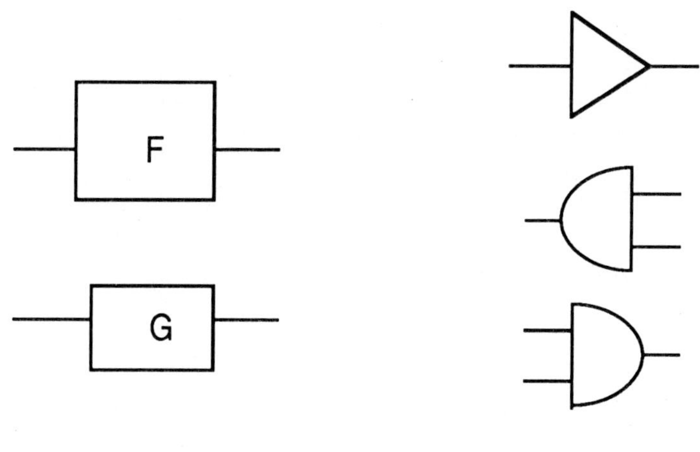

Figure 6.3. [G, F] $[And, And^{-1}, Not]$

Definition 7 $\langle a, b\rangle[F, G]\langle c, d\rangle =_{def} (a\ F\ c)\ \&\ (b\ G\ d)$

Par interacts with inverse and with composition according to Laws 4 and 5. The proofs follow easily from the definitions of par, inverse and composition.

Law 4 $[F, G]^{-1} = [F^{-1}, G^{-1}]$

Law 5 $[F1, G1]; [F2, G2] = [F1; F2, G1; G2]$

Some circuit forms appear so often that we introduce abbreviations for them.

Definition 8

$$\begin{aligned} \mathsf{fst}\ F &=_{def} [F, id] \\ \mathsf{snd}\ F &=_{def} [id, F] \end{aligned}$$

6.4 Map

Map places a circuit on each signal in a tuple of signals (or bus). It is a generic combining form, describing a class of circuits, rather than a single circuit of a particular size. Any diagram of necessity shows only one representative of the class (see Figure 6.4). This is one of the reasons for using a textual rather than a graphical language. Map is defined inductively. The algebraic laws which it obeys are again very simple.

Definition 9

$\langle\rangle(\mathsf{map}\ F)\langle\rangle$ ($=_{def}$ true is omitted here and in other base cases)

$a : b(\mathsf{map}\ F)c : d =_{def} a\ F\ c\ \&\ b(\mathsf{map}\ F)d$

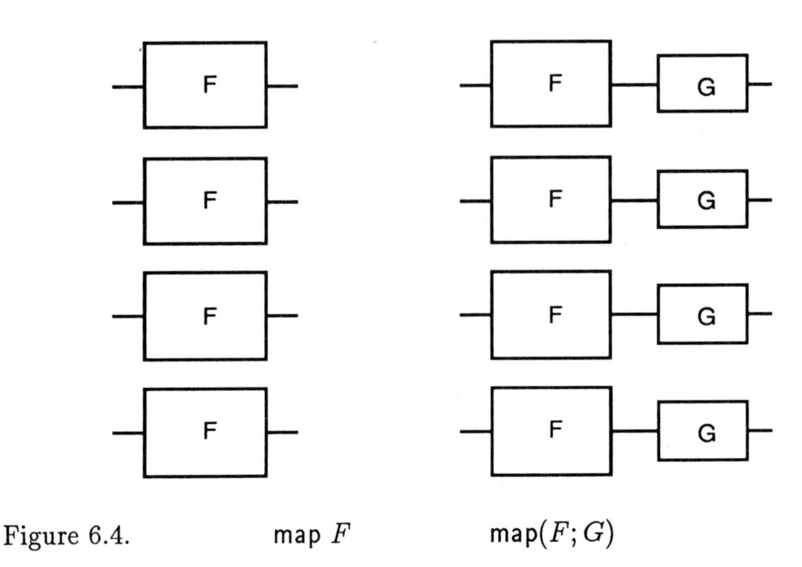

Figure 6.4. $\mathsf{map}\ F$ $\mathsf{map}(F;G)$

Law 6 $(\mathsf{map}\ F)^{-1} = \mathsf{map}\ F^{-1}$

Proof

Base Case:

$$\langle\rangle(\mathsf{map}\ F)^{-1}\langle\rangle \qquad \text{(definition of map, }^{-1})$$
$$\langle\rangle(\mathsf{map}\ F^{-1})\langle\rangle \qquad \text{(definition of map)}$$

Step:

$$
\begin{aligned}
a:b(\mathsf{map}\ F)^{-1}c:d &\Leftrightarrow c:d\ (\mathsf{map}\ F)\ a:b && \text{(definition of }^{-1})\\
&\Leftrightarrow (c\ F\ a)\ \&\ d\ (\mathsf{map}\ F)\ b && \text{(definition of map)}\\
&\Leftrightarrow (a\ F^{-1}\ c)\ \&\ b\ (\mathsf{map}\ F)^{-1}\ d && \text{(definition of }^{-1})\\
&\Leftrightarrow (a\ F^{-1}\ c)\ \&\ b\ (\mathsf{map}\ F^{-1})\ d && \text{(ind. hyp.)}\\
&\Leftrightarrow a:b\ (\mathsf{map}\ F^{-1})\ c:d && \text{(definition of map)}
\end{aligned}
$$

Law 7 $(\mathsf{map}\ F);(\mathsf{map}\ G) = \mathsf{map}(F\ ;\ G)$

7 DESCRIBING CIRCUIT WIRING

Wiring patterns are themselves circuits, so they are represented as stream relations. The most basic forms of plumbing are the selectors and constructors for signals. A new signal can be added to a tuple of signals either on the left or on the right. This gives us two constructors, each with two associated selectors.

Definition 10

$$\begin{array}{lll} \langle a,b\rangle\ app_L\ a:b & a:b\ first\ a & a\,\hat{}\,\langle b\rangle\ most\ a \\ \langle a,b\rangle\ app_R\ a\,\hat{}\,\langle b\rangle & a:b\ rest\ b & a\,\hat{}\,\langle b\rangle\ last\ b \end{array}$$

So, for example, *first* relates a tuple of streams (in the domain) to the first element of that tuple (in the range). So, *first* describes a whole class of circuits, just as map F does. Again, we can only draw diagrams of individual members of each class. In Figure 6.5 we illustrate the circuit *first* in the case when its domain is a 5-tuple.

$$\langle a,b,c,d,e\rangle\ first\ a$$

is true for all signals $a, b, \ldots, e$. We have labelled the signals in Figure 6.5 to illustrate this fact. We can only draw the diagram of a generic circuit when we have decided on a particular size and shape for its domain and range.

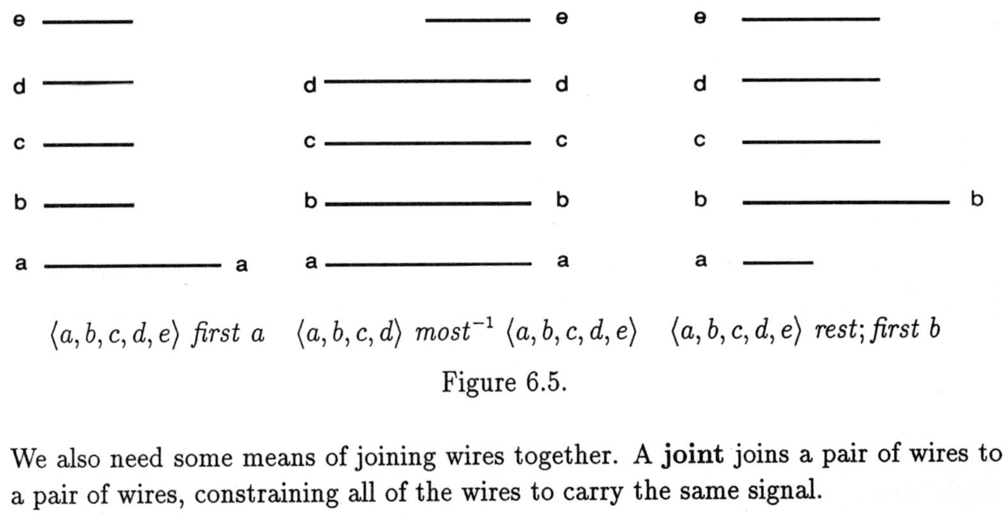

$\langle a,b,c,d,e\rangle\ first\ a$ $\quad\langle a,b,c,d\rangle\ most^{-1}\ \langle a,b,c,d,e\rangle$ $\quad\langle a,b,c,d,e\rangle\ rest; first\ b$

Figure 6.5.

We also need some means of joining wires together. A **joint** joins a pair of wires to a pair of wires, constraining all of the wires to carry the same signal.

Definition 11 $\quad\langle a,a\rangle\ join\ \langle a,a\rangle$

Join is introduced in this form for the sake of symmetry. It can be 'customised' using the selectors. For example,

$$\begin{array}{lll} \vdash & = join; first & \text{(a two-to-one joint)} \\ \dashv & = first^{-1}; join & \text{(a one-to-two joint).} \end{array}$$

Many-to-one and many-to-many joints can be constructed using the row and col combining forms introduced later.

It is useful to introduce abbreviations for the most common wiring patterns. For example, *dist* distributes a signal across a tuple of signals, to make a tuple of pairs of signals.

Definition 12

$$\langle a, \langle\rangle\rangle \ dist_L \ \langle\rangle$$
$$\langle a, b : c\rangle \ dist_L \ d : e \ =_{def} \ \langle a, b\rangle = d \ \ \& \ \ \langle a, c\rangle \ dist_L \ e$$
$$\langle\langle\rangle, c\rangle \ dist_R \ \langle\rangle$$
$$\langle a : b, c\rangle \ dist_R \ d : e \ =_{def} \ \langle a, c\rangle = d \ \ \& \ \ \langle b, c\rangle \ dist_R \ e$$

8 LOOP

Finally, we introduce feedback. Loop takes a circuit which relates a pair of signals to a pair of signals and joins the second of each of those pairs of signals together, to give a new circuit.

Definition 13 $\quad a \ (\mathsf{loop} \ H) \ b \ =_{def} \ \exists c.\langle a, c\rangle H \langle b, c\rangle$

This definition of loop contrasts sharply with the definition of μ in Sheeran [1983]. The new definition is much simpler and easier to understand. This is partly because we have chosen not to give the semantics by a translation to FP, and partly because we are content here to give a nonconstructive definition of loop. In general, the definitions of our combining forms are simplified because we introduce named signals into the definitions, and because we have chosen the notation for signal construction judiciously.

In loop, there is no constraint on the direction of data flow in the looped signal (c in the definition). Nor is there any requirement to place a unit delay on the fed back signal, as there was with the μ combining form in the functional case. So, loop is a more general combining form than the original μ. The user is now responsible for avoiding asynchronous cycles, that is cycles unbroken by delay. Loop is nicely symmetric.

Law 8 $\quad (\mathsf{loop} \ H)^{-1} = \mathsf{loop} \ H^{-1}$

We can now describe quite complicated wiring patterns (see Figure 6.8).

However, the combining forms introduced so far, composition, inverse, construction, par, and loop, tend to produce circuits with horizontal data flow. Our circuit 'tiles' have connections only on two sides.

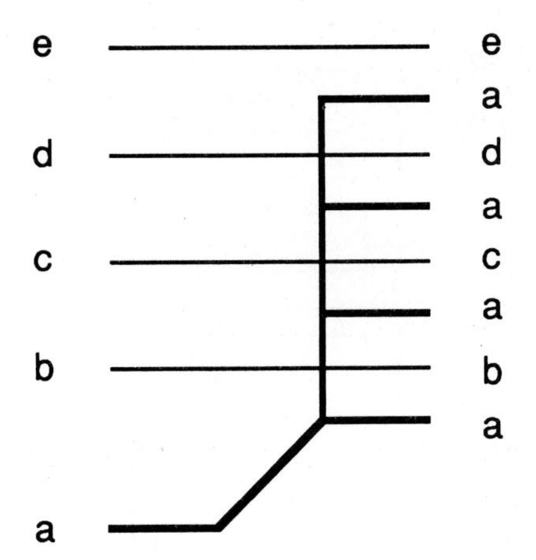

(i) $\langle a, \langle b, c, d, e\rangle\rangle$ $dist_L$ $\langle\langle a, b\rangle, \langle a, c\rangle, \langle a, d\rangle, \langle a, e\rangle\rangle$

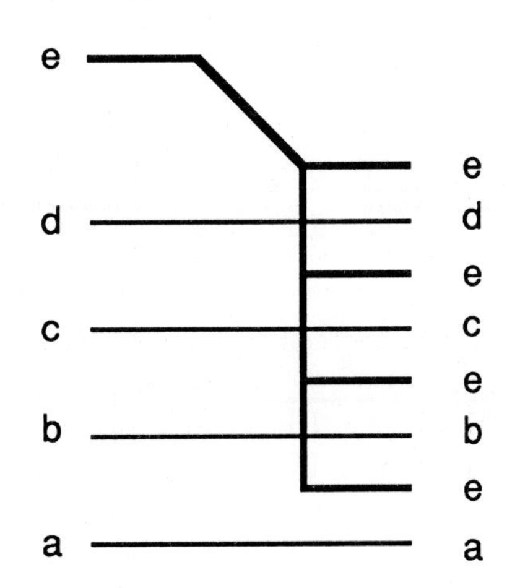

(ii) $\langle\langle a, b, c, d\rangle, e\rangle$ $dist_R$ $\langle\langle a, e\rangle, \langle b, e\rangle, \langle c, e\rangle, \langle d, e\rangle\rangle$

Figure 6.6.

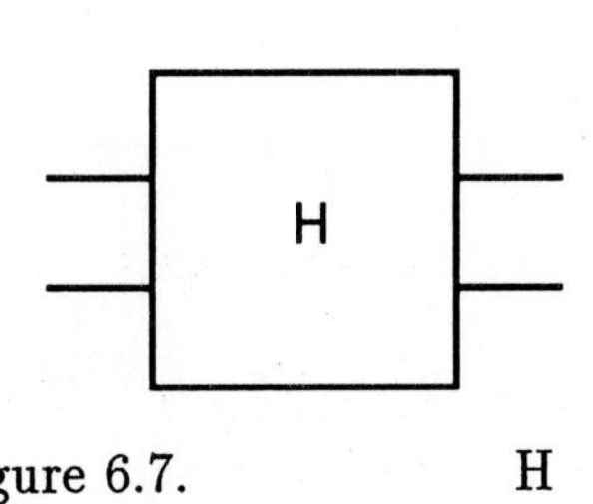

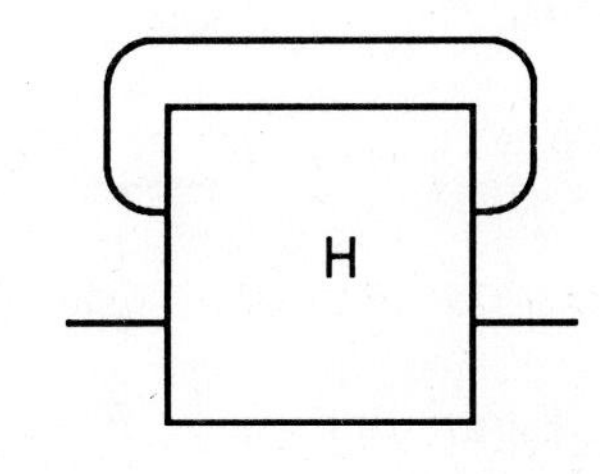

Figure 6.7. H loop H

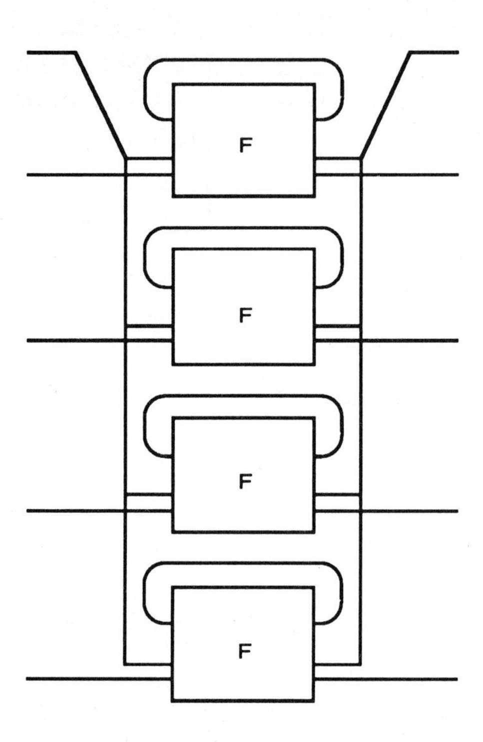

Figure 6.8. $dist_R$; map(loop F); $dist_R^{-1}$

9 INTRODUCING MORE COMPLICATED DATA FLOW

We can describe more general circuits by considering tiles with connections on all sides, and ways of plugging them together. We consider circuits of the form shown in Figure 6.9.

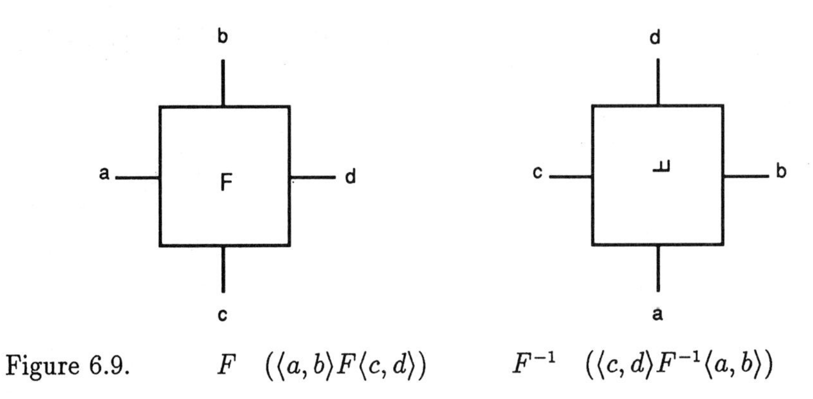

Figure 6.9. F $(\langle a,b\rangle F\langle c,d\rangle)$ F^{-1} $(\langle c,d\rangle F^{-1}\langle a,b\rangle)$

We make no change to the semantics of our language. We simply demand that circuits of this form must relate signals which are themselves pairs of signals. We associate the elements of those pairs with different edges of the circuit. By convention, we will associate the domain with the left and top edges and the range with the bottom and right edges. We read signals on vertical edges from bottom to top and those on horizontal edges from left to right. This gives an easy conversion between our two types of circuit. We need only bend some of the wires. It also means that relational inverse corresponds naturally to flipping a circuit about the bottom-left to top-right diagonal (see Figure 6.9). In the diagrams which follow, we will sometimes use two-sided and sometimes four-sided tiles. The choice will depend on the combining form used to connect the cell to its neighbours. For example, linear arrays of cells will always contain four-sided tiles, while triangles of cells will always be built from two-sided tiles.

9.1 Beside and above

Two natural combining forms for circuits of this form are **beside** ($\leftrightarrow$) and **above** ($\updownarrow$) (see Figure 6.10). We define **beside** directly, and then define **above** using **beside** and **inverse**.

Definition 14

$$\langle a, \langle b, c\rangle\rangle P \leftrightarrow Q\langle\langle d, e\rangle, f\rangle =_{def} \exists g. \langle a, b\rangle P\langle d, g\rangle \ \& \ \langle g, c\rangle Q\langle e, f\rangle$$
$$P \updownarrow Q =_{def} (Q^{-1} \leftrightarrow P^{-1})^{-1}$$

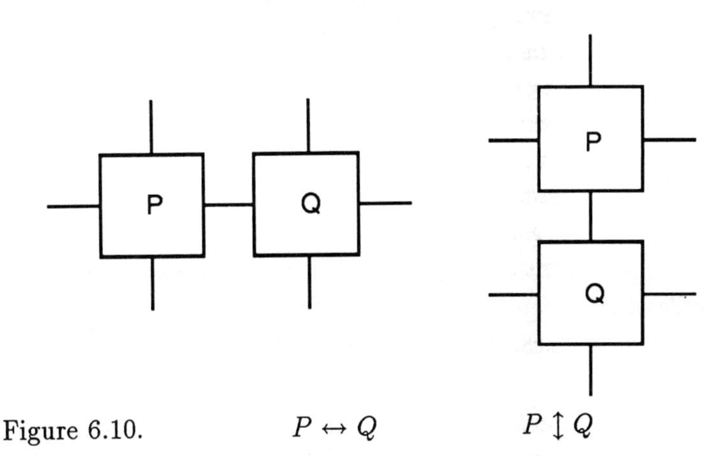

Figure 6.10. $P \leftrightarrow Q$ $P \updownarrow Q$

Law 9 $(P \leftrightarrow Q) \updownarrow (R \leftrightarrow S) = (P \updownarrow R) \leftrightarrow (Q \updownarrow S)$

9.2 Row and Col

The economics of VLSI encourages us to design regular circuits, so we introduce some suitable combining forms. row P plugs Ps together horizontally. Like map, it is a generic combining form; the number of elements in the linear array depends on the type of the signals in the domain and range. col P plugs Ps together vertically (see Figure 6.11).

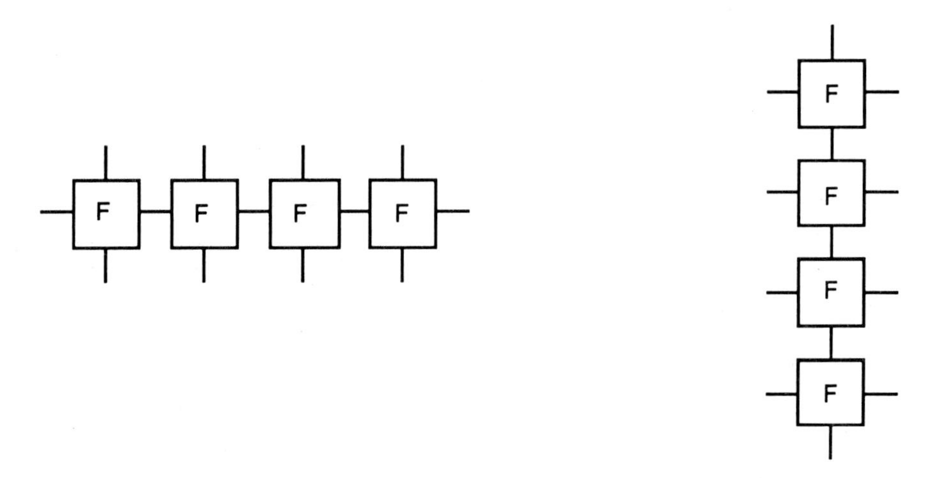

Figure 6.11. row F col F

Definition 15

$$\begin{array}{lll} \langle a, \langle\rangle\rangle \ \text{row}\ P\ \langle\langle\rangle, a\rangle & & \\ \langle a, b : c\rangle \ \text{row}\ P\ \langle d : e, f\rangle & =_{def} & \langle a, \langle b, c\rangle\rangle(P \leftrightarrow \ \text{row}\ P)\langle\langle d, e\rangle, f\rangle \\ \text{col}\ P & =_{def} & (\text{row}\ P^{-1})^{-1} \end{array}$$

Law 10 $(\text{row}\ P) \updownarrow (\text{row}\ Q) = \text{row}(P \updownarrow Q)$

Proof by induction, using the definition of row, Law 9.

By taking the inverse of both sides of this law, and replacing P^{-1} by P, and Q^{-1} by Q, we get a similar law relating col and **beside**.

Law 11 $\text{col}\ Q \leftrightarrow \text{col}\ P = \text{col}(Q \leftrightarrow P)$

We frequently use this style of developing new laws by 'inverting' old ones. It reduces considerably the number of proofs that must be done.

9.3 Triangles

When we consider transformation rules for row and col, we find that 'triangles' like those shown in Figure 6.12 arise often enough to justify the introduction of two new combining forms, upΔ and downΔ. We define downΔ in terms of upΔ and *rev*, the relation which reverses a tuple of signals.

Definition 16

$$\begin{array}{lll} \langle\rangle(\mathsf{up}\Delta\ F)\langle\rangle & & \\ a:b\ (\mathsf{up}\Delta\ F)\ c:d & =_{def} & (a=c)\ \&\ b\ (\mathsf{map}\ F;(\mathsf{up}\Delta\ F))\ d \\ \mathsf{down}\Delta\ F & =_{def} & rev;\ (\mathsf{up}\Delta\ F)\ ;rev^{-1} \end{array}$$

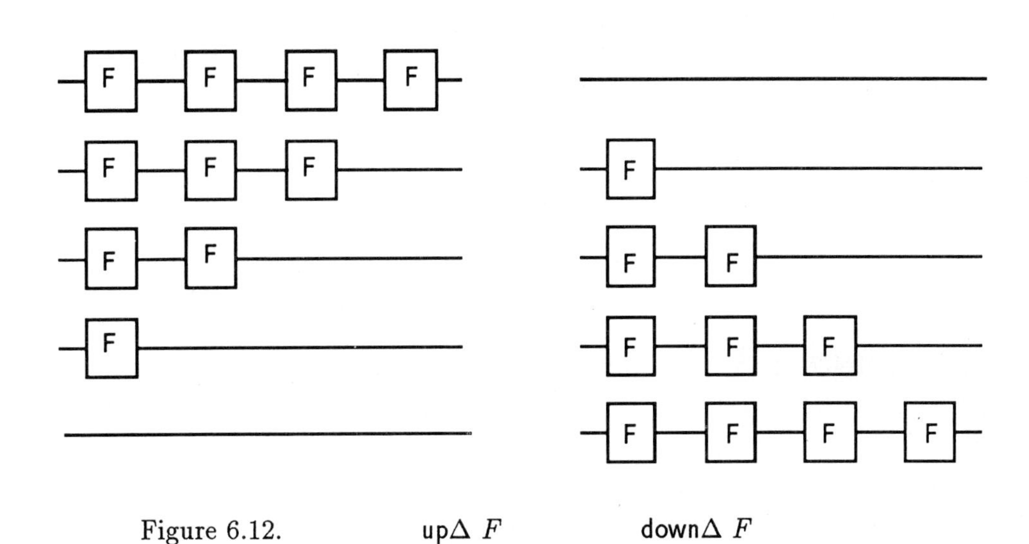

Figure 6.12. upΔ F downΔ F

It is useful to define a second slightly more complicated form of triangle. upΔ2 (F, G) relates a pair to a pair. The first elements of the pairs, which are tuples of length n, are related by upΔ F. The second elements are related by G^n. Figure 6.13 shows upΔ2 (F, G) for domain $\langle\langle a, b, c, d, e\rangle, f\rangle$ and range $\langle\langle g, h, i, j, k\rangle, l\rangle$, so in this case n is 5.

Definition 17

$$\langle\langle\rangle, a\rangle\ \mathsf{up}\Delta 2(F,G)\langle\langle\rangle, a\rangle$$
$$\langle a:b, c\rangle\ \mathsf{up}\Delta 2(F,G)\langle d:e, f\rangle$$
$$=_{def}\ (a=d)\ \&\ \langle b,c\rangle \mathsf{fst}(\mathsf{map}\ F); \mathsf{snd}\ G; \mathsf{up}\Delta 2(F,G)\langle e,f\rangle$$

downΔ2 is defined similarly. We will not use it.

Instead of introducing upΔ2, we could use subscripts to indicate the number of cells required in rows, columns etc. The first draft of this chapter used subscripts, and confirmed our view that subscripts should be avoided where possible.

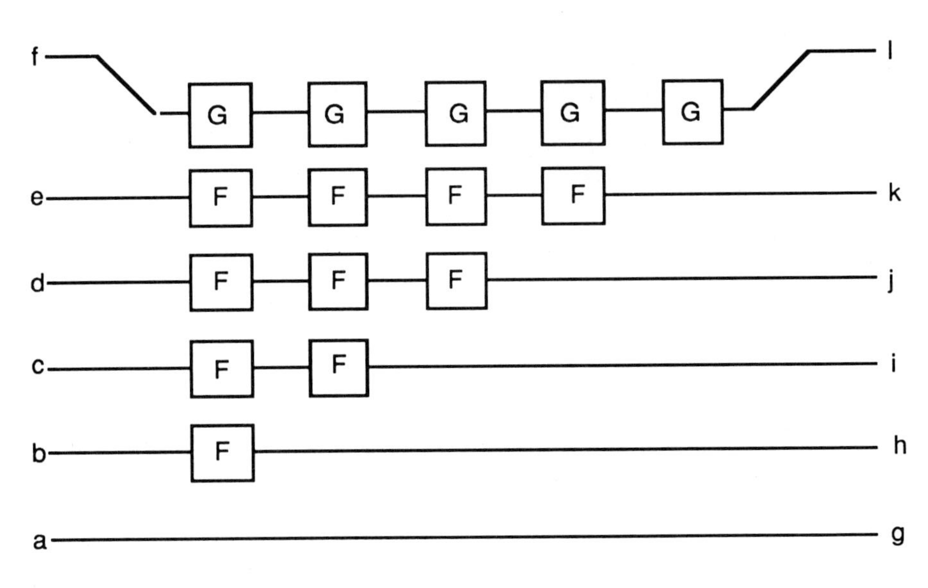

Figure 6.13. $\mathsf{up}\Delta2\ (F, G)$

10 RELATING LINEAR ARRAYS AND TRIANGLES

Next, we need to know how linear arrays and triangles interact. We will introduce a single law relating row, upΔ and upΔ2, and we will use it to derive several other useful laws. The law is in the form 'if G can be transformed in this way, then row G can be transformed as shown in Figure 6.14'.

Law 12

$$\frac{G = [A, B]; G; [C, A^{-1}] \qquad A^{-1}; A = id}{\mathsf{row}\ G = \mathsf{snd}(\mathsf{up}\Delta\ B); \mathsf{row}(G; \mathsf{snd}\ A); \mathsf{up}\Delta2(C, A^{-1})}$$

Law 12 tells us how to replace the cell G in a row by G with A 'attached' to its right hand edge (i.e., $G; \mathsf{snd}\ A$). First, we must find a transformation of G which leaves it with A on its left hand edge and A^{-1} on its right hand edge. Circuits B and C are introduced to ensure that the new circuit $([A, B]; G; [C, A^{-1}])$ has the same behaviour as the original G. Next, we apply this transformation zero times to the leftmost G, once to the next G, twice to the next and so on. This leaves $A^{-k}; A^{k+1} (= A)$ between the kth and $(k+1)$th Gs. Thus, we have replaced each G by $(G; \mathsf{snd}\ A)$. The triangles of Bs and Cs and the row of A^{-1} circuits on the right must be introduced to ensure that the new circuit has the same behaviour as the old. The proof, which is left as an exercise for the reader, is by induction on the length of the row. It uses the definition of upΔ and Law 13, which shows what happens if we apply our transformation once to each element of a row.

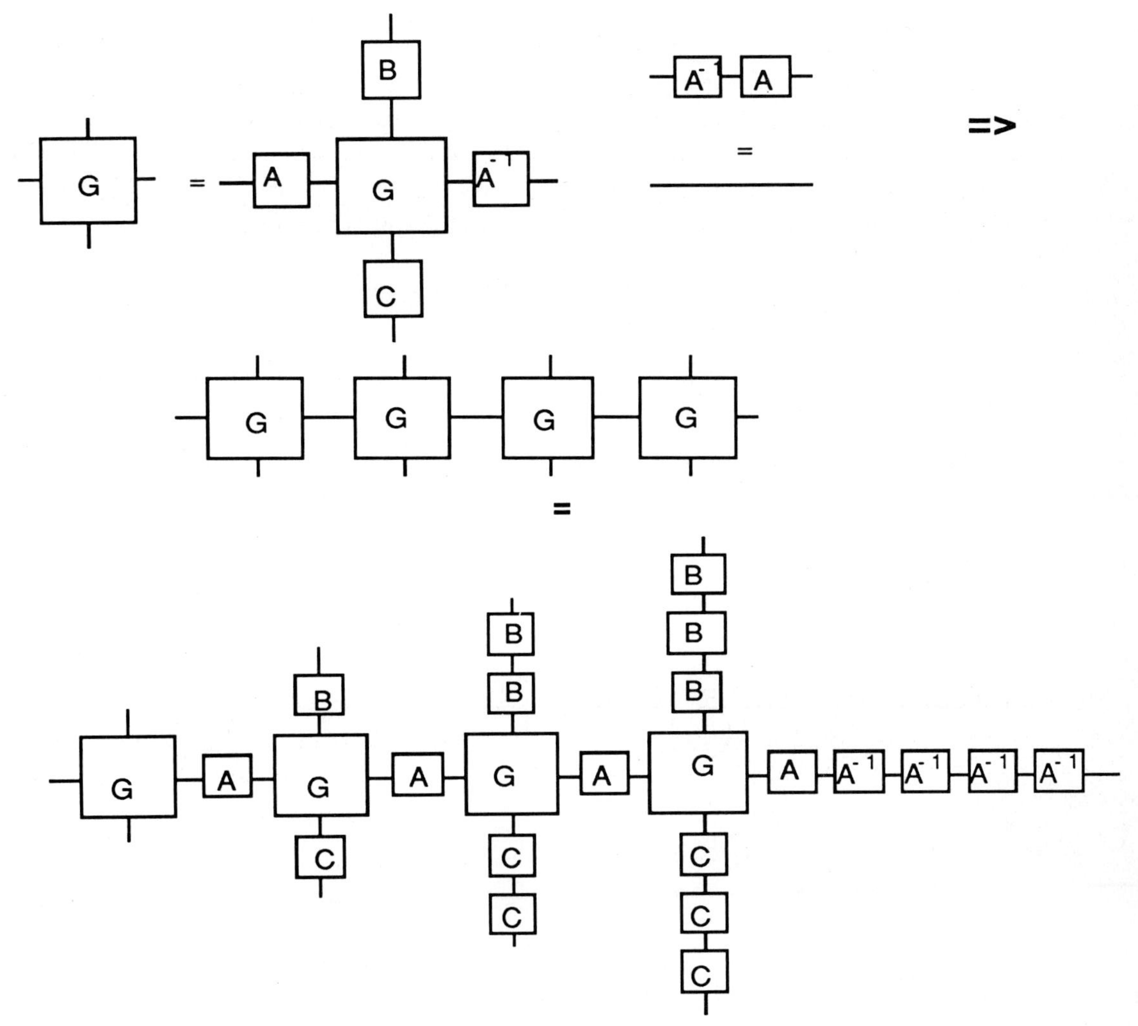

Figure 6.14. Law 12 illustrated.

Law 13

$$\frac{G = [A, B]; G; [C, A^{-1}] \qquad A^{-1}; A = id}{\mathsf{row}\ G = [A,\ \mathsf{map}\ B];\ \mathsf{row}\ G; [\mathsf{map}\ C, A^{-1}]}$$

Again, we can invert Law 12, which relates row and triangle, to give a law which relates col and triangle.

Law 14

$$\frac{G = [C, A^{-1}]; G; [A, B] \qquad A; A^{-1} = id^{-1} = id}{\mathsf{col}\ G = \mathsf{up}\Delta 2(C, A^{-1}); \mathsf{col}(\mathsf{snd}\ A; G); \mathsf{snd}(\mathsf{up}\Delta\ B)}$$

Laws 12 and 14 have many uses. We can use them to work out how to place latches on the internal arcs of linear arrays. For instance, Law 12.1 is an **instance** of Law 12. We have chosen the values of the variables A, B and C so that the premise of the law is guaranteed to hold. We appeal to the retiming law (Law 3) in making this choice. Then, substituting those same values into the conclusion gives us a true statement about the placing of latches on the internal arcs of a row.

Law 12.1 $\qquad$ $\mathsf{row}\ G = \mathsf{snd}(\mathsf{up}\Delta\ D); \mathsf{row}(G; \mathsf{snd}\ D); \mathsf{up}\Delta 2(D^{-1}, D^{-1})$

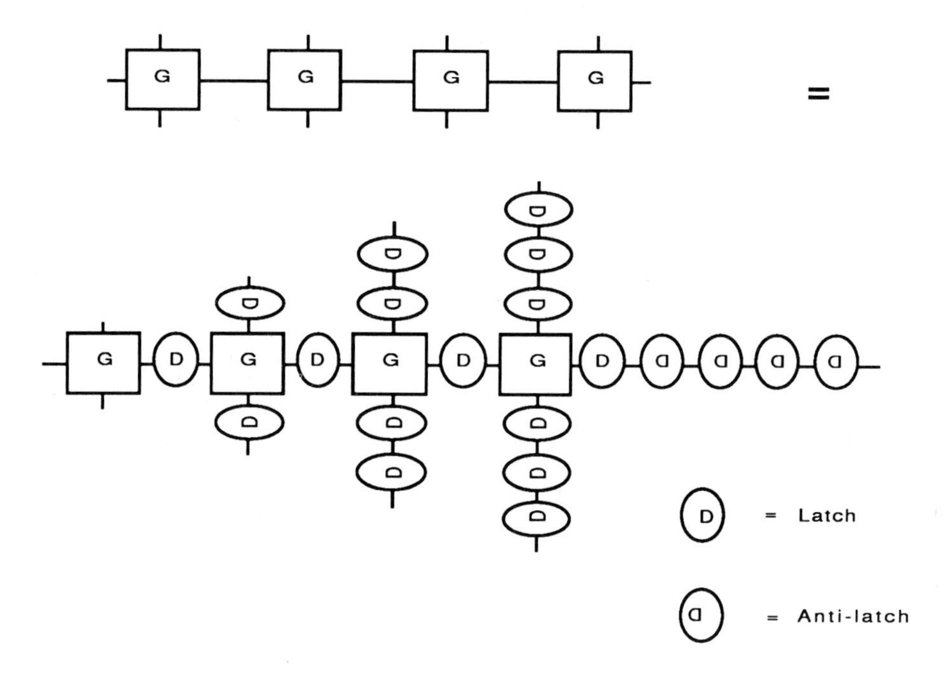

Figure 6.15. Putting latches on the internal arcs of a row (Law 12.1)

The circuit shown in Figure 6.15 will only make sense if all the horizontal data flow is from left to right. Otherwise we would be trying to drive some latches 'backwards' and this is physically impossible. In any example, we must choose the value of A so that arcs with left to right flow have Ds placed on them, and arcs with right to left flow have D^{-1}s placed on them. This guarantees that all latches are used correctly, and time is not abused. For example, if we have a pair of signals , the first flowing left to right and the second right to left, then the required value of A is $[D, D^{-1}]$.

It is important to note that Law 12 covers many different possibilities, depending on the data flow through the circuit. This is in contrast to our earlier functional approach in which it was necessary to have a separate law for each possibility. Arrays with counterflowing data presented particular problems in μFP and the fact that we can now deal with them so simply is an important advantage of the relational approach. If, given A, we can find B and C so that the premise of Law 12 holds, then we simply apply the law by pattern matching.

Law 12.1 can be thought of as an application of the cut theorem (Kung [1987]) in which we have made vertical cuts through the circuit and placed latches on those cuts. However, our law gives us more because it forces us to determine the consequences for the behaviour of the circuit of placing those latches. It shows us the data skew which results, while the cut theorem does not.

11 GRIDS

By using row and col together, we can describe rectangular grids. There are two ways of building a grid, as a row of columns or as a column of rows. Law 15 follows directly from this and from the properties of row and col.

Definition 18 $\quad \mathsf{grid}\ F \ =_{def}\ \mathsf{row}(\mathsf{col}\ F) = \mathsf{col}(\mathsf{row}\ F)$

Law 15 $\quad (\mathsf{grid}\ F)^{-1} = \mathsf{grid}(F^{-1})$

To derive the law relating grid and triangle, we first combine Laws 13 and 14 to give us a rule which tells us how to place a circuit on the top edge of each cell in a **grid**. Law 14 tells us how to replace each G by ($\mathsf{snd}\ B\ \ ;\ G$) in a column and we use Law 13 to apply this transformation to every colum in the grid. So, we replace each G in the grid by ($\mathsf{snd}\ B\ \ ;\ G$)).

$$\frac{G = [A, B^{-1}]; G; [B, A^{-1}] \qquad A^{-1}; A = id \qquad B; B^{-1} = id}{\mathsf{grid}\ G = \mathsf{up}\Delta 2(A, \mathsf{map}\ B^{-1}); \mathsf{grid}(\mathsf{snd}\ B\ ; G); \mathsf{snd}(\mathsf{up}\Delta\ A^{-1})}$$

If we swap A and B and then perform law inversion, we find that the premise remains unchanged, but we get a new conclusion:

$$\frac{G = [A, B^{-1}]; G; [B, A^{-1}] \qquad A^{-1}; A = id \qquad B; B^{-1} = id}{\mathsf{grid}\ G = \mathsf{snd}(\mathsf{up}\Delta\ B^{-1}); \mathsf{grid}(G; \mathsf{snd}\ A); \mathsf{up}\Delta 2(B, \mathsf{map}\ A^{-1})}$$

Next, we combine our two conclusions to give a single law which tells us how to place A's on the horizontal and B's on the vertical arcs of a grid (see Figure 6.16).

Law 16

$$\frac{G = [A, B^{-1}]; G; [B, A^{-1}] \qquad A^{-1}; A = id \qquad B; B^{-1} = id}{\mathsf{grid}\ G = \mathsf{skew}(A, B); \mathsf{grid}(\mathsf{snd}\ B; G; \mathsf{snd}\ A); \mathsf{skew}(B, A)}$$

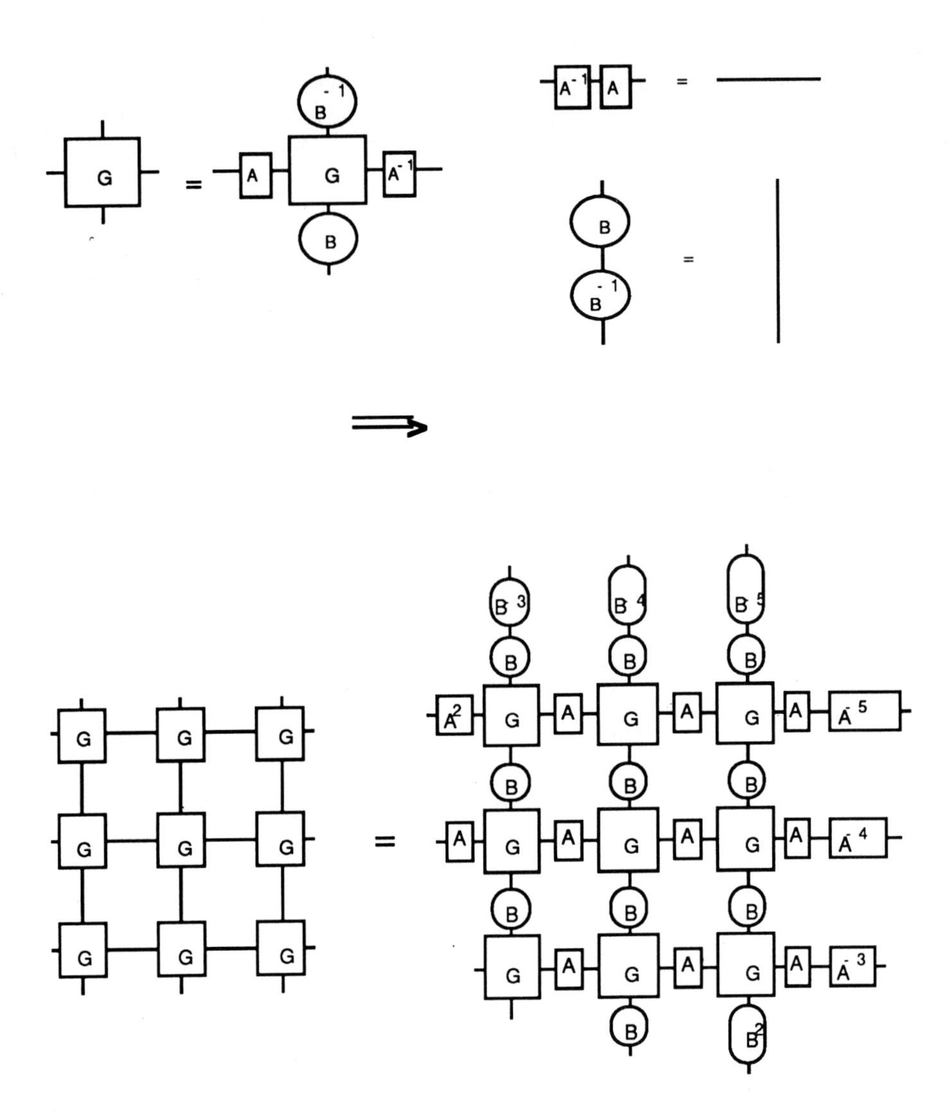

Figure 6.16. Placing circuits on the internal arcs of a grid (Law 16).

where

$$\mathsf{skew}(X, Y) = \mathsf{up}\Delta 2(X, \mathsf{map}\ Y^{-1}); \mathsf{snd}(\mathsf{up}\Delta\ Y^{-1})$$

We use Law 16 when making rectangular grids systolic. The 'A' and 'B' circuits are then combinations of latches and anti-latches, chosen to make the premise of the law true. The exact combinations depend on the directions of data flow and possibly on the data dependencies within the basic cell G.

12 DESCRIBING HEX-CONNECTED ARRAYS

We might now be expected to go on to provide a special combining form and related laws for describing hex-connected arrays. However, we can easily describe hex-connected arrays by adding a small amount of routing to a grid. We therefore avoid adding a new basic combining form. If F relates the signal $\langle a, b\rangle$ to the signal $\langle d, e\rangle$, then F^L relates the signal $\langle a, \langle b, c\rangle\rangle$ to the signal $\langle d, \langle e, c\rangle\rangle$. The signal c forms the 'bypass' shown in the top right hand corner. A grid built from circuits of this form is hex-connected (see Figure 6.17(c)).

$$\langle a, \langle b, c\rangle\rangle \; reorg \; \langle\langle a, b\rangle, c\rangle$$

$$F^L \;=_{def}\; reorg; \mathsf{fst}\ F; reorg^{-1}$$

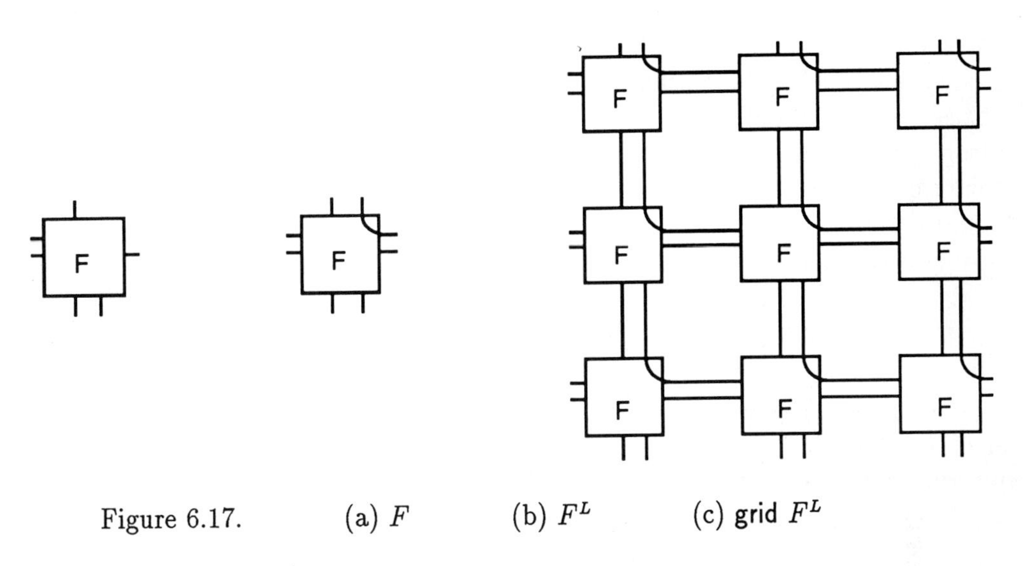

Figure 6.17. (a) F (b) F^L (c) grid F^L

We can now use our laws about grid to investigate ways of making hex-connected arrays systolic. In any example, our task is to find values of A and B which make the premise of Law 16 true, and which give a satisfactory systolic circuit when plugged into the conclusion. Because of the retiming property, the values $A = B^{-1} = D$ will work for all examples. If we set G to be F^L, our final grid is then made of cells of the form

$$\mathsf{snd}\ D^{-1}; F^L; \mathsf{snd}\ D$$

When we build the grid and cancel the latch/anti-latch pairs, we find that the bypass arcs end up without latches or anti-latches. This means that the final circuit is not even systolic.

If the data flow is such that $A = B = D$ makes the premise of Law 16 true, our final grid is then made of cells of the form

$$\mathsf{snd}\ D; F^L; \mathsf{snd}\ D$$

and the bypass arcs end up with two latches on them. From the point of view of circuit performance, it is preferable to have exactly one latch on every arc of a systolic circuit. If, however, the basic cell in the grid is combinational, we just have to accept two latches on some arcs. There are many such circuits in the literature (the systolic multiplier in Sheeran [1985] for example).

On the other hand, if the basic cell already contains some latches, it may be possible to ensure that the final circuit has one latch on each arc. A standard trick is to 'slow' the circuit by replacing each D in the basic cell by D^k (Leiserson [1983]) so that when the array is made systolic using Law 16, the offending arcs then end up with D on them, as required. The resulting k-slow circuit computes the same function as before, but it must operate on k separate interleaved streams of data. In the case of hex-connected arrays, k is usually 3. Many hex-connected circuits are 3-slow, some, such as the matrix multiplier in Mead & Conway [1980], unnecessarily so. For similar reasons, many linear arrays with counterflowing data are 2-slow (Sheeran [1986]). Using the relational approach, we have developed a theory of 'slow' circuits (Sheeran [1988]). It allows us to reason formally about a design technique which has often been used in a very *ad hoc* way.

13 OTHER COMBINING FORMS

We will not define any more combining forms here. The final language is likely to include some versions of the reduce (or fold) combining form, which is much loved by functional programmers. Reduce can be regarded as a degenerate form of col, in which the cells relate a pair of signals to a single signal. So, some of the ports in col are not present. Some form of conditional is also necessary. Finally, it is useful to introduce a 'flip about the top-left to bottom-right diagonal' combining form. Rotations can then be defined using this and inverse.

14 DATA ABSTRACTION

As an unexpected bonus, we have found that using relations facilitates reasoning about data abstraction. Circuits are usually developed top down. At some stage in the development, it is necessary to make a 'jump' between levels of abstraction. Often, this is the switch from the word level to the bit level. The specification will refer to integers (up to a certain size) and the implementation will refer to bits. In presentations of circuit developments, such switches between levels of abstraction are usually done informally (e.g., Sheeran [1985]). However, the whole process can be made more formal by the use of abstraction and representation relations. For example, if *bits* relates (streams of) integers to (streams of) twos complement bit sequences, then we can express integer addition in terms of bit level operations as follows.

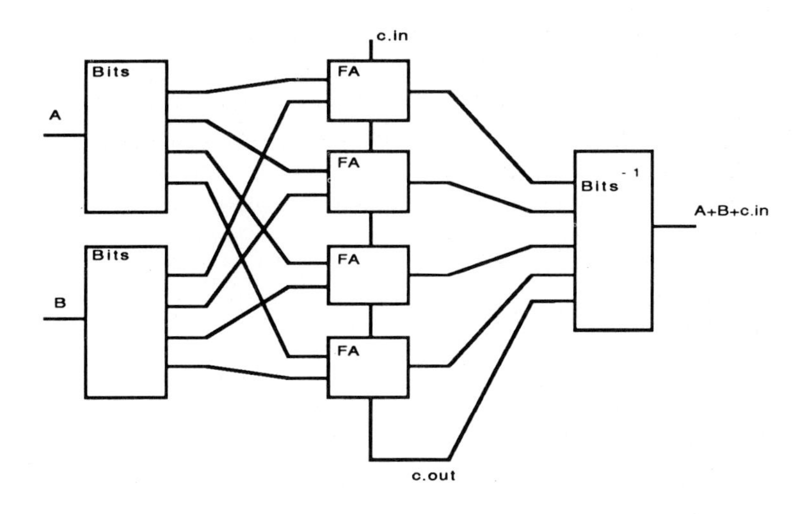

Figure 6.18. Relating integer- and bit-addition

$$int+ = \mathsf{fst}([bits, bits]; zip); (\mathsf{col}\ FA); app_L; bits^{-1}$$

int+ relates a pair of integers and a carry-in bit to an integer. The column of full adders, on the other hand, relates two interleaved bit sequences and a carry-in to a carry-out and a bit sequence. The carry-out is the most significant bit of the final result, so it is appended to the sum sequence, using app_L. The resulting sequence is then converted back into integer form, by $bits^{-1}$. In a circuit development, we can replace any occurrence of *int+* by the whole of the right hand side. We have 'wrapped up' the right hand side in abstraction relations so that, internally, only bits are manipulated. The conversion relations show how our higher level data types are represented using bits. The presence of the inverse combining form facilitates this.

It is often useful to make the change to the bit level gradually. Thus, a partially developed design may manipulate both integers and bits. Such circuits are easily represented in this formalism. Having moved part of the way to the bit level, it may be easier to choose the bit level representations for the remaining signals. Some circuits mix more than one bit level representation. For example, the systolic bit-level multiplier in McCanny & McWhirter [1982] represents integers in both twos complement and carry save form. Our relational notation allows us to specify data abstraction and representation precisely. This is an improvement on the informal techniques that we used previously.

15 ALTERNATIVE INTERPRETATIONS

Sometimes, we are not interested in a complete description of a circuit's behaviour but only in a certain aspect of that behaviour. We can get this kind of abstract information by plugging abstract versions of the circuit primitives into the circuit description. In our case, one useful abstract view of the circuit is the one which is concerned only with the direction of flow along the wires. We are not attempting to deal with true bidirectionality, so, in any final implementation, we want every wire to have a unique direction. The abstract directions are 'left-to-right' ($>>$) and 'right-to-left' ($<<$). Inputs in the domain of a relation and outputs in the range are $>>$. Outputs in the domain and inputs in the range are $<<$.

Each primitive circuit will have a corresponding abstract relation giving the set of allowed directions of flow for domain and range signals. For some primitive circuits, there is only one choice available. For instance, the abstract version of *And* is the singleton relation $\{((>>,>>),>>)\}$ since it must have both its inputs in the domain and its output in the range. We will write this as

$$And^* \;=\; \{((>>,>>),>>)\}$$

A slightly more complicated *And*1, defined as follows

$$< a,b > \;\; And1 \; c \;\; =_{def} \; < a,c > \;\; And \; b$$

would have the assignment

$$And1^* \;=\; \{((>>,<<),<<)\}.$$

The inputs are the first element of the domain, and the range. The output is the second element of the domain.

Not, which is its own inverse, allows two possible choices, giving

$$Not^* = \{(>>,>>),(<<,<<)\}.$$

The identity function (on signals), which corresponds to a wire or bus, has as abstract value the identity function on $>>, <<$ and tuples involving them. A wire or bus has no influence on the direction of signals; it merely passes them on. Selector functions now operate on tuples rather than signals, but are otherwise unchanged. A latch, D, must be replaced by the identity on $>>$ and on tuples containing only $>>$. This expresses the requirement that it must be run 'forwards'.

As well as replacing all primitive circuits and wires by abstract versions, we must also replace the combining forms by their abstract versions. The abstract version of relational inverse must reverse the directions of all arrows as well as swapping domain and range. Inverting a relation causes the directions of flow along all wires to be reversed. For example, And^* is $\{((>>,>>),>>)\}$ and $(And^{-1})^*$ is $\{(<<,(<<,<<))\}$. So,

$$(F^{-1})^* = rev\text{-}dir; (F^*)^{-1}; rev\text{-}dir^{-1}$$

where $rev\text{-}dir (= rev\text{-}dir^{-1})$ reverses the direction of a single arrow , or of all arrows in a tuple. All combining forms defined using inverse will have abstract values which reflect this definition of $(F^{-1})^*$.

We have chosen our representations for direction of flow so that the abstract value of composition is still ordinary relational composition. The relations composed are relations between tuples rather than signals, but that is the only difference. Thus,

$$(F;G)^* = F^*;G^*.$$

For example,

$$\begin{aligned}(And; Not)^* &= And^*; Not^* \\ &= \{((>>,>>),>>)\}; \{(>>,>>),(<<,<<)\} \\ &= \{((>>,>>),>>)\}\end{aligned}$$

The second possible direction for the Not is ruled out because the $>>$ in the range of And^* and the $<<$ in the domain of Not^* do not match. So, a clash of directions along the wire between And and Not is disallowed. The par combining form behaves similarly.

$$[F, G]^* = [F^*, G^*]$$

So,

$$\begin{aligned}[Not, Not]^* &= [Not^*, Not^*] \\ &= [\{(>>,>>),(<<,<<)\}, \{(>>,>>),(<<,<<)\}] \\ &= \{((>>,>>),(>>,>>)),((<<,<<),(<<,<<)), \\ &\quad ((<<,>>),(<<,>>)),((>>,<<),(>>,<<))\}.\end{aligned}$$

In this case, there are four allowed assignments of direction, corresponding to all combinations of the two for each individual Not.

In fact, inverse is the only combining form which has any effect on the directions of signals. * can be distributed through all other combining forms. For example,

$$\begin{aligned}(\mathsf{map}\ F)^* &= \mathsf{map}\ F^* \\ (\mathsf{row}\ F)^* &= \mathsf{row}\ F^* \\ (\mathsf{up}\Delta\ F)^* &= \mathsf{up}\Delta\ F^*.\end{aligned}$$

We know (Definition 15) that

$$\text{col } P = (\text{row } P^{-1})^{-1}$$

By taking the abstract values of both sides, we find that

$$(\text{col } P)^* = \text{col } P^*.$$

Given a relation describing a circuit, we can now calculate the abstract value of that relation, giving the set of allowed assignments of directions to the signals in the domain and range of the relation. The expected assignment of directions should be an element of that set. If that set is empty, as in the case of $And; And^{-1}$, then there are no allowed assignments and the circuit cannot be implemented. It seems likely that the best way to implement a simulator for Ruby would be to use this kind of analysis to find directions of flow and to 'run' the result as a functional program. A similar approach is used in Clocksin & Leeser [1986] to determine signal flow through MOS transistor networks.

An alternative approach is to assign to each primitive circuit a singleton relation containing one of its allowed assignments. This is a way of checking that we know which way round to plug any circuit primitive which has more than one allowed assignment.

Checking of directions of data flow in a circuit is important. However, it is not sufficient to guarantee that the circuit described is realistic. Any useful circuit will accept any value of the correct type on its input ports and produce, in response, a unique value of the correct type on its output ports. So, a proper circuit is a total function from its inputs to its outputs. All of our primitive circuits are proper. Unfortunately, plugging together two proper circuits correctly, using relational composition, is no guarantee that the resulting circuit will be proper. The resulting relation need not correspond to a function of the inputs, and if it is a function, it need not be total. The problem arises when the resulting composition creates a cycle of dependencies. Such a cycle can never be cut so that information flows in one direction across the cut. So, the resulting circuit might not be a function of its inputs. So, whenever we describe a circuit, we must be careful to avoid unresolved (asynchronous) loops. This is similar to the need to avoid deadlock in concurrent programs.

16 RELATED WORK

The use of relational circuit descriptions is not new. Mike Gordon and others at Cambridge (Gordon [1986]) have been using Higher Order Logic (HOL) for describing and reasoning about circuits for some time. Their work is inspired by that of Keith Hanna, who uses a formalism based on type theory (Hanna & Daeche [1986]).

In HOL, a circuit is represented not by a relation but by a predicate characterising the relation which the circuit maintains on its ports. Circuits are composed by *And*ing their predicates, with equally named ports joined. Internal wires can be hidden using existential quantification. This use of & and ∃ corresponds exactly to our use of relational composition. The main motivation behind the use of relations in HOL is the desire to capture truly bidirectional behaviour. For example, simple transistor models which capture bidirectionality are among the standard HOL examples (Camilleri, Gordon & Melham [1986]). We have made no effort to capture such true bidirectionality. Instead, we have demonstrated that relations can ease the description and design of circuits, even when functional descriptions would have been sufficient. HOL is a much more powerful and general purpose formalism than Ruby. Ruby provides combining forms for describing the most common circuit structures, and related laws to help in the design of those structures. The fact that a circuit layout must eventually be produced has influenced the design of the language. The emphasis is on correct design rather than on *post hoc* verification. Ruby could perhaps be thought of as a specialised subset of HOL.

Boute has presented a generalisation of the lambda calculus designed to capture bidirectional data flow (see Boute [1986] for a survey of work on sematics of digital systems). As in Sheeran [1983] and in this chapter, the usefulness of having various different interpretations of circuit expressions is emphasised. Boute chooses to name 'channels', citing the unreadability of the original μFP, with its myriad projection functions or selectors as one of the reasons. Here, we have largely solved that problem by using signal variables in the definitions of primitives and combining forms and in the proofs of laws. Where necessary, we move to the object level, yet we retain the advantages of using combinators which obey nice algebraic laws. It is interesting to compare the description of a full-adder taken from Sheeran [1983] with the Ruby description (Figure 6.19).

A proliferation of projection functions should obviously be avoided. Our solution is to choose our combining forms and wiring relations carefully. Boute's is to introduce variables into his bidirectional β calculus. He also introduces a universal combinator, and translations between the two description methods. He has demonstrated the use of his calculus by defining some simple combining forms for cascading circuits and proving some properties of these combining forms. He gives an example of the use of the combining forms in the description of an analogue circuit. It seems to have been the desire to describe and reason about analogue circuits which prompted the use of relations in this case.

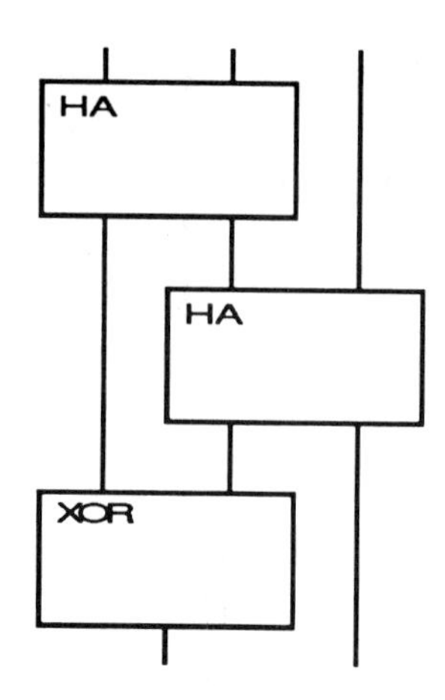

$$FA = [Xor \circ [1, 1 \circ 2], 2 \circ 2] \circ [1 \circ 1, HA \circ [2 \circ 1, 2]] \circ [HA \circ 1, 2] \quad (\mu FP)$$

$$FA = \mathsf{fst}\ HA; reorg^{-1}; \mathsf{snd}\ HA; reorg; \mathsf{fst} Xor$$

$$\text{where } \langle a, \langle b, c\rangle\rangle \ reorg \ \langle\langle a, b\rangle, c\rangle \qquad (see\ Fig.\ 6.18) \quad (Ruby)$$

Figure 6.19. Comparing circuit description styles: the original μFP vs Ruby

17 CONCLUSION AND FUTURE WORK

We have introduced a simple language for describing circuits. A circuit is viewed as a transformer of synchronous streams or signals, and circuits are plugged together using combining forms which have both structural and semantic interpretations. We have abstracted away from the notions of input and output. A circuit is simply a binary relation between signals, not a function from input signals to output signals. Relational composition is now the most natural combining form. The set of combining forms which arises when one makes this switch to relations is slightly different from the functional set. The new combining forms exhibit some pleasing symmetries.

We have defined the combining forms of Ruby directly, whereas in the functional case we gave the semantics by translating into FP. We have simplified the definitions by introducing a notation for signal construction and by allowing references to signals in the definitions. This is a move away from the FP style of allowing only 'function-level' reasoning. It has simplified the inductive proofs of the laws which the combining forms obey. Yet we retain the ability to describe circuits in a 'higher order' style, without naming of channels. This makes it easier to reason about those circuits, by applying the algebraic laws. A comparison with the definitions of the combining forms in Sheeran [1983] shows the advantages of the new style of definition.

The move to relational descriptions simplifies reasoning about circuits with complicated data flow. We have demonstrated this by showing how one reasons about the placing of latches on the internal arcs of linear and hexagonal arrays. Anti-latches, which fit nicely into the relational setting, are useful here. We have not attempted to describe circuits in which wires can carry signals in both directions. We have demonstrated that there are advantages in using relations for describing and reasoning about circuits, even in the absence of true bidirectionality. We have also illustrated the use of alternative 'abstract' interpretations of our primitives and combining forms as a means of analysing circuit behaviour.

One of our motivations in moving to relational descriptions was a desire to move away from the notion of design by equivalence transformation (as presented in this chapter) towards the more general design by refinement. Designers tend to start with a fairly abstract specification of the required behaviour, just as programmers do. Gradually, the specification is refined into a circuit which is made of components, which we 'know how to build'. Some of the design steps will not be true equivalence transformations. For example, the initial specification may be nondeterministic, but the final circuit will probably be deterministic. We would like to investigate this more general view of design further, and the move to relational descriptions is a step in that direction.

Acknowledgements
I am grateful to John Williams and John Hughes, who provided constructive criticism of a very poor first draft. Geraint Jones and Wayne Luk provided comments on a later draft. It was a pleasure to take into account the perceptive comments and suggestions of an anonymous referee.

Several other papers about the relational approach to hardware description have appeared in print since I wrote this one. Some representative examples are Sheeran & Jones [1987], Sheeran [1988], Luk & Jones [1988b, 1989] and Sheeran [1989].

REFERENCES

Boute [1986]
R. T. Boute (1986) 'Current work on the semantics of digital systems', *Formal Aspects of VLSI Design*, ed. G. Milne & P. A. Sumrahmanyam, North-Holland.

Camillieri, Gordon & Melham [1986]
A. Camillieri, M. Gordon and T. Melham (1986) 'Hardware verification using higher order logic', in *Proceedings of the IFIP International WG10.2 Working Conference: From HDL Descriptions to Guaranteed Correct Circuit Designs*, North-Holland.

Clocksin & Leeser [1986]
W. F. Clocksin and M. E. Leesor (1986) 'Automatic determination of signal flow through MOS transistor networks', *Integration, the VLSI Journal*, **4**, no. 4.

Gordon [1986]
M. J. C. Gordon (1986) 'Why higher order logic is a good formalism for specifying and verifying hardware', in *Formal Aspects of VLSI Design*, ed. G. Milne & P. A. Sumrahmanyam, North-Holland.

Hanna & Daeche [1986]
F. K. Hanna and N. Daeche (1986) 'Specification and verification using higher order logic', in *Formal Aspects of VLSI Design*, ed. G. Milne & P. A. Sumrahmanyam, North-Holland.

Kahn [1974]
G. Kahn (1974) 'The semantics of a simple language for parallel programming', *Information Processing 74*, ed. J. L. Rosenfield, North-Holland, Amsterdam.

Kung [1987]
S. Y. Kung (1987) *VLSI Array Processors*, Prentice-Hall.

Luk & Jones [1988a]
W. Luk and G. Jones (1988) 'The derivation of regular synchronous circuits', in *Proceedings of the International Conference on Systolic Arrays*, ed. K. Bromley. S-Y. Kung & E.Swartzlander, IEEE Computer Society Press.

Luk & Jones [1988b]
W. Luk and G. Jones (1988) 'From specification to parameterised architectures', *The fusion of hardware design and verification*, ed. G. Milne, North-Holland.

Luk & Jones [1989]
W. Luk and G. Jones (1989) 'Parametrized retiming of regular computational arrays', *Parallel processing for computer vision and display*, ed. P. M. Dew, R. A. Earnshaw & T. R. Heywood, Addison-Wesley.

Leiserson [1983]
C. E. Leiserson (1983) 'Systolic and semisystolic design' [Extended Abstract], in *Proceedings of the IEEE International Conference on Computer Design.*

McCanny & McWhirter [1982]
J. V. McCanny and J. G. McWhirter (1982) 'Completely iterative pipelined multiplier array suitable for VLSI', in *IEE Proceedings*, **129**, Pt. G, No. 2.

Mead & Conway [1980]
C. A. Mead and L. Conway (1980) *Introduction to VLSI Systems*, Addison-Wesley.

Milne [1983]
G. J. Milne (1983) 'CIRCAL, A calculus for circuit description', *Integration, the VLSI Journal*, **1**, nos. 2 & 3.

Sheeran [1983]
M. Sheeran (1983) 'μFP, an algebraic VLSI design language', D. Phil thesis, Programming Research Group, Oxford University (appears as Tech. Monograph PRG-39).

Sheeran [1984]
M. Sheeran (1984) 'μFP, a language for VLSI design', in *Proceedings of the ACM Symposium on LISP and Functional Programming.*

Sheeran [1985]
M. Sheeran (1985) 'Designing regular array architectures using higher order functions', in *Proceedings of the International Conference on Functional Programming Languages and Computer Architecture*, Springer-Verlag LNCS 201.

Sheeran [1986]
M. Sheeran (1986) 'Design and verification of regular synchronous circuits', *IEE Proceedings*, **133**, Pt. E, No. 5.

Sheeran [1988]
M. Sheeran (1988) 'Retiming and slowdown in Ruby', in *The fusion of hardware design and verification*, ed. G. Milne, North-Holland.

Sheeran [1989]
M. Sheeran (1989) 'Describing hardware algorithms in Ruby', in *Proceedings IFIP TC10/WG10.1 Workshop on Concepts and Characteristics of Declarative Systems*, ed. G. David, R. Boute & B. Shriver, North-Holland.

Sheeran & Jones [1987]
M. Sheeran and G. Jones (1987) 'Relations + higher order functions = hardware descriptions', in *Proceedings, CompEuro 87*, IEEE.

Weiser & Davis [1981]
U. C. Weiser and A. L. Davis (1981) 'A wavefront notation tool for VLSI array design', in *VLSI Systems and Computations*, ed. H. T. Kung, R. F. Sproull & G. L. Steele, Computer Science Press.

7 The synthesis of VLSI signal processors: theory and example

H. C. YUNG

ABSTRACT
The theme of this chapter centres on the automatic synthesis of cost effective and highly parallel digital signal processors suitable for VLSI implementation. The proposed synthesis model is studied in detail and the concepts of signal modelling and data flow analysis are discussed. This is further illustrated by the COSPRO (COnfigurable Signal PROcessor) simulator – a primitive version of the automatic synthesis concept developed at the Department of Electrical & Electronic Engineering, University of Newcastle Upon Tyne. Binary addition is chosen as a case study to demonstrate the application of the concept.

1 INTRODUCTION

1.1 Digital signal processing

Digital signal processing (DSP), a counterpart of analog signal processing, began to blossom in the mid 1960s when semiconductor and computer technologies were able to offer a massive increase in flexibility and reliability. Within the short period of twenty years, this field has matured rapidly in both theory and applications, and contributed significantly to the understanding in many diverse areas of science and technology. The range of applications has grown to include almost every part of our lives, from microprocessor controlled domestic appliances to computerised banking systems and highly sophisticated missile guidance systems. Many other areas such as biomedical engineering, seismic research, radar and sonar detection and countermeasures, acoustics and speech, telecommunications, image processing and understanding, thermography, office automation and computer graphics employ DSP to a great extent, and are heavily applied in military, intelligence, industrial and commercial environments. In this way DSP is becoming an integral part of our world (Oppenheim & Schafer [1975], Rabiner & Gold [1975], Peled & Liu [1976], Gonzalez & Wintz [1987]).

In principle, DSP may be considered as a branch of computational mathematics concerning computer agorithms or as an engineering discipline concerning analog filter and transform theories. In practice, DSP deals with the sampling of a continuous (analog) process to generate a set of 'raw' discrete-time signals. This set of discrete-time signals may either be used to evaluate the characteristics of the sampled process, or be transformed into a desirable form by various deterministic and/or stochastic processes. The aims of these processes are mainly towards extracting, improving and enhancing the input information so as to provide a better understanding of the information by human beings, and to use it to control other electronic/mechanical devices (Rabiner & Gold [1975]).

DSP can be classified into two main categories: filtering and spectral analysis. Filtering can be further sub-divided into finite impulse response (FIR) and infinite impulse response (IIR); spectral analysis can be partitioned into spectral computation using transform techniuqes (i.e., DFT, FFT, convolution, etc.) and statistical or prediction techniques as in the case of random signals. Based on the numerical analytic techniques developed in the seventeenth and eighteenth centuries, the theory of discrete-time linear time-invariant systems is now the corner stone of the entire discipline (Rabiner & Gold [1975]).

Most of the past effort has been concentrated on one-dimensional signal processing because of its computation simplicity. Multi-dimensional signal processing such as stereo vision processing is now receiving much attention as a result of the increase in computing resources and the decrease in computing cost, both in hardware and software. Although the basis of these multi-dimensional systems is still the discrete-time linear time-invariant system, the theory and implementation is very much overshadowed by the overall system complexity and input complexity.

1.2 Signal processing implementation

Within its many applications, techniques in signal processing implementation have developed into many different forms. Using general purpose computers is a cheap and flexible way to implement algorithms in software, but is usually slow in performance compared with dedicated signal processors. Dedicated signal processors using bit-slice Transistor-Transitor Logic (TTL) or Emitter-Coupled Logic (ECL) devices on printed circuit board (PCB) are much faster, but the eventual system is always too specific and rigid, imposing great difficulties in system modification and expansion. Since the advent of VLSI, more signal processors can now be integrated onto a single piece of silicon (including both general purpose and dedicated systems) (Fairbairn [1982], Kinniment [1982]). However the coefficients and other parameters in the signal processors are normally represented by finite register lengths disregarding the implementation styles (in dedcated digital hardware or as a computer algorithm) (Oppenheim & Weinstein [1972]). The major errors introduced by having finite register lengths are the analog-to-digital conversion noise, uncorrelated roundoff noise, inaccuracies in process response due to coefficient quantization and correlated roundoff noise or limit cycles. The overall DSP system performance such as the dynamic range, stability and noise content etc is affected by the type of arithmetic used in the processing algorithm, the type of quantization and the exact DSP structure (Peled & Liu [1976]). Also the information extraction process in signal processing often requires the execution of millions or even billions of complex computations in milliseconds (Allen [1985], Kung [1987]). Such complexity imposes many restrictions on the design and development of signal processors. Since VLSI technology promises a solution to this problem of complexity, our ability to manage the theoretical complexity and technological promise becomes vital in future signal processor implementation (Trimberger [1983]).

1.3 An intelligent autonomous system

The synthesis of signal processors has always been an intuitive process by which designers create new designs based purely on their expertise – experience, knowledge, insight and creativity. Some of these techniques can be learnt but not others, and so the standard of processor system design varies. If tools to assist designers in their thinking and decision processes are not developed then this variation in standards will remain, and the task of creating larger and more complex processor systems will become unmanagable. The features of these tools should be intelligent and autonomous – that is intelligent enough to trace back and enhance the designers thinking and decision process, and capable of handling all the complexity of the implementation on its own, requiring minimum interaction from the designer. This much anticipated intelligent automomous signal processing design system will then establish the standard for the construction of signal processing systems – from concept to product (Denyer & Renshaw [1983]).

1.4 Chapter organization

The overall organization of this chapter is as follows: Section 2 outlines the theory of signal processing, which includes an overview of the fundamental concepts, the importance of discrete-time linear time-invariant (LTI) systems and a brief discussion on two-dimensional signal processing. The basic system synthesis concept is presented in Section 3. The system will be described in detail, and various limitations and constraints will be discussed. Section 4 details the idea of data flow representation, which is the major component of the system. Different analytic techniques and manipulations of data flow graphs (DFG) will be covered. Section 5 discusses the preliminary form of an intelligent autonomous tool which runs as a software simulator. The simulator is capable of emulating a realistic signal input condition, a COnfigurable Signal PROcessing (COSPRO) array and hardware descriptions for silicon compilation. Section 6 describes a simple but practical example with some detailed analysis. It is hoped that the example will illustrate the main concept described in Section 3.

2 SIGNAL PROCESSING THEORY

2.1 Discrete-time linear time-invariant systems

While analogue signals are represented in a continuous-time domain, digital signals are represented in a discrete-time domain. These so-called *discrete-time signals* are defined only for discrete values of the independent variable, time. Generally time is quantized uniformly; i.e., $t = nT$, where T is the interval between time samples. Discrete-time signals are represented mathematically as sequences of numbers whose amplitude may take on a continuum of values. One way of generating a sequence of discrete-time signals is by uniformly sampling a continuous-time waveform. In practice, this is done by using an analog-to-digital converter. Some important sequences frequently used in digital signal processing are given in Figure 7.1, including a unit impulse, a unit step, a decaying exponential and a sinusodial. Arbitrary sequences may be represented in terms of delayed and scaled impulses.

A *discrete-time system* may be viewed as an algorithm for converting one sequence (input) into another sequence (output). Such representation is given in Figure 7.2. The input is called $x(n)$ and the output is called $y(n)$, and the output is functionally related to the input by the relation

$$y(n) = f[x(n)] \tag{1}$$

where $f[.]$ is determined by the specific system.

A system is said to be *linear* if $x_1(n)$ and $x_2(n)$ are specific inputs to the system and $y_1(n)$ and $y_2(n)$ are the respective outputs, then if the sequence $ax_1(n) + bx_2(n)$ is applied at the input, the sequence $ay_1(n) + by_2(n)$ is obtained at the output, where a and b are arbitrary constants. In a *time-invariant system*, if the input sequence $x(n)$ produces an output sequence $y(n)$, then the input sequence $x(n - n_0)$ produces the output sequence $y(n - n_0)$ for all n_0.

From the above definitions of linear and time-invariant system, we can derive a convolutional relation between the input and the output sequences. Let $x(n)$ be the input to an LTI system and $y(n)$ be the output of the system; and let $h(n)$ be the response of the system to a digital impulse $u_0(n)$ as shown in Figure 7.1. As $x(n)$ is an arbitrary sequence which can be represented in terms of delayed and scaled impulses, it can be expressed as in equation (2):

$$x(n) = \sum_{m=-\infty}^{\infty} x(m)u_0(n - m). \tag{2}$$

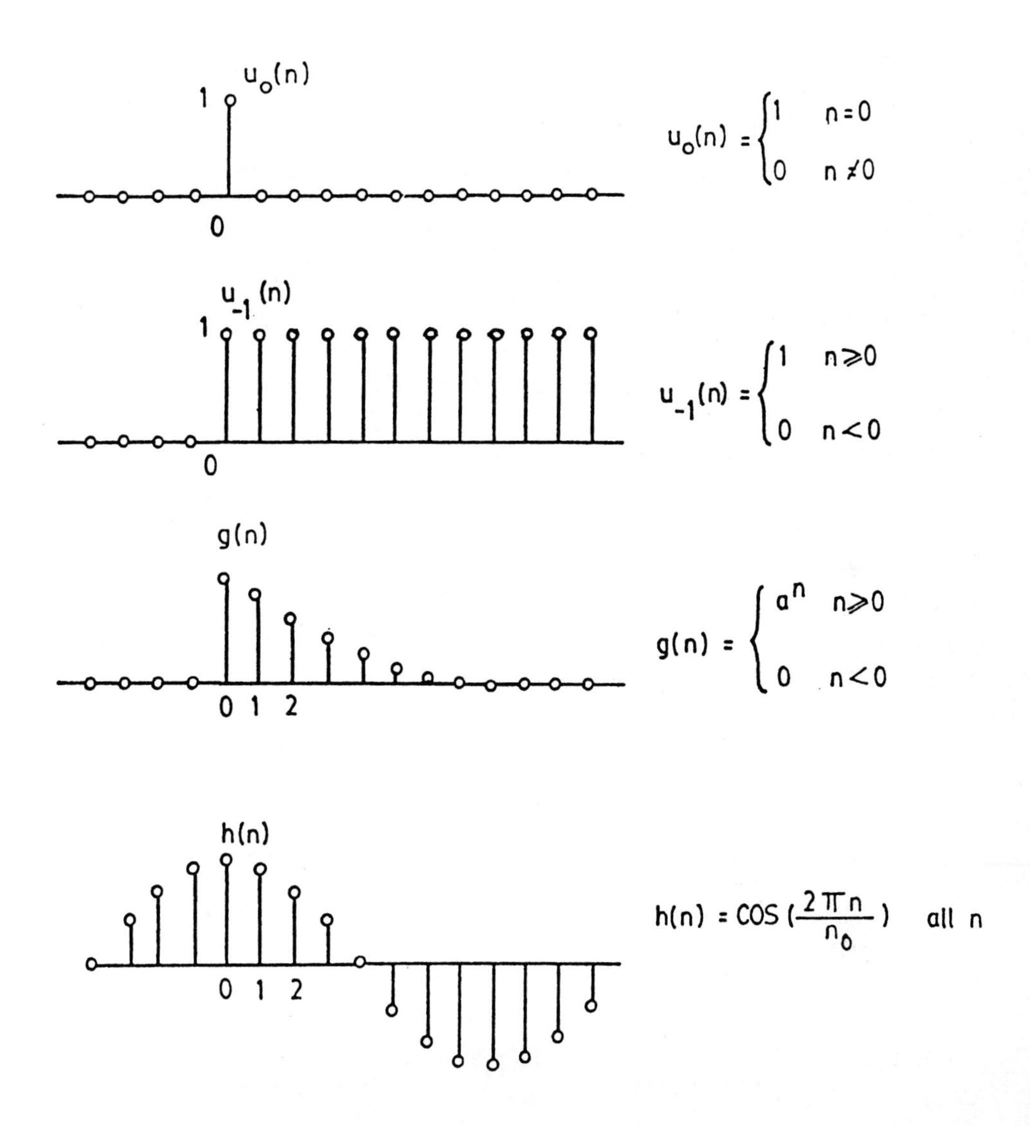

Figure 7.1. Some important sequences used in DSP

Since $h(n)$ is the response to the sequence $u_0(n)$, then by time invariance we can say that $h(n-m)$ is the response to the sequence $u_0(n-m)$. Similarly, by linearity, the response to the sequence $x(m)u_0(n-m)$ must be $x(m)h(n-m)$. Thus, the response to $x(n)$ must be

$$y(n) = \sum_{m=-\infty}^{\infty} x(m)h(n-m) \tag{3}$$

which is the desired convolution relation. Equivalently, by a simple change of vari-

Figure 7.2. Discrete-time system

ables, equation (3) can be written in the form

$$y(n) = \sum_{m=-\infty}^{\infty} h(m)x(n-m). \tag{4}$$

Thus, in the case of LTI systems, the sequence $h(n)$ completely characterizes the system as shown in Figure 7.3. A graphical representation of the convolution process (equations (2) and (3)) is given in Figure 7.4.

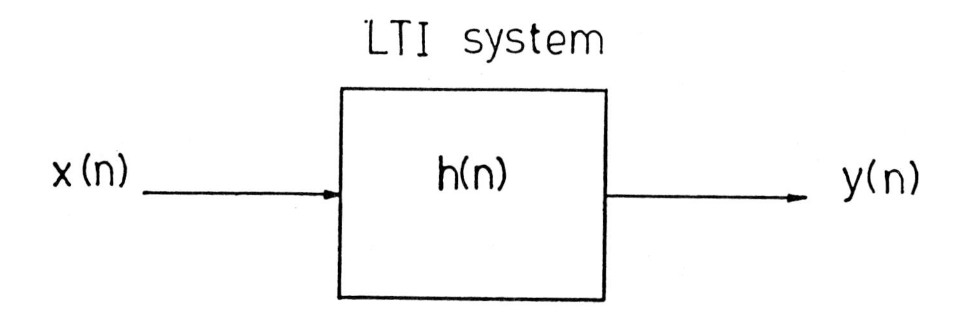

Figure 7.3. Discrete-time LTI system

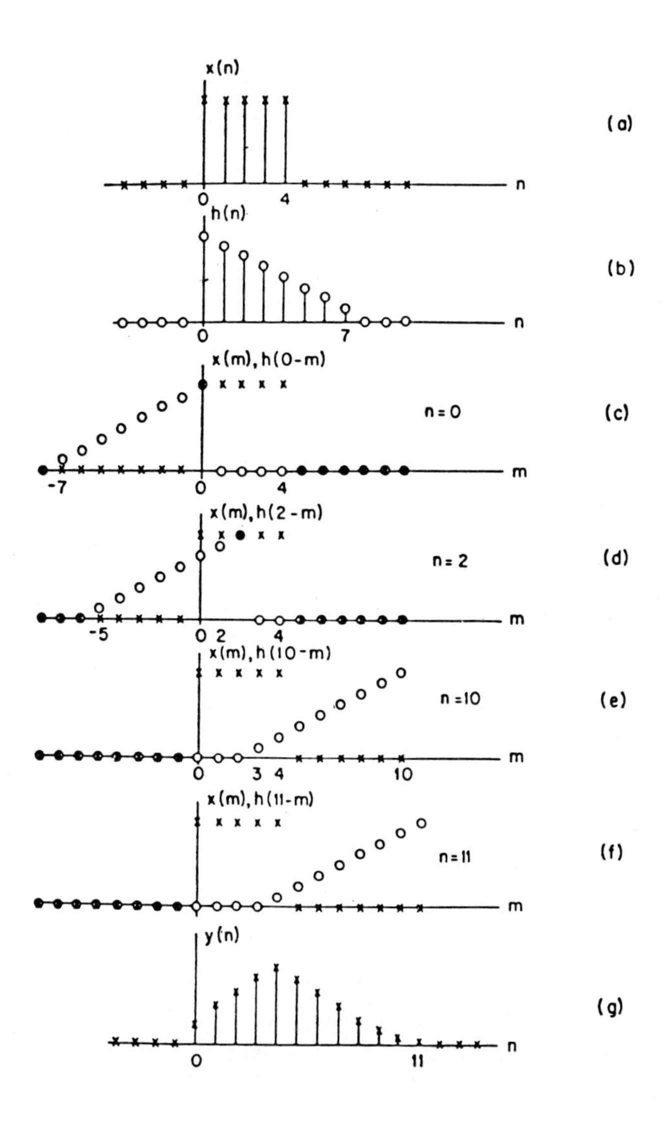

Figure 7.4. Graphical representation of a convolution process

A major sub-class of linear time-invariant discrete systems are those where the input and output sequences $x(n)$ and $y(n)$ are related via a constant coefficient linear difference equation. The difference equation representation of LTI systems is extremely important because it often offers insights into efficient ways of realizing these systems. Furthermore, from the appropriate difference equations one can learn a great deal about the characteristics of the specific system under investigation, e.g., the natural frequencies of the system and their multiplicity, the order of the system, the frequencies at which there is zero transmission, etc.

The most general expression for an Mth-order linear, constant coefficient, difference equation of a causal system is of the form (an LTI system is said to be causal or realizable if the output at $n = n_0$ is dependent only on values of the input for $n \leq n_0$)

$$y(n) = \sum_{i=0}^{M} b_i x(n-i) - \sum_{i=1}^{M} a_i y(n-i) \tag{5}$$

where M is the order of the system and $\{b_i\}$ and $\{a_i\}$ characterise the system with $n \geq 0$ and a_M is non-zero.

Equation (5) is in a useful form for solution by direct substitution. Given the set of initial conditions and the input sequence $x(n)$, the output sequence $y(n)$ for $n \geq 0$ may be computed directly from equation (5). For closed form solution, one may obtain two sets of solutions to the difference equation, a homogeneous solution and a particular solution. The homogeneous solution is obtained by setting terms involving the input $x(n)$ to zero and finding outputs that are possible with zero inputs. It is this class of solution that essentially characterizes the specific system. The particular solution is obtained by guessing a sequence $y(n)$ that would be obtained with the given input sequence $x(n)$. The initial conditions are used to determine the arbitrary coefficients in the homogeneous solution. Hence the importance of difference equation in DSP is its realizable simplicity. The most general first and second order difference equations are given below:

$$y(n) = -a_1 y(n-1) + b_0 x(n) + b_1 x(n-1), \tag{6}$$

$$y(n) = -a_1 y(n-1) - a_2 y(n-2) + b_0 x(n) + b_1 x(n-1) + b_2 x(n-2). \tag{7}$$

Another important aspect of equations (6) and (7) is that higher-order systems can be decomposed into a cascade or a parallel combination of first and second order systems, indicating the possibility of realising the hardware system in modular form and employing multiply and add operations in each module.

An important technique in studying and realising discrete-time systems as described above is the z-transform, where z is a complex number. The importance of z-transform is analogous to the Fourier transform in continuous-time systems. It offers a powerful means to determine the transfer function of a LTI system, and often gives insight into the realization possibility. For example, a LTI system can be realized as in Figure 7.5 directly from the transformed expression given by equation (8).

$$Y(z) = [\sum_{i=0}^{M} b_i z^{-k}]X(z) - [\sum_{i=1}^{M} a_i z^{-k}]Y(z) \tag{8}$$

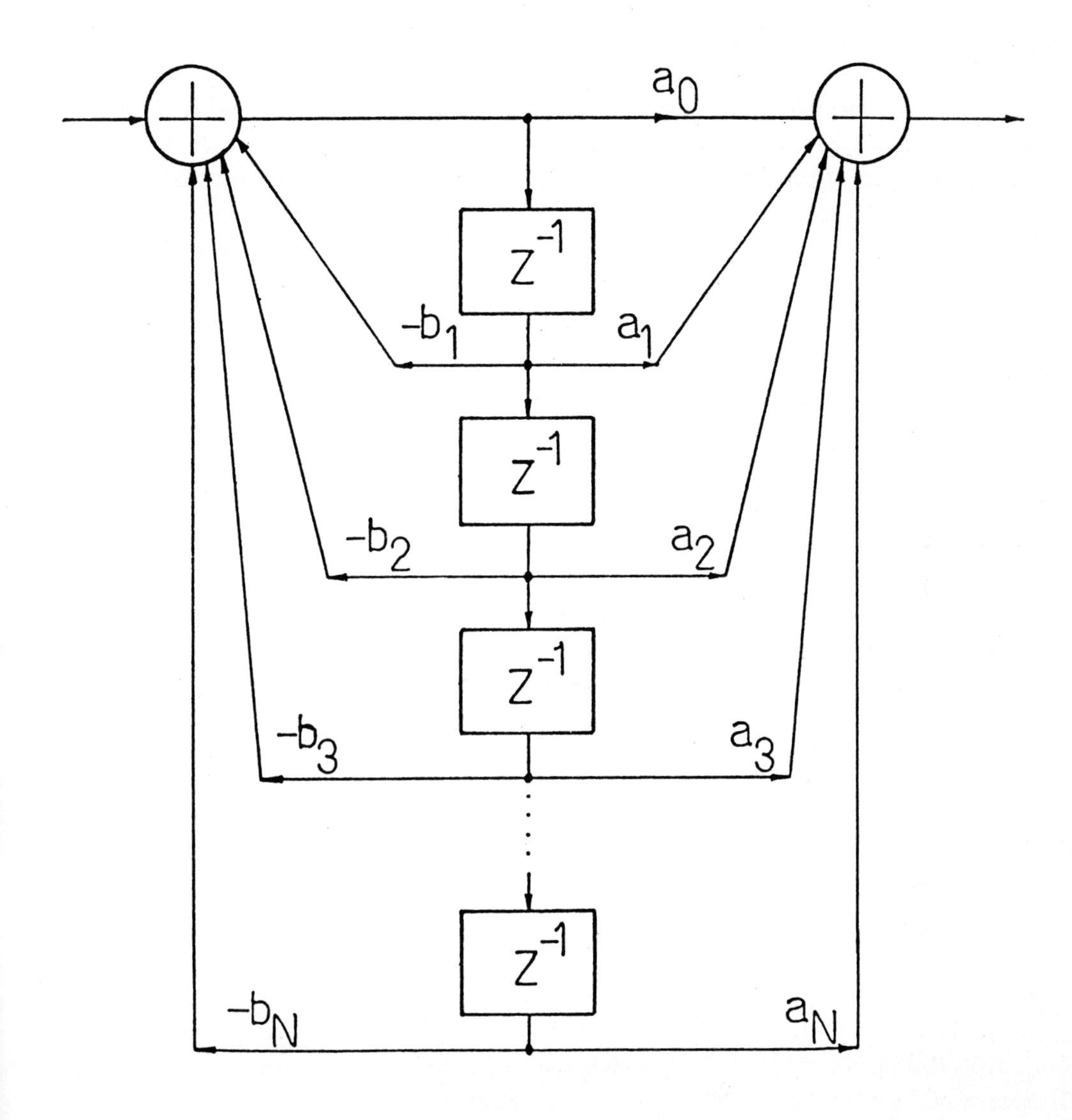

Figure 7.5. Direct realization of LTI system

where $Y(z)$ and $X(z)$ are the z-transform of $\{y(n)\}$ and $\{x(n-i)\}$ respectively. This encompasses a multitude of DSP algorithms ranging from recursive and non-recursive filters (low-pass, high-pass, bandpass and bandstop) to spectrum analysis on the z-plane (Oppenheim & Schafer [1975], Rabiner & Gold [1975], Peled & Liu [1976], Chen [1984]). The parallel nature of these algorithms indicates a great scope for VLSI implementation (Allen [1985], Kung [1987]).

2.2 Two-dimensional signal processing

Signals acquired from photographic inputs such as CCD camera, radar, medical X-ray scanner, tomographic machine, etc are inherently two-dimensional. Although the majority of the one-dimensional signal processing algorithms are still applicable in two dimensions, it is more appropriate to use two-dimensional concepts and systems, especially some of the important one-dimensional concepts are not readily expandable to two dimensions. Let $x(n_1, n_2)$, $y(n_1, n_2)$ and $h(m_1, m_2)$ be the two-dimensional input, output and response sequences where n_1 and n_2 are integer variables. The convolution relation given in equation (3) for linear time-invariant systems may be written as equation (9) for two-dimensional sequence:

$$y(n_1, n_2) = \sum_{m_1=-\infty}^{\infty} \sum_{m_2=-\infty}^{\infty} h(m_1, m_2)x(n_1 - m_1, n_2 - m_2). \tag{9}$$

As in the one-dimensional case, linear time-invariant two-dimensional filters can often be described by a constant coefficient, linear difference equation relating the output of the filter to its input. The most general form of the difference equation for a realisable filter is given by

$$\sum_{m_1=0}^{M_1} \sum_{m_2=0}^{M_2} \alpha_{m_1,m_2} y(n_1 - m_1, n_2 - m_2) = \sum_{l_1=0}^{L_1} \sum_{l_2=0}^{L_2} \beta_{l_1,l_2} x(n_1 - l_1, n_2 - l_2) \tag{10}$$

where $\{\alpha_{m_1,m_2}\}$ and $\{\beta_{l_1,l_2}\}$ are the set of constant coefficients that characterize the particular filter. Given a set of initial conditions, equation (10) can be written into a recursive form and solved by techniques similar to the one-dimensional case.

Some commonly used convolution relations include image convolution, correlation, averaging, contour finding and tracing, compression and coding. From a computational mathematical point of view, these expressions are simply the representation of matrix multiplication in general. This highlights the fact that discrete-time LTI systems have a unifying influence on the entire field of DSP (Oppenheim & Schafer [1975], Rabiner & Gold [1975]), which also indicates that an autonomous design system would be feasible under this well-defined environment.

3 SIGNAL PROCESSING SYSTEM MODELLING

3.1 VLSI signal processing system

The basic characteristics of VLSI systems can be simply viewed as cost, performance, power dissipation and complexity. Normally the cost of constructing a VLSI system includes the research and development cost, the man-hours spent on designing it, the fabrication/packaging cost, sales and marketing, etc. All these are related to one common criterion of the system – the size of the chip. Therefore it is possible to estimate the cost of the target VLSI system by simply considering the size of the system on silicon. The performance of the system is defined as the time taken between the input signals entering the system and the output signals leaving the system. And the power dissipation of the system is defined as both the static and dynamic power consumed by the whole system. Thus the complexity of the system is defined as the product of the cost and performance of the system. Mathematically, if A is the area of the chip, and T is the time for a signal to propagate from the input to the output of the chip, then the complexity of the chip is defined as

$$Complexity = (A/A_0)(T/T_0)^{2\alpha} \tag{11}$$

where α is either 0 or 1 and A_0 and T_0 are positive constants which depend on the technology (Brent & Kung [1979]).

All of these have been treated formally by many researchers (e.g., Thompson [1980]) at different degrees, some are more rigorous than others. In many cases, the complexity of VLSI systems overwhelms all other problems in I. C. design and introduces enormous difficulties in creating and debugging new designs. There are however three key considerations in the efficient management of the complexity of VLSI systems. The architectural issues of decomposition and regularity were discussed in Yung & Allen [1986a] and Yung [1985], leaving the third issue of automatic synthesis to be treated here. As indicated in Yung & Allen [1986a] and Yung [1985], future VLSI systems will be highly concurrent functionally and structurally with complexity decomposition and regularity as their main features. To deal with this concurrency at the layout level could be highly demanding, and since correctness-by-construction is necessary, an autonomous design system is need. There are many approaches to silicon compilation, some begin with a functional description, and others with a structural description. So the choice of silicon compiler is still very much in the hand of the designer, which as discussed in section 1.3, should be autonomous as well. This emphasises the reason for using signal modelling – the concept of designing a VLSI signal processing system from specifications, not from algorithms.

Some off-the-shelf examples of VLSI signal processors are INMOS IMSA100 (INMOS [1987]), Motorola DSP56000 (Kloher [1986]), NEC upd77230 (NEC [1986]), and Texas TMS320C25 (Texas [1986]). These devices showed a high degree of functional decomposition and architectural regularity, indicating the use of advanced CAD tools or silicon compilors in their development. Furthermore, the combination of VLSI and advanced DSP architectures offers outstanding computational precision, fast instruction cycle time and application versatility (Aliphas & Feldman [1987]).

3.2 Synthesis flow chart overview

The principle of signal modelling is to drive the synthesis of dedicated VLSI signal processors by the application specifications and the signal features of the desired system rather than some general algorithms or operations which may not closely match the required speed performance under a certain cost level, power consumption or volume limitation.

As shown in Figure 7.6, the approach of establishing a system signal model from the target application, signal parameters and information content identifies the appropriate processing algorithms for the specific types of applications and input conditions. Once the algorithm(s) is identified detailed classification of the arithmetics used can also be carried out. The operations involved in these algorithms are mostly standard arithmetic and logical operators (Section 2). But there are specific operations required for differentiation, integration and interpolation, etc. However all these operations are considered as the composition of the structure of the VLSI processor. The flow of the signals through the structure describes the interconnection pattern of the structure. Thus the processor architecture can be determined by detailed knowledge of the flow of signals through the structure and the composition of the structure. At this point, complexity analysis can be performed on the prospective architecture which would then reflect the cost-performance, the technology constraints and the application specification of the system. This may eventually indicate a modification of the algorithms, the signal flow graphs or the architectures. Optimization factors may be tightened or loosened in order to meet the initial specification. When the synthesis system or the designer interacting with the system is satisfied with the optimization procedure and the architecture, hardware implementation is the next step. This process involves the use of a rule-based floor-planner which will layout the processing system according to the data flow graph. Any physical impossibilities at this stage will result in modifying the data flow graph again to suit the format of the floor-planner. If a decision between VLSI masking and PCB production is necessary then this will be determined by the eventual complexity of the system.

3.3 Modelling procedure details

3.3.1 Signal parameters and information content In the design of VLSI signal processors, process operation is always considered the major issue, determining the overall structure of the final system. However one area which has not received as much attention as it should, is that of the signal parameters, which are directly related to the applications. Basically governed by the application itself, the study of signal parameters includes the understanding of the signal source, the nature of the noise, signal to noise ratio, dynamic range, transmission medium, frequency, transducer

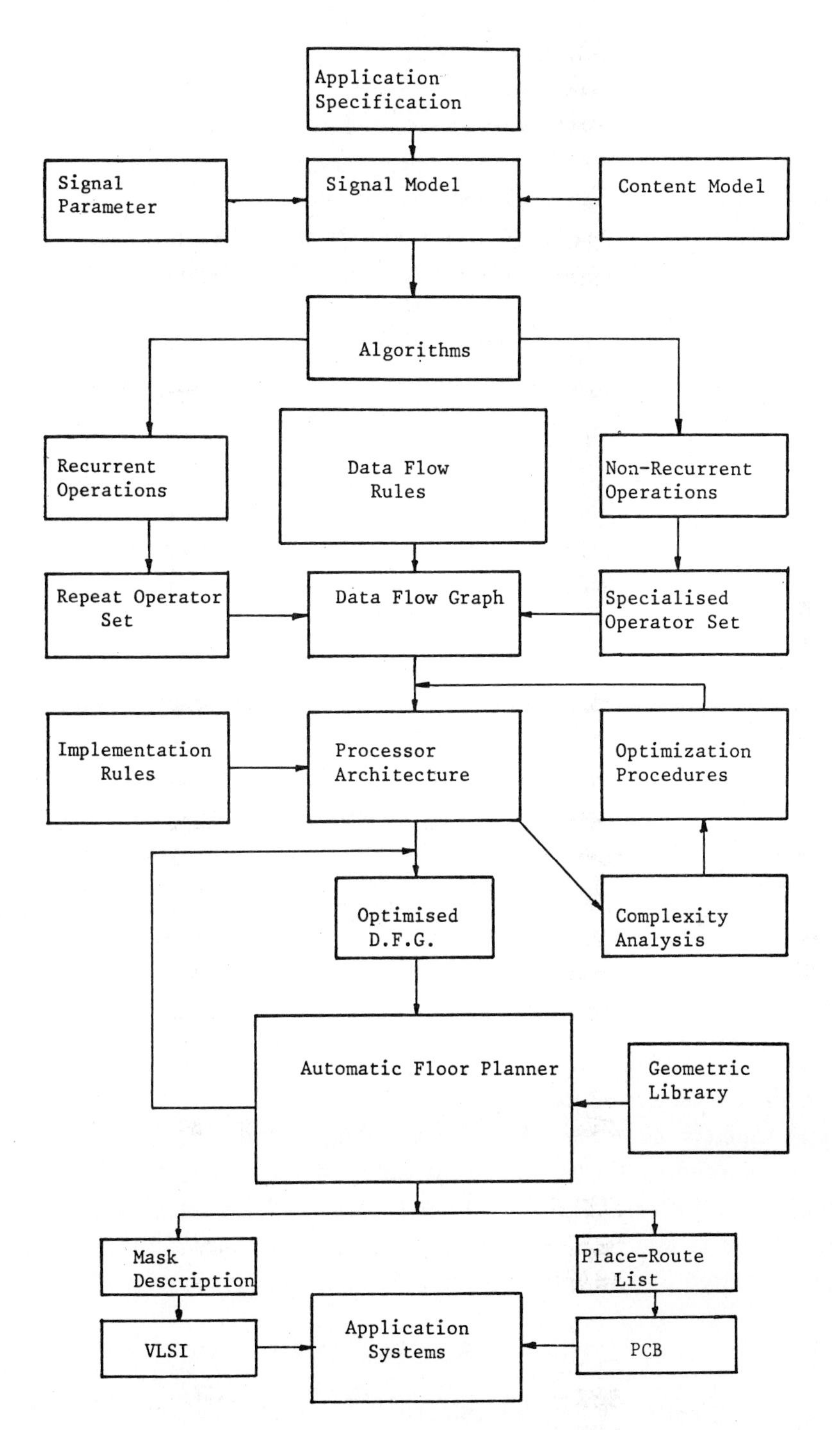

Figure 7.6. Signal processing synthesis flow-chart

characteristics, sampling rate and quantization.

For example, the signal parameters anticipated in audio communication channels normally have 3 kHz source, Gaussian noise, low SNR and 8 kHz sampling rate. Research in the past has indicated that the technique of 'inverse filtering' or linear prediction is appropriate for formant extraction. However there are two different ways of defining the correlation which in turn leads to two different methods of solution for the set of predictor coefficients. The auto-correlation method minimises the mean squared error over the infinite interval with the assumption that the speech signal is zero outside a finite interval as it would be in practice. The predictor coefficient matrix in this case is Toeplitz in nature and may be solved by methods such as Gaussian reduction and elimination, or the Levinson and Durbin recursive procedure. The covariance method minimises the mean squared error over a fixed (finite) interval. The predictor coefficient matrix in this case is symmetric but not Toeplitz, and has the properties of a covariance matrix. Therefore the signal parameters determine which method is best to extract the formant (Witten [1982], Ayer [1987]).

In the case of two-dimensional image processing, the input image may suffer from sampling and quantization error, imaging contortion, and other errors due to inaccurate camera calibration. Quality enhancement of the image is essentially a problem-oriented process. For example, a method that is useful for enhancing X-ray images may not necessarily be the best method for enhancing images of Mars transmitted by a space probe. Thus the input image determines whether a Fourier transform of the image is appropriate (frequency-domain methods), or a direct manipulation of the pixels is required (spatial-domain methods). Some of the signal types are illustrated in Figure 7.7 (Adams & Yung [1987]).

In theory, the application system may be considered as a 'black box' with p input terminals and q output terminals. If the inputs and outputs are denoted by a $p \times 1$ and a $q \times 1$ column vector respectively, the input function $\mathbf{f}(n)$ may be denoted by the value of $\mathbf{f}$ at time $t = nT$ (where T is the sampling period), and the output function $\mathbf{g}(n)$ may be denoted by the value of $\mathbf{g}$ at time $t = nT$, where $\mathbf{f}$ is the $p \times 1$ vector and $\mathbf{g}$ is the $q \times 1$ vector defined over $[0, \infty)$. Hence, if the system is assumed to be relaxed (a system is said to be relaxed at time t_0 if and only if the output is solely and uniquely excited by the input over $[t_0, \infty)$), it is legitimate to write

$$\mathbf{g} = \mathrm{H}\mathbf{f} \tag{12}$$

where H is some operator or function (algorithm) that specifies uniquely the ouput $\mathbf{g}$ in terms of the input $\mathbf{f}$ of the system. However, if the system is not initially relaxed, then $\mathbf{g}$ in equation (12) depends not only on $\mathbf{f}$ but also on the initial conditions at t_0.

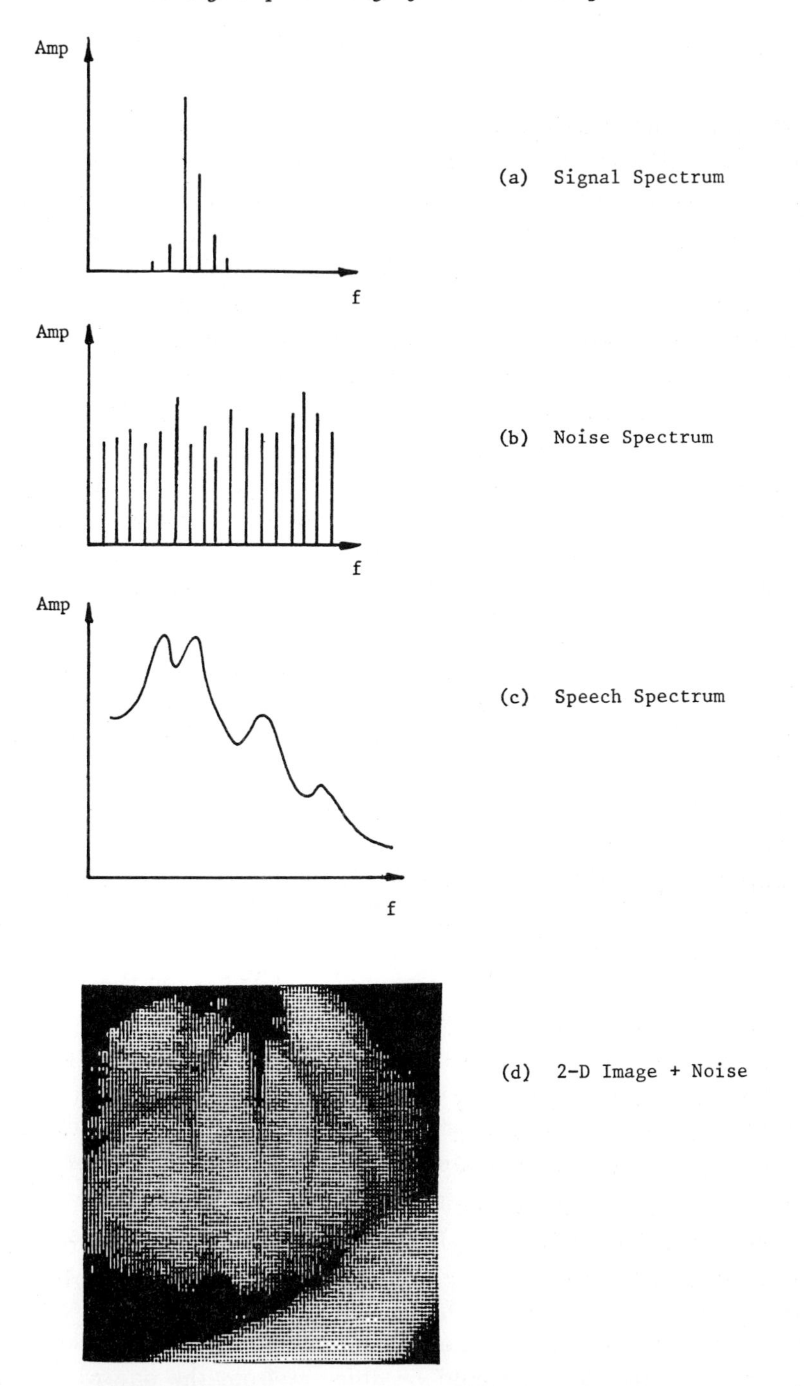

Figure 7.7. Signal parameter types (a) signal spectrum; (b) noise spectrum; (c) speech spectrum; (d) two-dimensional image and noise

The initial conditions at t_0 are called the *state*. Hence the state at t_0 is the information that together with the input, determines uniquely the output. In linear system theory, the state of a system can be represented by a finite-dimensional column vector $\mathbf{x}$, called the state vector where its components are called state variables. As the input function may be represented by a mathematical equation (Figure 7.1) which may be a composite expression of a real input condition and the state may be determined from the information content, the output function is determinable if H is known. In this case, H is the set of equations that describes the unique relations between the input, output, and state (equation 13):

$$\mathbf{x}(\mathrm{n}) = \mathbf{u}(\mathbf{x}(\mathrm{n}), \mathbf{f}(\mathrm{n}), \mathrm{n}), \qquad (13a)$$
$$\mathbf{g}(\mathrm{n}) = \mathbf{v}(\mathbf{x}(\mathrm{n}), \mathbf{f}(\mathrm{n}), \mathrm{n}), \qquad (13b)$$

where $\mathbf{x}$ is the state, $\mathbf{g}$ is the output, and $\mathbf{f}$ is the input [12]. There is no loss of generality in using equation (13b) to describe the output of the system because by the definition of the state, the knowledge of $\mathbf{x}(n)$ and $\mathbf{f}(n)$ suffices to determine $\mathbf{g}(n)$. Furthermore, different input conditions imply different $\mathbf{u}$ and $\mathbf{v}$ if the same output function is to be expected. This identifies the significance of understanding the input signal features in deriving the transfer function to acheive a defined output. Detailed discussion of this is given in the following section.

3.3.2 Algorithm specification The next important step in the procedure is to derive the mathematical description of the system. This often ends up with different equations to describe the same system when both the input-output description and the state-variable description are studied (see the previous section).

Let us consider equation (5) given in Section 2. A LTI system is essentially characterised by the coefficients of this Mth-order difference equation, which are determined by the homogeneous solution of the equation and the initial conditions of the system. Hence the output $y(n)$ of the system is determined by the input $x(n-i)$ and the previous history of the system $(y(n-i))$. For a constant coefficient system, if the input is a composite of the information content and a noise distribution $(x(n-i)+s(n-i))$ as in practice, then the output may deviate from the theoretical expectation due to the noise component. The worst case effect of this component may cause the system to oscillate, therefore failing to achieve the required specification. In order to produce a specified output, the noise component has to be quantified when the arbitrary coefficients of the difference equation are being calculated. In this case, the resultant equation describing the system would be different from the one without considering the noise components. Moreover the nature of the noise would also be different under different physical environments. As a matter of fact, noise from signal paths involving atmospheric conditions are often difficult to predict and control. Under such environment, the coefficients may have to be re-calculated after every change of condition so that the resultant description of the system (difference equation) would always produce a similar performance.

In other cases, for example, feature extraction in robot vision comprise a sequence of sub-processes describing the complete process. Under a well lit assembly environment, the features of a workpiece under the vision camera may be extracted after edge detection, dilation, erosion, and object measurements. However, if the visual input is from a badly lit and noisy environment, extra sub-processes such as noise cleaning, histogramming, and grey scale transformation are required to improve the noisy and under-exposed image in addition to the usual sequence of sub-processes. Therefore, in terms of finding the complete set of equations characterising the system, the input signal parameters and conditions play an important role, and should be clearly identified before the set of equations is selected.

3.3.3 Operation representation The representation of operations in the system description perhaps may be studied at two levels. The first level represents the primary operators which are standard arithmetic operators $(+, -, *, /)$, and standard logical operators (and, or, complement, shift, rotate, compare). In the case of arithmetic operators, the number system used and the methodology employed determine the accuracy, dynamic range and stability or the whole system. The choice of number system is basically restricted by the hardware register length and the potential numeric range of the results. For example, the floating point system is often used to cope with algorithms which are iterative on large numeric values. The choice of methodology is a matter of the performance specification. It may be broken down into greater details such as and/or gates and full adders.

The second level represents the composite primary operators which may be simply $A+B$ or the more complicated equations (6) and (7). In general it may be divided into the *recurrent* type and *non-recurrent* type. Most of the algorithms in DSP comprise of recurrent operations involving recursions or iterations; for example, a discrete-time LTI system is in itself recursive (equation (5)). However a very small number of the operations are non-recurrent. The required operator set is specialised; for example, differentiator, integrator, logarithmic operator, histogrammer, etc, which can also be presented by the primary operators in detail. Since the computation nodes in a DFG are a direct realization of the operations, the nature of the operators determines the composition of the DFG.

3.3.4 Data flow concept and processor architecture Once the system is decomposed into a set of operators, a simple data flow concept may be used to further analyse the system, and from there, to identify possible realizable architecture. This concept essentially converts operators into 'computational nodes', and data communications into 'computational paths'. The resultant graphical representation enables the designer to visualise the flow of signals which is directly related to the structure and behaviour of the system. Similar to logic diagrams, this representation allows the study of observability and controllability of the system, in addition to critical paths and race round conditions. Problematic structures and unreliable behaviours may thus be identified and eliminated easily, giving a higher chance of success. It also indicates whether the system is physically realizable, and whether there exist possibilities for optimization. The transfer from a conceptual and mathematical model to a physical structure is a major step and will be further investigated in Section 4. Figure 7.8 shows two examples of data flow graphs.

3.3.5 Complexity analysis It is becoming a normal code of practice in system design to estimate the cost and performance of the system before committing the design to hardware prototyping. There are two objectives in this. The first is to extract the technical details of the system, and the second is to optimise the realisable structure to a better cost-performance figure. The complexity of the system thus represents the cost – the silicon area occupied by the VLSI processor; and the performance – the longest time taken for an input signal to propagate across the entire system to the output. The analysis of these two issues will therefore give the designer a good picture of how fast and expensive the system is going to be. As long as the initial specification is met within the cost constraint; then optimization analysis can be used to improve the cost-performance figure.

Complexity analysis as described by Thompson [1980] and many other researchers uses the upper bound (or lower bound) concept to define the cost and performance figure. Unfortunately the asymptotic results of complexity theory do not always carry much practical importance. In practise, this big 'O' theory is not necessarily practical. Therefore the resulting cost-performance figure may deviate from the actual measured figure (actual cost spent and experimental performance) if only bounds are considered. It would be more accurate to consider the complete mathematical expression instead of ignoring the proportionality constant. This will be illustrated in Section 6 by an example.

3.3.6 From paper design to silicon Silicon Compilation is now a familar term, and has appeared in many publications in the CAD area in the last few years (Denyer & Renshaw [1983], Southard [1983]). Design Automation has been advocated for many years but its achievement has been made possible only by the power of VLSI processors, thus allowing even more powerful VLSI processors to be designed in shorter

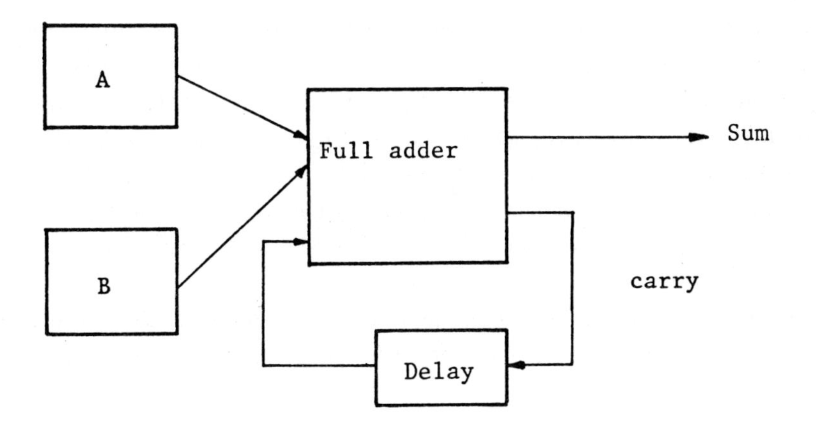

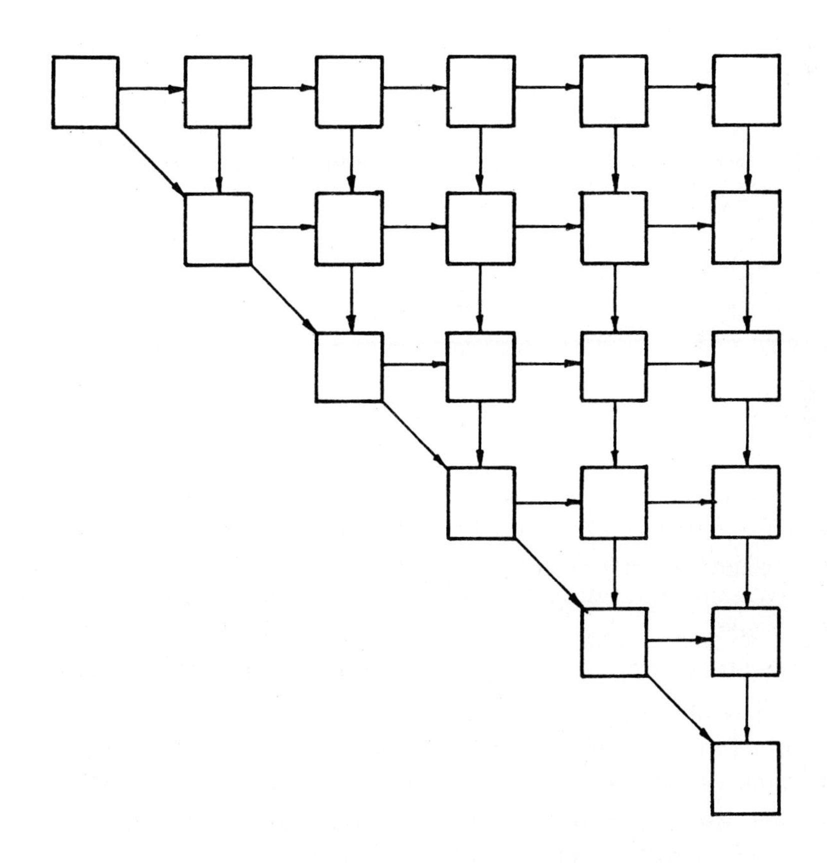

Figure 7.8. Data flow graphs

time with less effort. Although many tasks in CAD can be handled smoothly by software tools such as simulators, layout editors, circuit extractors and other more specific and sophisticated packages, a considerable amount of human interaction is still needed. If human interaction is to be reduced to a minimum, the process of VLSI implementation – transferring paper design to silicon – still requires much attention. A brief attempt is made in Section 5 to try to identify the requirements of such tool, and its possible future development.

4 DATA FLOW REPRESENTATION

4.1 Data flow analysis

Data flow representation was first proposed by Keller *et al.* [1980], and since then it has been followed up by many other reseachers (e.g., Kung [1987] and Bowen & Brown [1985]). The concept of data flow is of representing computational elements, logical elements and delay elements by means of distinct nodes. Each node identifies part of the functional behaviour of the whole system. Data movement is indicated by interconnecting paths between nodes with arrow-heads showing the direction of movement. Hence the complete functionality of a system may be decomposed into managable functional nodes interconnected by data movement paths. With the knowledge of the structural and physical behaviours of the nodes, the implementation of the data flow representation may be a direct realisable architecture. The beauty of the data flow approach is that it bears a natural relationship to the behaviour of digital systems and allows rigorous design verification at a high level of abstraction. Figure 7.9 illustrates the idea of data flow: the computation node is represented by a box with a label indicating the function of the node, and the arrows connecting the boxes represent the direction of data movement.

Therefore the main aim of *data flow analysis* is to create a consistent representation of the hierarchical decomposition of each system node and level in terms of the logical flow of data elements between logical processing functions. The purpose of this is to allow the overall complexity of logical system design to be handled in tractable stages. This is of course an important criterior in managing VLSI systems, especially in assisting the designer to design effectively. The complexity of the system may also be quantified and optimised. Moreover, such logical layout of data flow also implies lesser efforts in generating test patterns for fault identification. After the logical data flow requirements are established, the control flow of the system should also be considered in the same way so as to obtain a detailed insight into the system complexity. In general DFGs provide a mechanism to support general decomposition of major system functions and the data movement between them. This directed graph methodology also provides a basis for the initial quantitative analysis of performance

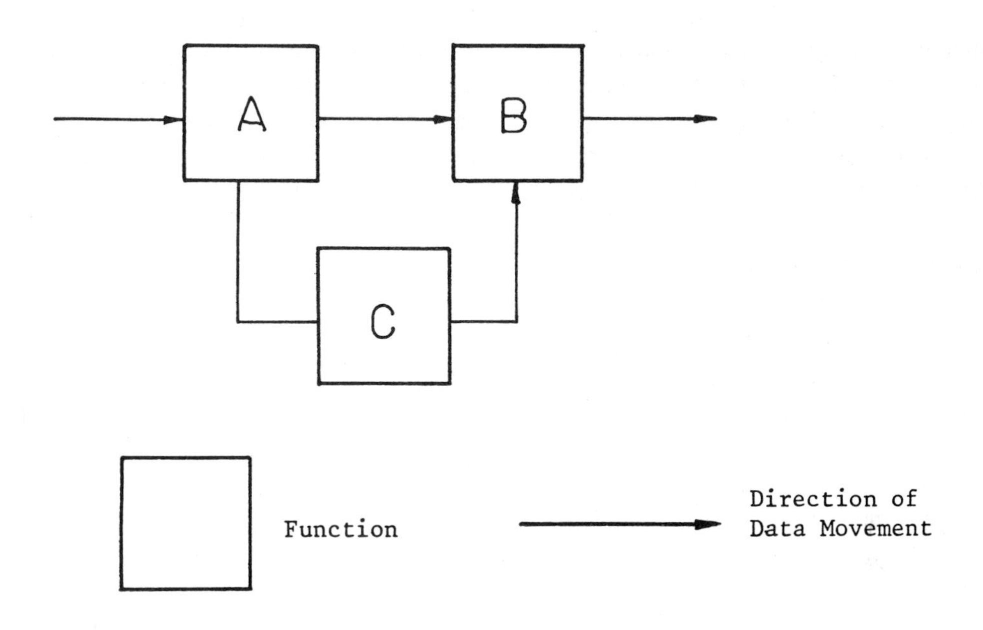

Figure 7.9. Illustration of data flow concept

requirements, and the derivation of initial system loading parameters. There are four system parameters which can be identified on data flow graphs:

(i) data transfer rate (flow rate through the DFG);
(ii) function execution time budgets (computation rate plus flow rate);
(iii) computational loading estimates (operations/iteration/node/time budget);
(iv) memory requirement for the storage of input/output data.

The data transfer rate is calculated by assuming a zero delay in all the nodes, and a unit delay along a path. This rate is proportional to the number of path sections on the longest data path. In practice, each processing node contributes a certain delay, according to the function of the node, to the overall execution time of the system. This is called the function execution time budget – the sum of the computation rate through nodes and the flow rate through paths. From this time budget (the system throughput), the computational loading at each node may be estimated, and the overall distribution of computational loading may also be analysed. This is useful to know if overloading or uneven loading are to be avoided. Perhaps the implicit element in DSP system design is the memory requirement. Although memory ICs are inexpensive, powerful and dense nowadays, the amount of storage space for input/output data has always been overlooked. The data flow method gives the designer a fuller picture of the implementation commitment, and detailed issues as described above may be studied in depth separately. Hence, and initial estimate of the system bandwidth, latency and best/worst case delay may be deduced from these parameters. It may also be possible to identify race conditions, and unobservable or uncontrollable nodes in the system – such conditions may reduce the testability of the system.

The *data flow description* of a system basically consists of two parts: a hierarchical set of data flow graphs and a corresponding data dictionary. The set of DFGs represents both the logical and control flow of the system, whereas the data dictionary serves as the record of house keeping such as node details, node order, connectivity, edge directions, input data nature, etc. Figure 7.10 shows the use of data flow graphs to represent the functional decomposition of an FFT butterfly operation. Data inputs are represented by rectangular boxes labelled by the data values such as the real component of the absolute value of X. It should be noted that these are not nodes. Each computational node is identified by a label $(a.b.c)$ and the function of the node (e.g., real X means multiplication on real numbers).

Some basic principles for creating data flow graphs are listed below (Bowen & Brown [1985]).

(i) First, identify the data elements and data structures, then identify the processing functions.
(ii) Each node should represent a well defined function with precisely defined input and output data structures.
(iii) Explicit data storage nodes may be included and are usually desirable when the data being stored are to be accessed by more than one function.
(iv) Every node and path of the data flow graph should be concisely and unambiguously labelled on the data flow graph.
(v) Data flow graphs should have data flowing in one direction only on each path.
(vi) Number each node to show its location in the hierarchical decomposition of the system.

4.2 Partitioning of algorithms and DFGs

If a VLSI signal processing system is to be represented by a hierarchical set of DFGs, then one must develop a systematic way of partitioning an algorithm so as to reduce it into a set of sub-algorithms in such a way that each of these sub-algorithms could be represented by a physically realisable DFG. Theoretically, this set of DFGs can

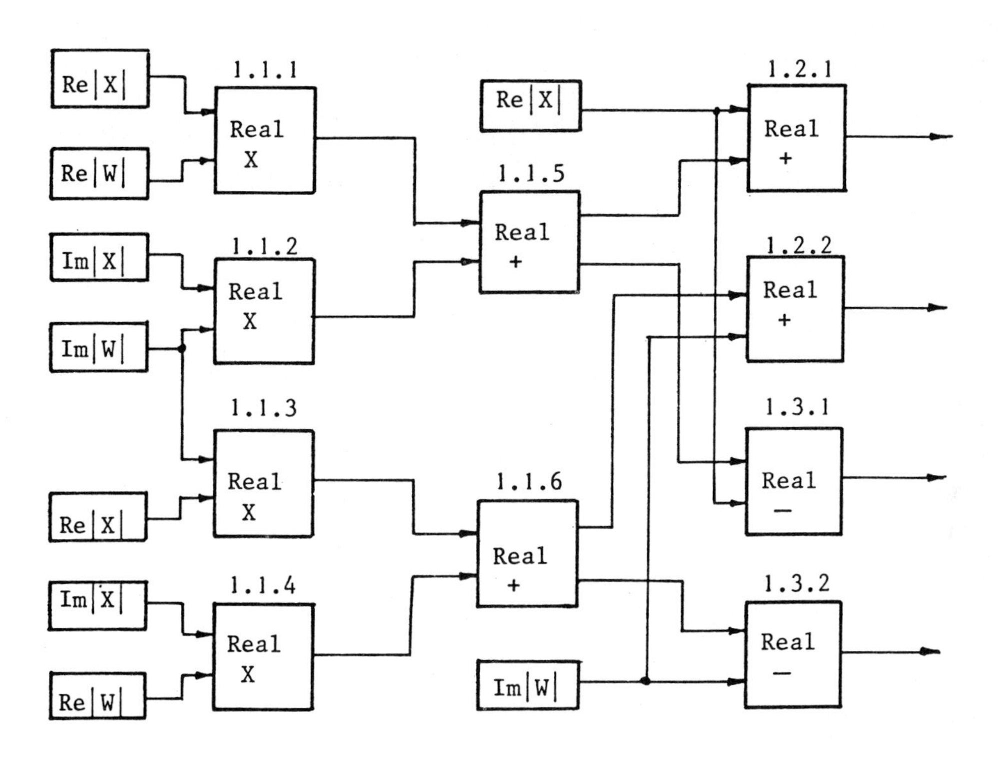

Figure 7.10. An FFT data flow graph

be implemented on an array architecture in which the number of processors is equal to the number of computation nodes. The other extreme is to use one processor serially by mapping all the computation nodes on it. But it is often more practical to have a multiplexed architecture which has fewer processors than the number of computation nodes, so that some of the computations can be performed in parallel but not all (see Figure 7.11). In an architecture of this kind, the partitioning of the algorithm should lead to a hierarchical set of DFGs which represents the logical, data and control flow of the system, and at the same time, can be mapped onto the size of the architecture. By selecting the number of processors in the architecture, this approach allows a higher degree of flexibility to adjust to a certain cost–performance level. In the case of signal processing algorithms as described in Sections 2 and 3, the convolution process is essentially based on multiplication and addition in an iterative way. This feature makes the mapping of a DFG onto a smaller architecture easier. However it is not sure that the same principle will apply to more general algorithms.

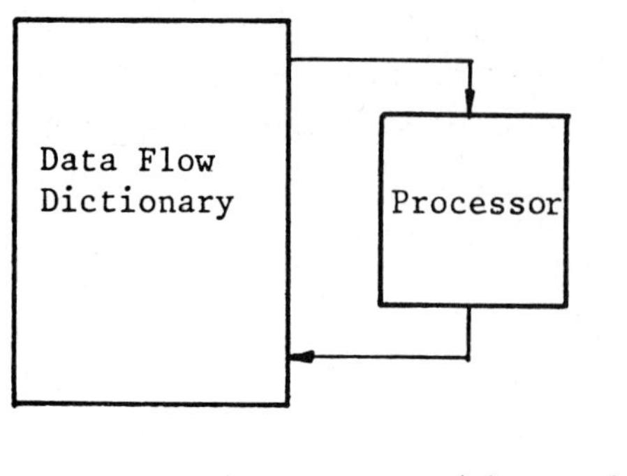

(a) Serial

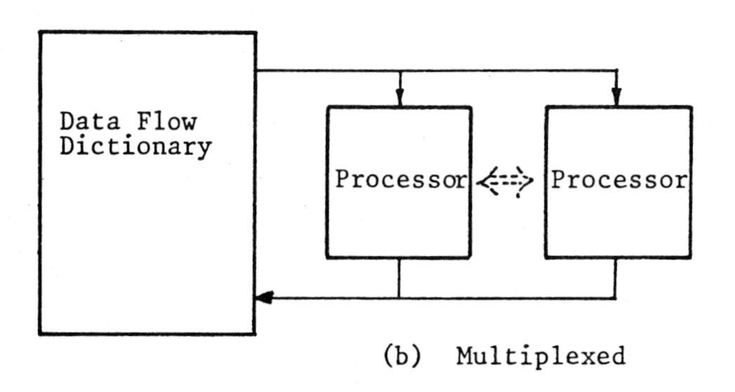

(b) Multiplexed

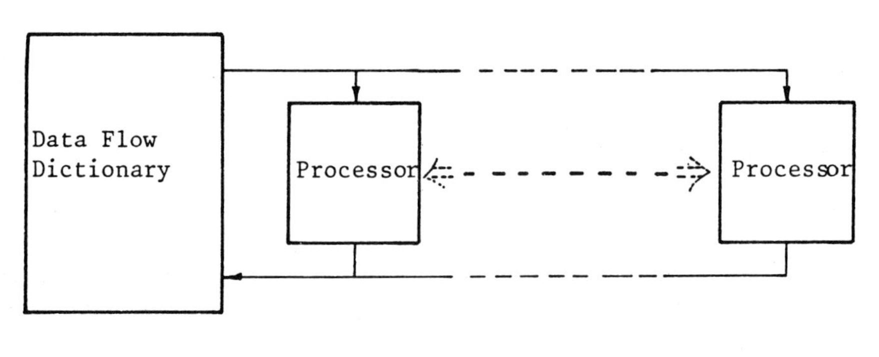

(c) Array

Figure 7.11. Some processor architectures: (a) serial; (b) multiplexed; (c) array

A few points should be noted and closely observed in partitioning the algorithm and DFG. Firstly, the accuracy of computation should not be affected by the partitioning process, i.e., the process should not reduce the number of nodes in the graph resulting in a high percentage of error in the computation. Secondly, no additional delays should be caused by the partitioning process. So the partitioning process should not introduce extra nodes in the graph. Furthermore, the insertion of extra buffers to ensure synchronization should also be avoided. Thirdly, the process should not require any increase in the complexity of the VLSI signal processor, and the number of overheads in external hardware and external communication caused by partitioning should be as small as possible. This should be studied carefully because it is often true that when a multiplexed architecture is used, critical sequencing of data with feedback paths may become dominant, causing an increase in interconnection complexity and extra overheads to deal with the sequencing.

4.3 Types of data flow graphs

In general the structure of a graph is determined by the algorithm, which does not necessary conform to a particular type of structure. However in signal processing, the well-defined mathematical background does indicate that most structures in this field can be represented by a combination of linear graphs and multiplexed graphs. This statement is illustrated in Figure 7.12, showing how a DFT sequence may be interpreted by a few different forms and bounded by a totally linear graph and a totally multiplexed graph. This knowledge also enables the designer to decide upon the architecture most appropriate for his system. For example, a linear graph presents an array structure whilst the rest present the different degrees of multiplexing and parallelism that are achievable.

5 A DATA FLOW GRAPH SIMULATOR – COSPRO

5.1 Configurable signal processor – the concept

Most signal processors are inherently software polymorphic in nature and the construction of software programs has no limit depending only on the application tasks. Although generality and flexibility are achieved by software polymorphism, system throughput is the price to pay, hence real time signal processing becomes doubly difficult. As there are many algorithmically specialised processing architectures for specific problems, hardware configurability becomes an appealing alternative for parallel signal processing at an economic cost (Snyder [1982], Yalamanchili & Aggarwal [1985]). The constraints of this configurable strategy are firstly, a pool of processing elements should be available for interconnection; and secondly, a limited variation of configurations are allowed so that the change over of configurations can be controlled by a computing machine. Thus the definition of *hardware configurability* is

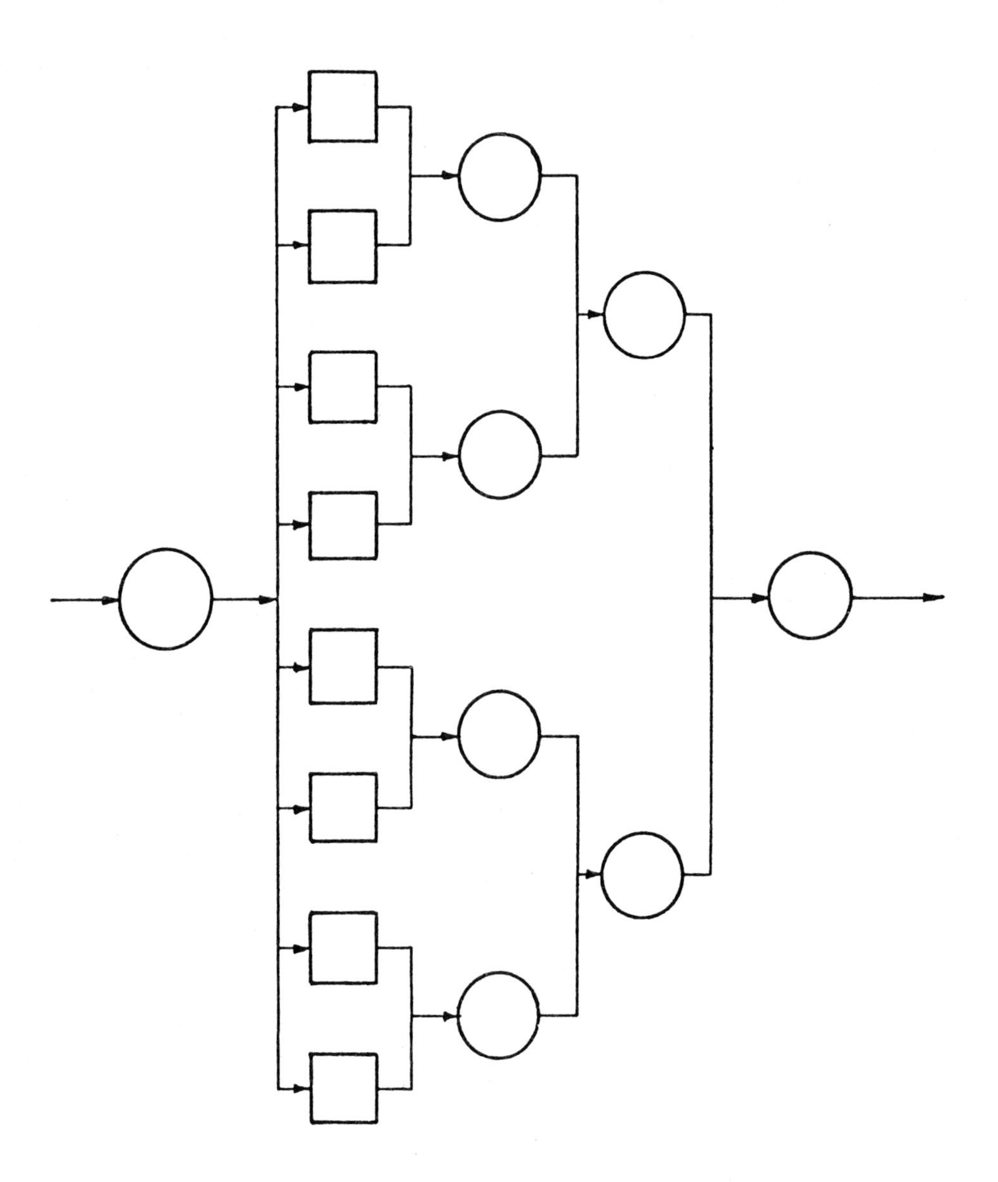

Figure 7.12. Types of DFG: (a) (b) fully multiplexed DFG

on interconnecting the processing elements according to a set of defined structures (e.g., linear arrays, two-dimensional mesh, etc.) centrally or locally controlled by a controller (Snyder [1982], Smith [1987]). This concept is illustrated in Figure 7.13. The pool of processing elements (in circle) is arranged in a mesh structure, with the interconnection pattern being determined locally by the Configuration Control Code (CCC), allowing four different configurations to be implemented on the same mesh. This concept of hardware configurability was implemented in PASCAL on an Amdahl machine.

5.2 System organization

The simulator organization is shown in Figure 7.14 of which a preliminary version is in operation at the University of Newcastle. The core of the simulator is a two dimensional mesh array capable of configurating itself into at least five different types of systolic arrays. The data communication is on a self-timed, data-driven basis of which a handshaking sequence is required to ensure a complete and successful data transfer. A number of support modules exist to assist data entry into the simulator, an output graphic display. The menu-driven input module consists of five levels of commands allowing the user to input signal parameters and construct composite real signals. It also allows user to specify a filter specifications say, and generates a transfer function describing the filter. The root extractor determines the roots of the function and generates a coefficient vector. An output response formatter was written to reformat the output data from the simulator and plot this set of data in time domain or frequency domain on a graphic terminal. Hardcopying and line printing are inherent in the formatter. This ultility enables the user to study both the input and output waveforms in great details including the numeric matrices, the time functions and frequency responses.

A number of modules are under development at the moment. These are the data flow graph generator, the technology library, the rule library, diagnosis output, performance evaluator and the hardware description generator. The data flow graph generator is designed to provide configuration information to the simulator every time a new transfer function is received. The technology library provides the implementation information which will be used by the performance evaluator and the diagnosis module. Similarly, the rule library provides a set of rules on the configuration, the implementation limitations, etc and will be used by the performance evaluator and the diagnosis module. Thus the accuracy of the evaluation will depend on the support of both libraries. The hardware description is expected to be a fully functional and optimised description of the designed system.

5.3 Processing array

The processing array is an 8×8 mesh which can be expanded to 256×256. As shown in Figure 7.15 the tightly coupled array has a complete interconnection topology (i.e., each element has communication paths with each of its 8 neighbours) to allow maximum flexibility. Input/output data to and from the array resides in the

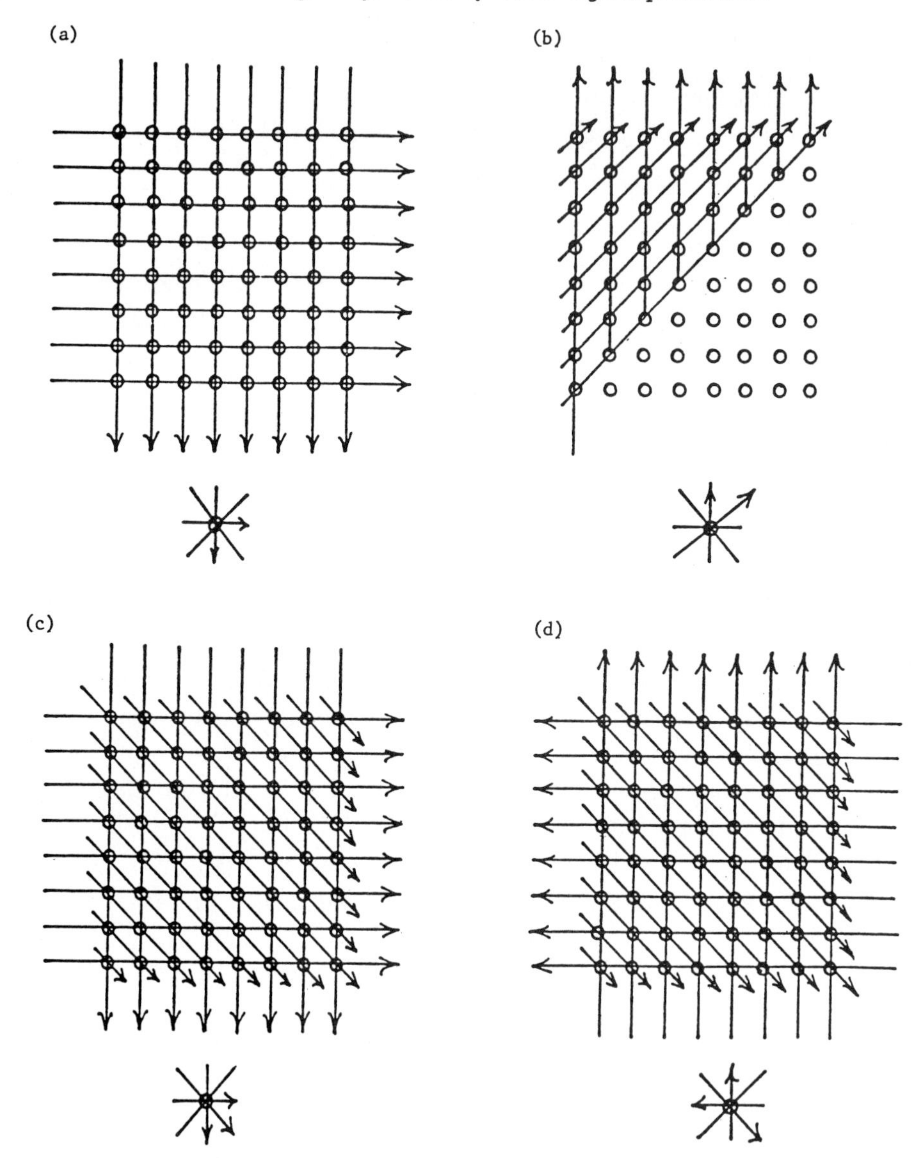

Figure 7.13. Some of the possible configurations and their CCC.

(a) Orthogonal configuration
CCC: 10100000
Application: 2-D convolution

(b) Triangular configuration
CCC: 00000011
Application: matrix triangularisation

(c) Wavefront configuration
CCC: 11100000
Application: matrix multiplication

(d) Hexagonal configuration
CCC: 01001010
Application: L-U decomposition

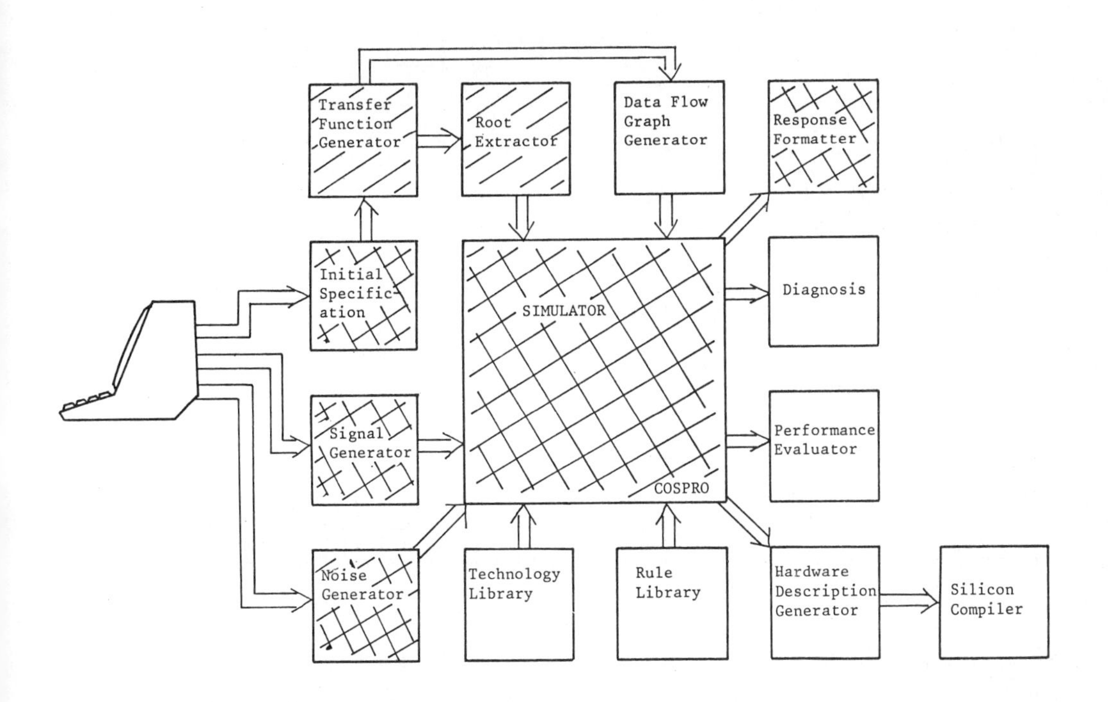

Figure 7.14. A data flow graph simulator

memory pages connected to the *boundary* elements. The processor at the corner (1,1) interfaces to a control processor (not shown) and serves as an initiator of new processes. Subsequent processors are activated upon receiving the required set of data and handshaking signals. Although each element is connected to eight neighbours, not all neighbours are in communication at the same time. The communication is determined by the control word which is propagated just one cycle before the data is sent. This CCC contains the configuration information and sets the communication paths up in each processor. For configurations such as the systolic arrays, the CCC

Figure 7.15. Array organisation – tightly coupled

remain the same throughout. But for other types of arrays, the CCC may need to be updated while it is passed on to the neighbours. The processing element is a simple multiply-and-add arithmetic cell with multiplexors and control logics for handling neighbouring communications and CCC setting up and transfer.

```
PROCESSING ELEMENT;
WHILE ON DO
BEGIN
    REPEAT
        WAIT UNTIL RECEIVE CALL FROM A NEIGHBOUR;
        OBTAIN DATA FROM THAT NEIGHBOUR;
    UNTIL ALL DATA(AIN, BIN AND CIN) IS OBTAINED;
    COBEGIN
        DECODE CCC;
        EFFECT COMMUNICATION PATHS ACCORDINGLY;
    COEND
    COBEGIN
        COUT:= CIN+(AIN*BIN);
    COEND
    REPEAT
        SEND SIGNALS TO CCC SPECIFIED NEIGHBOURS;
    UNTIL RESULTS ARE READ BY ALL SPECIFIED NEIGHBOURS;
END;
```

Figure 7.16. Behavioural level algorithm of the processing element

With this configuration technique at the processor level, there is no loss of generality of the system's configurability, and also there is no loss of computation speed on the reduction of system hardware (Snyder [1982]). A behavioural description of the processing element is given in Figure 7.16.

5.4 Input module

The first entry level into the simulator is through a menu-driven module supporting a signal generator and a noise generator. The signal generator offers two possible choices: sine or square waves with user defined amplitude and frequency. These two parameters are entered via a 'question-and-answer' section during the signal generation stage. More than one set of amplitude and frequency is allowed, and all the different sets of amplitude and frequency (different signal components) will be compiled into the expression signal $= (g(f_1, A_1)+g(f_2, A_2)+\cdots+g(f_i, A_i))$ where this signal will be sampled and formatted into a matrix ready for input into the COSPRO simulator. By compiling a composite periodic wave, almost any periodic wave can be generated (Smith [1987]).

In practice, noise signal is always present and superimposed with the ideal input signal. This noise may take different forms. The two common types are Gaussian (or normal) distributions, and uniform distribution. Although non-Gaussian noise may appear in telecommunication channels, it is sufficient to approximate Gaussian noise to be the main source of interference. The noise generator in the input module is based on the Gaussian pseudo-random number sequence with a sufficiently large N, where N is the length of the sequence. This component is added to the ideal signal before the sampling of the signal expression, upon request by the user. The Signal-to-Noise Ratio (SNR) may also be defined by the user via the menu which will generate an input matrix consisting of a noisy signal.

Generating a noisy input signal is a simple task as compared with generating the vector coefficients which is to characterise the behaviours of the system. The first approach taken in the input module was to use a look-up table for storing the coefficients. The second approach taken was to generate the coefficients from a specification of the signal processing application. For example, the parameters required for a bandpass filtering are the cut-off frequencies. From these parameters, the transfer function of the algorithm is generated (partially working), and the roots of the function are extracted using a standard root extractor. Hence, the coefficients of the algorithm are compiled into a vector ready to be used by the simulator.

5.5 Output module

The output module was built around a graphics package available at Newcastle University. Apart from handling the bulk of output data, either for archiving or printing, the output module manages mainly the graphic display of the input and output waveforms at both time and frequency domains. Figure 7.17 shows the input signal in time domain. The signal is a 3 Hz square wave with a peak-to-peak amplitude equals 100V, SNR=3 and a d.c. offset. Figure 7.18 shows the fifth and seventh order Butterworth filter responses in frequency domain, with all the values calculated by the simulator.

5.6 Further development

The future development of the simulator will concentrate on three issues: data flow graph generation, library support and diagnosis/evaluation support. Data flow graph generation is essentially the formation of graphical outputs from a transfer function, and determine the configuration control code from the graph. This includes the conversion of arithematic and logical operators and their interrelationships into nodes and connecting patterns.

The development of the library support and the diagnosis/evaluation support is heavily related. The technology library will provide information related to the type of target technology if the data flow graph is implemented (not just on the simulator). From this information, the performance evaluator will be able to estimate the cost

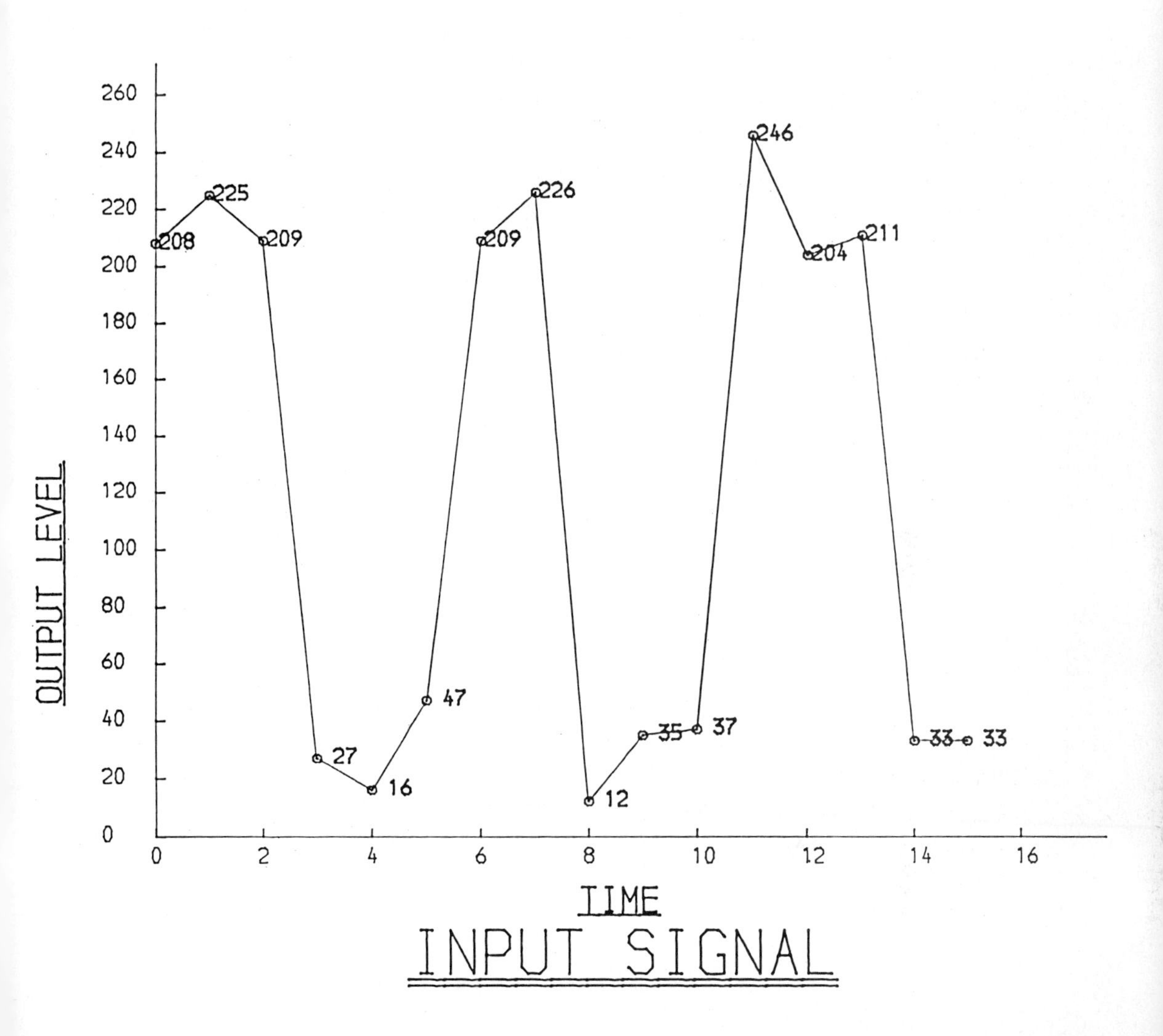

Figure 7.17. Input signal plot in time domain

and performance of the system. The rule library will set out rules of configuration and interact with the data flow graph generator. Also, the results of a configuration will be analysed and diagnostics given. This will be fed back to the transfer function generator for reference and the data flow graph generator for further analysis if a new data flow graph is required.

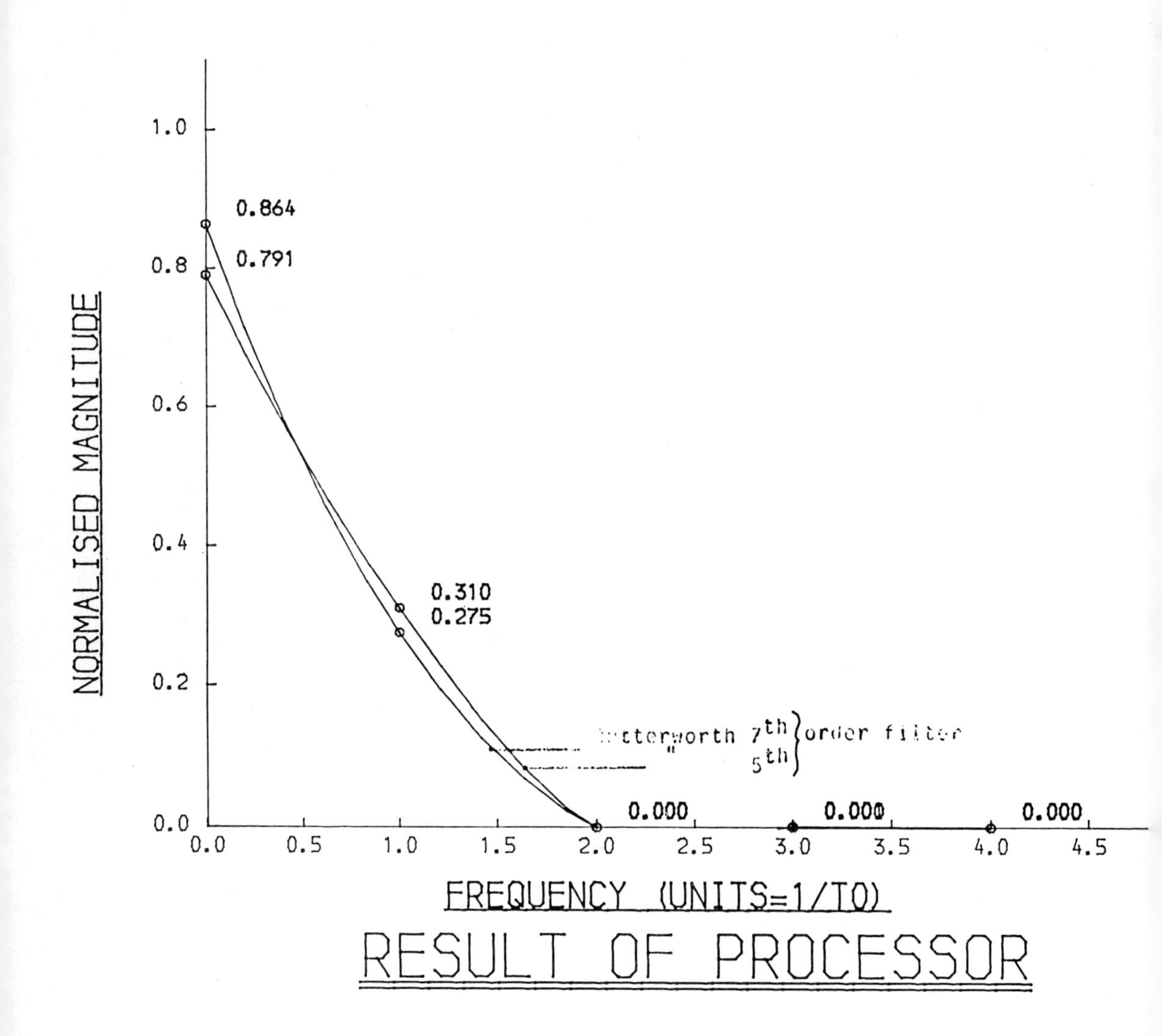

Figure 7.18. Output signal plot in frequency domain

6 BINARY ADDITION – AN EXAMPLE

6.1 The input conditions and signal parameters

A simple example is chosen deliberately to show the overall conceptual flow of the synthesis methodology (Yung & Allen [1985, 1986b]). The application of this is for image processing in the area of industrial robotics. In this type of application, the main number crunching tasks are feature extraction and segmentation. The signal parameters defining this application are a moderate dynamic range using fixed number representation of 8 bit, a 13 MHz video rate, high SNR (because the lighting condition is reasonably clean) and the signal content is in two dimensions. The main operations are logical operations and bit-serial additions (full or half addition).

6.2 Algorithm and its decomposition

Let A (augend) and B (addend) be two n-bit unsigned binary operands given by equation (14):

$$A = a_{n-1}a_{n-2}\ldots a_1a_0, \qquad (14a)$$

$$B = b_{n-1}b_{n-2}\ldots b_1b_0, \qquad (14b)$$

where a_i and b_i are the bit values of weighting 2^{i-1} for $i = 1,\ldots,n$ and can be either one or zero. Theoretical addition of A and B starts from the least significant bits (a_0, b_0), assuming a zero initial carry. A bit-by-bit addition is performed until the most significant bits (a_{n-1}, b_{n-1}) are reached. One algorithm which offers high concurrency is given in equation (15) (Yung & Allen [1986b]).

$$A + B = 2^{n/2}(C + E) + (D + F) \qquad (15)$$

where A and B are n-bit operands; and C, E are the most significant $(n/2)$-bits of A, B respectively, and D, F are the least significant $(n/2)$-bits of A, B respectively. From equation (15), $A + B$ is achieved after performing two $(n/2)$-bit additions and one possible $(n/2)$-bit carry propagation if there is a carry generated from $D + F$. If we consider $(C + E)$ and $(D + F)$ individually, each of these $(n/2)$-bit operands may also be decomposed into two equal halves, i.e., $(n/4)$ bits each. Further decomposition is possible until the lowest level leaf cell is reached which is a simple one-bit addition without carry (half addition). The whole addition algorithm may thus be treated as an addition hierarchy having $\log_2 n$ levels, each may be derived from the components taken from the next lower level. In general the ith addition at the ith level of the hierarchy may be expressed as a composite of the $(i - 1)$th additions.

6.3 DFG and processor architecture

The data flow graph of equation (15) is given in Figure 7.19 with the corresponding architecture shown in Figure 7.20. Figure 19 shows the full flow of signals from the top level of the hierarchy to the lowest level. From the DFG, the flow rate may be determined (five units for an eight-bit addition). Similarly the computation rate ($\sim$ 14 full adder delay units) and the function execution time budgets ($\sim$ 14 FA delay units + 5 communication units) may be determined in the same way. Therefore the loading estimates would be 1/15 for the top left hand node. Memory requirement in this case is only for the two operands and the addition result. This requires two n-bit locations and one $(n+1)$-bit location. For this simple example, the whole algorithm could be implemented onto VLSI. However, more complicated algorithm may need to be partitioned.

Figure 7.20 shows the n-bit level of the hardware architecture. As discussed in the previous section, further expansion of the $n/2$ blocks is possible. The $(n/2)$-bit level shares the same block diagram except the word size is now half of the n-bit level diagram.

6.4 Complexity analysis

The complexity analysis of the addition architecture is based on the chip area, the computational time and the overall area-time complexity given in Section 3.3.5. Owing to the inherent nature of the design the adder complexity is determined by the branching ratio as well as the word size (Smith [1987], Yung & Allen [1986b]). The chip area and computation time are given in equation (16):

$$\frac{A_N}{N} = [1 + N(N-1)(\log_2 N - 2) + \log_2^2 N - 3\frac{\log_2^2 N}{N} + 4\frac{\log_2 N}{N} + \frac{4}{N} + \frac{5}{4}], \quad (16a)$$

$$T_N = 2 + (\frac{N}{4} + 2)\log_2 N, \quad (16b)$$

where A_N/N is the normalised area, N is the word size and T_N is the computational time. Equations (16a), (16b) and their product are plotted in Figures 7.21, 7.22 and 7.23 respectively. These are shown in Figures 7.21, 7.22 and 7.23 together with the complexity figures of other popular addition architectures, some are linear (i.e., carry propagate addition) and some are multiplexed (i.e., conditional sum addition). These figures are useful to locate the complexity of a particular architecture and make comparison with each other or with the initial specification if necessary. The detailed layouts of the architecture can be referred to in Yung & Allen [1986b].

7 DISCUSSIONS/CONCLUSIONS

This chapter outlined the theory of a signal processing synthesis methodology applicable in both VLSI or PCB technologies. It identifies the possibility of optimising the cost and performance of the designed system, and matching the final system with the application requirements. The theory began with analysing the signal parameters, application specification and information content. From these three features

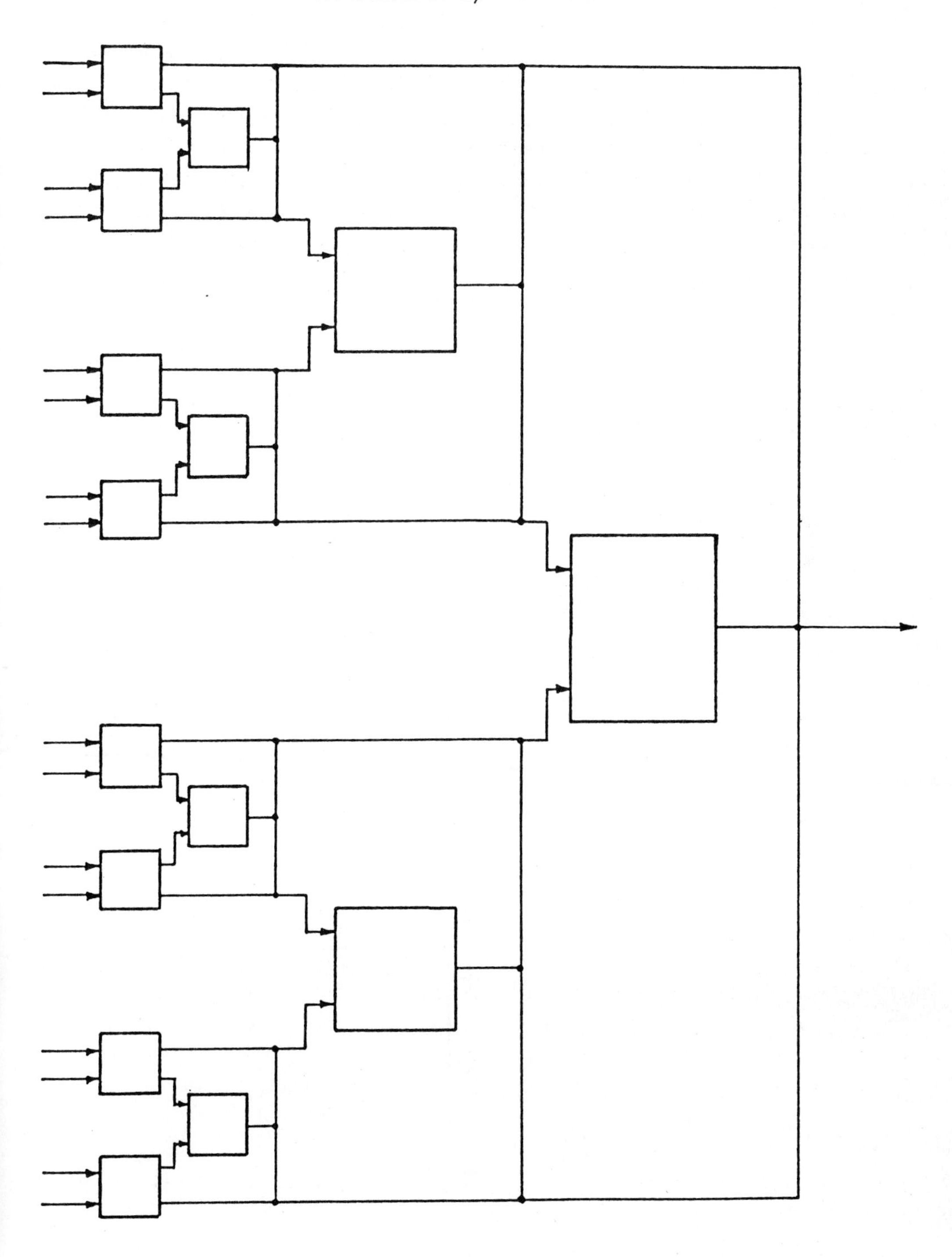

Figure 7.19. Recursive DFG for addition

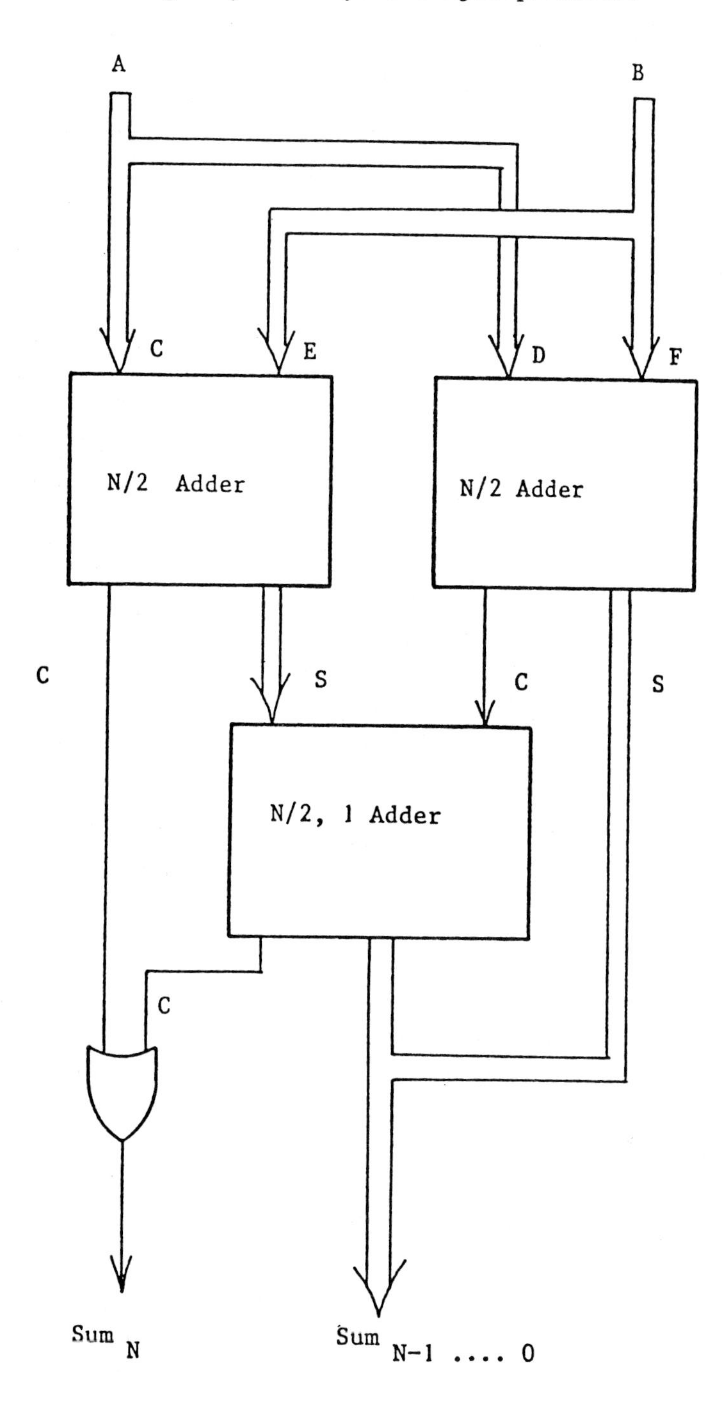

Figure 7.20. Recursive addition architecture

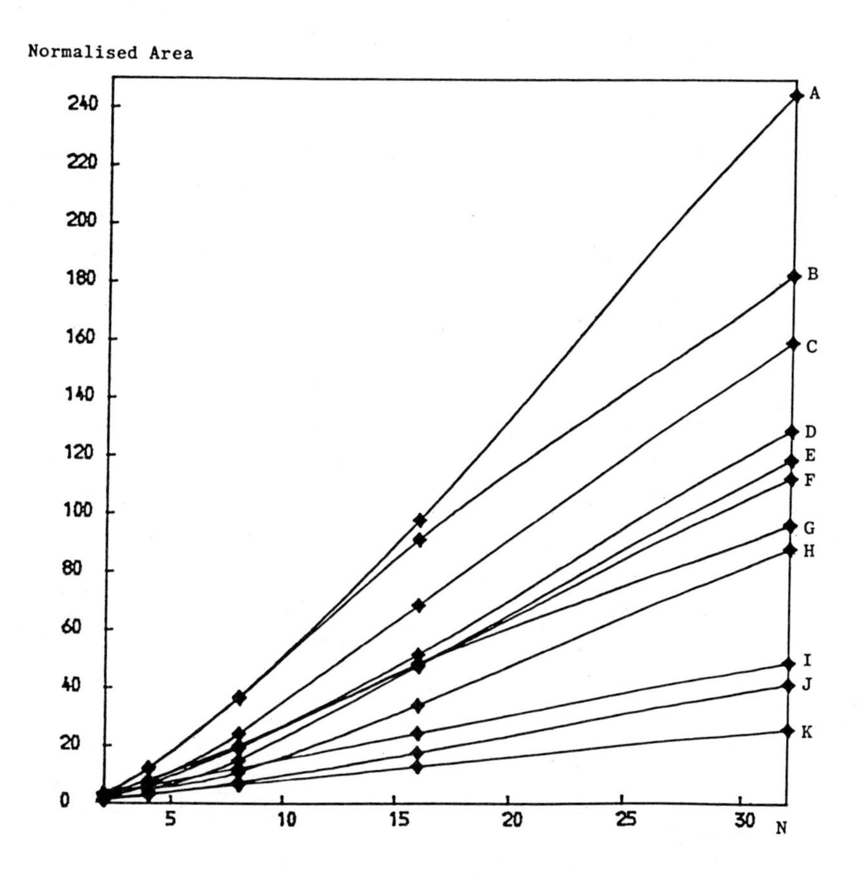

Figure 7.21. Adder layout area vs N

A = CLA (ML, PLA), B = CLA(SL, PLA), C = Carry select,
D = CLA(ML, RL), E = Conditional sum, F = RA (PLA),
G = CLA (SL, RL), H = Brent/Kung CLA, I = CPA (PLA),
J = RA (RL), K = CPA (RL),

CPA = carry propagate adder, CLA = carry lookahead adder,
ML = multi-level, SL = single level,
PLA = programmable logic array, RL = random logic.

of the system, an appropriate algorithm may be selected or derived. For complex systems, the partitioning of the algorithm at the functional and architectural level will generate a set of realizable data flow graphs. The graphical description of the system function enables the system details to be understood and analysed, and this will lead to a processor architecture. Once the processor architecture is known, complexity analysis will indicate the optimality of the processor and the feasibility of the

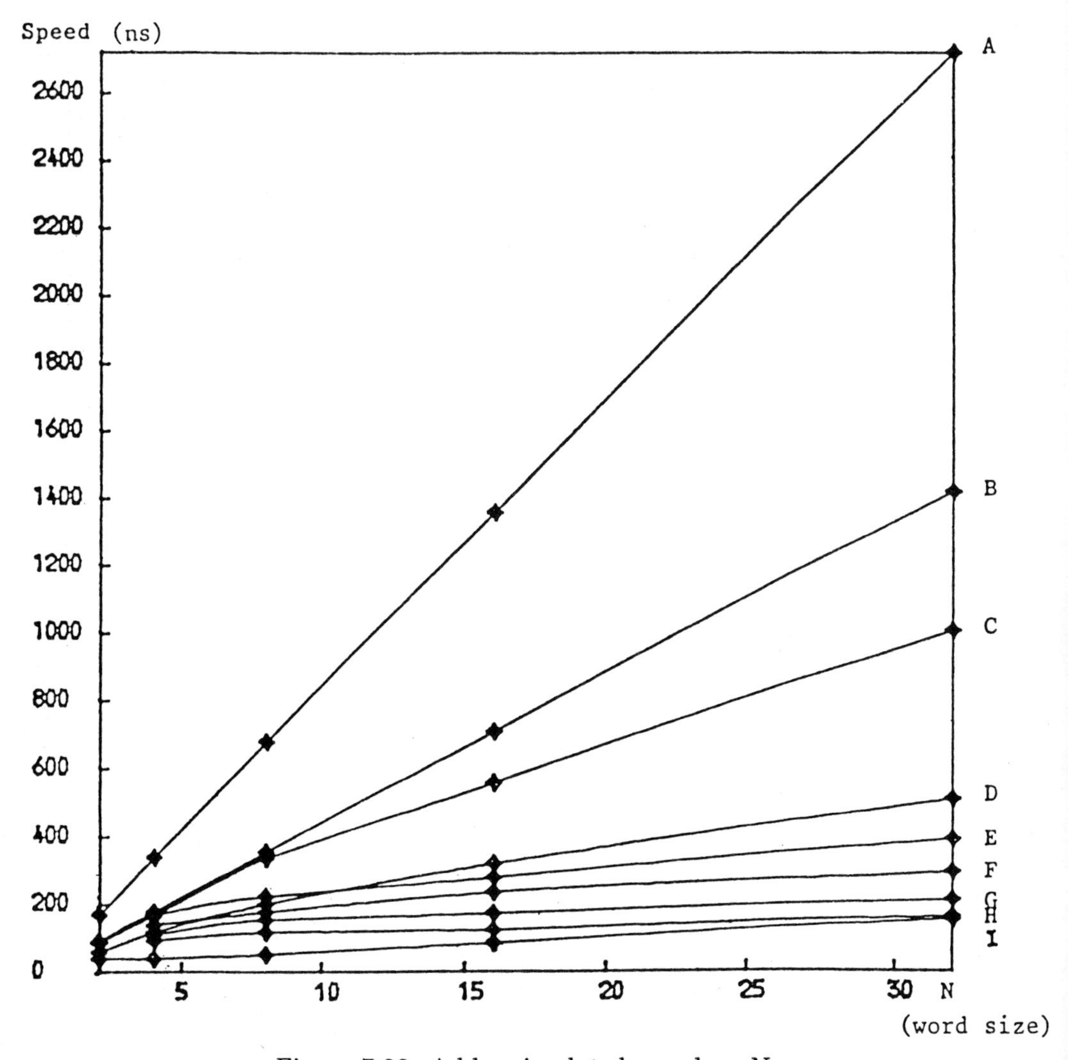

Figure 7.22. Adder simulated speed vs N

A = CPA (PLA), B = CPA(RL), C = CLA (SL, RL),
D = RA(RL), E = CLA (ML, RL), F = Brent/Kung CLA,
G = Carry select, H = Conditional sum, I = RA (pipelined).

final system satisfying the application requirements. A preliminary version of such synthesis tool is in operation which can assist designers in defining an input condition, the designed system characteristics and in studying the processing output on graphic display. It is invisaged that such a tool will eventually take over most of the designers' workload. It also provides assistance in the decision making process, and hopefully in the future, refine and improve such decisions on its own. The example of

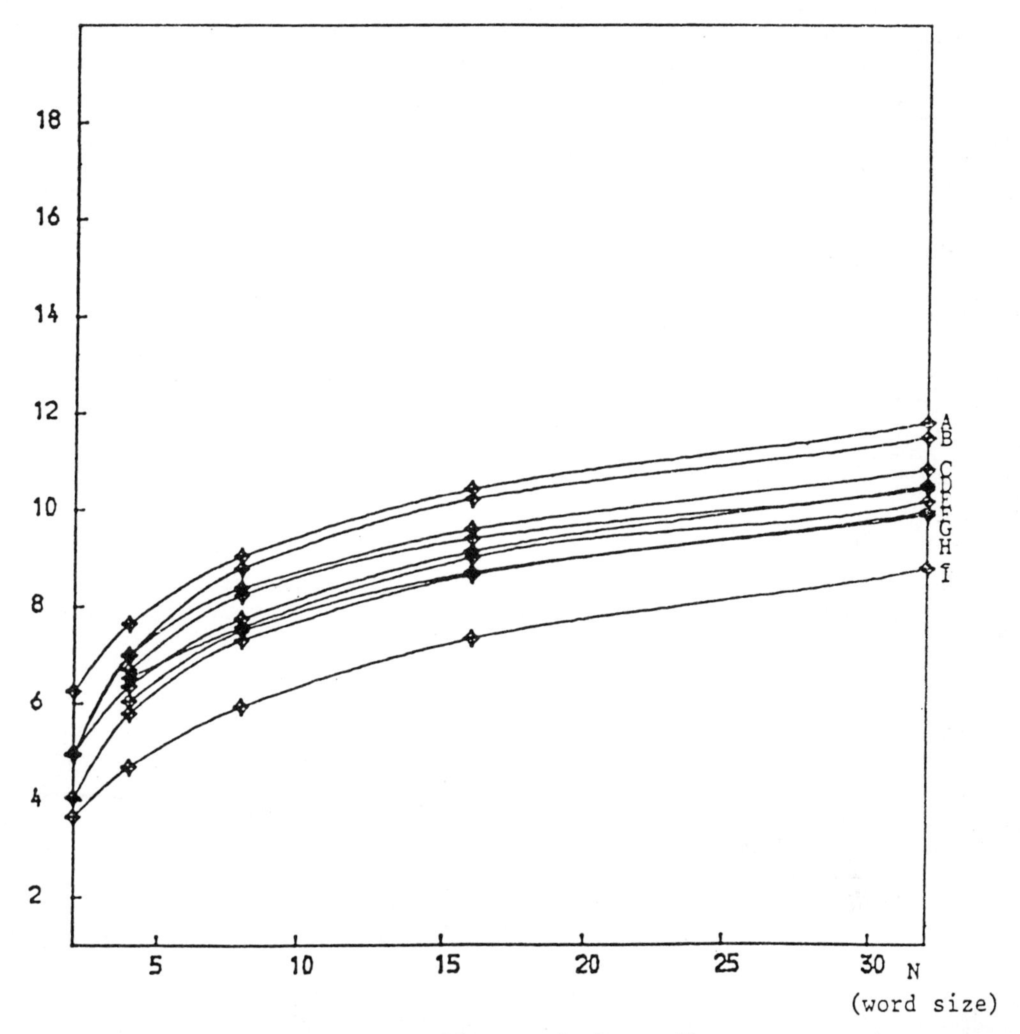

Figure 7.23. Adder complexity vs N

A = CPA (PLA), B = CLA(SL, RL), C = CLA(ML, RL),
D = CPA (RL), E = Carry select, F = Brent/Kung,
G = RA (RL), H = Conditional sum, I = RA (pipelined)(bandwidth).

binary addition used in image processing demonstrates simply the feasibility of this methodology, and how easy it is to design a signal processor in VLSI. Based on this synthesis methdology, rule-based support may be interfaced to the processing module to enhance the decision making process in the system, and this will eventually lead to a signal processing synthesis system capable of making major design decisions and producing the final hardware without any human interaction.

ACKNOWLEDGEMENT

The author would like to acknowledge the contributions from K. Sehat and G. Smith in the Simulator development, and the helpful suggestions of the referees. Part of this work was supported by the EEC ESPRIT 278 project.

REFERENCES

Adams & Yung [1987]
A. E. Adams and H. C. Yung (1987) 'NIPS – a flexible image processing system', 169–73 in *Proceedings of IEEE Asian Electronics Conference.*

Aliphas & Feldman [1987]
A. Aliphas and J. A. Feldman (1987) 'The versatility of digital signal processing chips', *IEEE Spectrum*, **24**, No.6, 40–5.

Allen [1985]
J. Allen (1985) 'Computer architecture for digital signal processing', *Proceedings of IEEE*, **73**, No.5, 852–73.

Ayer [1987]
C. M. Ayer (1987) 'Extraction of formant data from speech using linear predictive coding techniques', Technical Report, Department of Electrical and Electronic Engineering, University of Newcastle-upon-Tyne.

Bowen & Brown [1985]
B. A. Bowen and W. R. Brown (1985) *VLSI Systems Design for Digital Signal Processing*, vol. II, *System Design*, Prentice-Hall.

Brent & Kung [1979]
R. P. Brent and H. T. Kung (1979) 'The area-time complexity of binary multiplication', Technical Report, CMU.

Chen [1984]
C.-T. Chen (1984) *Linear System Theory and Design*, 2nd edn., Holt, Rinehart and Winston.

Denyer & Renshaw [1983]
P. B. Denyer and D. Renshaw (1983) 'Case studies in VLSI signal processing using a silicon compiler', pp. 939–42 in *Proceedings of ICASSP.*

Fairbairn [1982]
D. G. Fairbairn (1982) 'VLSI: a new frontier for system designers', *IEEE Computer*, **15**, No.1, 87–96.

Gonzalez & Wintz [1987]
R. C. Gonzalez and P. Wintz (1987) *Digital Image Processing*, 2nd edn., Addison-Wesley.

INMOS [1987]
INMOS (1987) *IMSA100 Application Manual.*

Keller *et al.* [1980]
R. M. Keller, G. Lindstrom and S. Patyil (1980) 'Data flow concepts for hardware design', pp. 105–11 in *COMPCON '80, 22nd IEEE Computer Society Conference.*

Kinniment [1982]
D. J. Kinniment (1982) 'VLSI and machine architecture', pp. 24–33 in *VLSI Architectures*, ed. B. Randell & P. Treleaven, Prentice-Hall.

Kloher [1986]
K. L. Kloher (1986) 'Architectural features of the Motorola DSP56000 digital signal processor', in *VLSI Signal Processing II*, IEEE Press.

Kung [1987]
S. Y. Kung (1987) *VLSI Array Processors*, Prentice-Hall, Englewood Cliffs, N.J.

NEC [1986]
NEC Electronics (1986) *uPD77230 User's Manual*, Mountain View.

Oppenheim & Schafer [1975]
A. V. Oppenheim and R. W. Schafer (1975) *Digital Signal Processing*, Prentice-Hall, Englewood Cliffs, N.J.

Oppenheim & Weinstein [1972]
A. V. Oppenheim and C. W. Weinstein (1972) 'Effects of finite register length in digital filters and the FFT', *Proceedings IEEE*, **60**, No.8, 957–76.

Peled & Liu [1976]
A. Peled and B. Liu (1976) *Digital Signal Processing – Theory, Design, and Implementation*, Wiley.

Rabiner & Gold [1975]
L. R. Rabiner and B. Gold (1975) *Theory and Applicaton of Digital Signal Processing*, Prentice-Hall, Englewood Cliffs, N.J.

Smith [1987]
G. Smith (1987) 'I/O interface for COSPRO simulator', Technical Report, Department of Electrical & Electronic Engineering, University of Newcastle-upon-Tyne.

Snyder [1982]
L. Snyder (1982) 'Introduction to the configurable, highly parallel computer', *IEEE Computer*, **15**, No.1, pp. 47–56.

Southard [1983]
J. P. Southard (1983) 'MacPitts : an approach to silicon compilation', *IEEE Computer*, **16**, No.12, pp. 74–82.

Texas [1986]
Texas Instruments Inc. (1986) *TMS320C25 User's Guide.*

Thompson [1980]
C. D. Thompson (1980) 'A complexity theory for VLSI', Ph.D. Thesis, Department of Computer Science, CMU.

Trimberger [1983]
S. Trimberger (1983) 'Reaching for the million-transistor chip', *IEEE Spectrum*, **20**, No.11, pp. 100–2.

Witten [982]
I. H. Witten (1982) *Principles of Computer Speech*, Academic Press.

Yalamanchili & Aggarwal [1985]
S. Yalamanchili and J. K. Aggarwal (1985) 'Reconfiguration strategies for parallel architectures', *IEEE Computer*, **18**, No.12, pp. 44–61.

Yung [1985]
H. C. Yung (1985) 'Recursive and concurrent VLSI architectures for digital signal processing', Ph.D. Thesis, University of Newcastle-upon-Tyne.

Yung & Allen [1985]
H. C. Yung and C. R. Allen (1985) 'Constrained decomposition of binary addition for VLSI signal/image processing', pp. 455–8 in *Proceedings of IEEE ISCAS.*

Yung & Allen [1986a]
H. C. Yung and C. R. Allen (1986) 'Recursive and concurrent VLSI architectures for signal processing', pp. 269–98 in *Advances in CAD for VLSI*, vol. 6, *Design Methodologies*, North-Holland.

Yung & Allen [1986b]
H. C. Yung and C. R. Allen (1986) 'Recursive addition and its parameterisation in VLSI', *IEEE Proceedings*, **133**, Pt. G, No. 5, pp. 256–64.

Part 3

Models of circuits and complexity theory

The chapters in this part examine the performance of VLSI systems from different viewpoints.

Chapter 8 looks at the use of discrete complexity models in VLSI design, both theoretically and practically. The results of an experiment on a basic hypothesis are reported.

The final chapter is a technical presentation of two recent innovative results in a field of complexity theory relevant to VLSI.

8 The prioritiser experiment: estimation and measurement of computation time in VLSI

P. M. DEW, E. S. KING, J. V. TUCKER AND A. WILLIAMS

1 INTRODUCTION

The potential complexity of VLSI circuits is such that a well-structured *design process* is necessary in order reliably to transform designs from the initial user specification into a VLSI device that fulfils the user requirements. A major part of that process is the *algorithm design stage*, which concerns the construction of formal specifications for the algorithms. These consist of logical specifications in a formal hardware specification language HSL, and performance specifications based upon a suitable algorithmic model of computation for VLSI.

At the *implementation stage*, during the process of laying out a VLSI circuit using a geometric description language GDL, software tools are required to simulate the design in order to test system logic, and to predict system performance, before fabrication. For this, physical models of circuit behaviour are developed, based upon electronic models that predict component characteristics.

We will study theoretically and empirically the relationships between formal hardware specification languages, the discrete VLSI models of computational complexity which they use to predict system performance, and performance estimation models.

Of particular interest are two questions.

(1) Is it possible to construct a *performance transformation* pt for a suitable model of complexity that would transform a high-level functional performance estimation c_{HSL} into a low-level circuit performance estimation c_{GDL} such that

$$pt(c_{HSL}) = k \cdot c_{GDL}.$$

The most important condition on pt is that it preserves 'performance correctness' (to within a constant factor k).

(2) How is the low-level performance estimate related to the performance of the fabricated VLSI device?

1.1 Algorithmic models of VLSI complexity

In the algorithm design process, a formal specification language must be equipped with a suitable complexity model, which can be used for performance estimation. Many complexity models presently used are based upon ideas originating in work by C. D. Thompson [1980] and Brent & Kung [1981]. There, a VLSI circuit implementation of an algorithm is represented by a *communication graph*, where the nodes model processors and the edges model communication channels transmitting data between the nodes. Processors transfer data synchronously, with only one data value being assigned to each edge in any given time-step. A set of technology-dependent constraints govern the physical layout of the communication graph. The model is used to establish complexity measures of *circuit area* A and *computation time* τ_C for a circuit C_f implementing a function f. Techniques based upon traditional complexity theory and graph theory are used to find asymptotic lower or upper bounds, by consideration of the *information flow* required by a particular algorithm, and the *band width* available in the communication graph.

Amongst the detailed assumptions used in proposed models lie certain discrepancies, especially when considering signal delay in communication channels. Of particular interest is the assumption that *the time required to transmit data through a series of processors is simply the sum of the times required by the individual processors in isolation.* This is called the *Linear Composition Rule*, and is used explicitly or implicitly in most analyses of algorithm performance in computer science.

1.2 Physical models for VLSI performance estimation

At the implementation stage in a design process, a VLSI circuit is represented using a *geometric description language* such as CIF (see Mead & Conway [1980]). This provides the patterning or mask information for each layer to be built up in the fabrication process required to produce the desired VLSI circuit in silicon. Before fabrication the VLSI circuit formed by the proposed layout must be checked for logical and performance correctness. In order to do this, the circuit is *extracted* from the layout, and is processed by a set of circuit simulators and analysers. The simulators employ models based upon physical laws governing the behaviour of the electronic components present in the circuit. The accuracy of the results produced is restricted by the sophistication of both the particular behavioural model, and of the circuit extraction process.

Several simulators and circuit timing analysers employ an *RC network model* of the circuit, which will be considered here to estimate the computation time. Each circuit element is replaced by a simple circuit network containing only resistances and capacitances of appropriate values. An estimate for the signal delay through the resulting

RC network can be found using simple delay formulae, and used as an estimate for the computation time of the circuit modelled by the network.

1.3 The prioritiser experiment

We wish to investigate the accuracy of the complexity models outlined above when tested upon simple digital devices, and to determine the validity of their underlying assumptions. A simple function called the *prioritiser* is studied (see Section 2 for a description). A formal logical specification of the prioritiser is produced using an algorithmic description language PR, and estimates of performance are made using the PR complexity model (Section 3).

The language PR is designed to reason about and simulate synchronous concurrent algorithms in general (for example, systolic algorithms and neural nets); for our purposes it represents a theoretically well-defined high-level algorithmic tool in which the role of the discrete complexity models is easily understood.

The problem of modelling a complete nMOS circuit using RC networks is investigated, followed by that of estimating the signal delay in the RC network models produced, so that low-level performance estimates for the prioritiser circuit can be made (Section 4). The RC network model is then adapted in Section 5 so that it can be used for simple nMOS circuits specified using hierarchical *computation graphs* (see McEvoy [1986]).

An experiment is described to investigate empirically the performance of VLSI devices implementing the prioritiser function; theoretical results are compared with those obtained empirically (Section 6). For a prioritiser circuit computing on input-word size n the algorithmic complexity model, used in the formal specification language, predicts an asymptotic computation time of $O(n)$ for each *carry-chain* circuit implementation used. However, from empirical results, the asymptotic computation time is found to be $O(n^2)$ for certain implementations, involving *non-level-restoring* carry-chains and long interconnecting wires. It is shown experimentally that the RC network model predicts the correct asymptotic computation times in each case.

From the results, the valid domain of application of each of the complexity models used can be determined, with the aim of developing a performance estimation methodology which is capable of being used with formal specification languages, but which also provides good estimates for circuit performance, that will compare well with empirical results.

1.4 Literature

On the subject of VLSI complexity theory useful surveys and contributions can be found in Baudet [1983], Bilardi *et al.* [1981] or Chazelle & Monier [1985]. An interesting comparison of complexity results for integer multiplication appears in Kühnel & Schmeck [1986] highlighting the effects caused by the use of different complexity model assumptions. Several circuit simulators and analysers have been developed based upon RC network models, and of particular note is work by Lin & Mead [1986], and Ousterhout [1983].

Work is currently in progress at Leeds on formal methods for algorithm specification (see for example Thompson & Tucker [1985] or Thompson [1987]) and the algorithm design process (see Harman & Tucker [1987], Eker & Tucker [1987], McEvoy [1986] or Martin & Tucker [1987]). In particular, studies of the formal treatment of performance have been undertaken.

2 DESCRIPTION OF THE PRIORITISER CIRCUIT

In this section, the basic architecture of a *register-to-register (R-R) transfer* circuit implementation of a boolean function f is given. A boolean function called the *prioritiser* will then be described, followed by two implementations of this function as modular circuits both based upon the *carry-chain*. The two implementations differ only in their use of *signal-level-restoring buffers* along the chain.

The purpose of using these particular methods of implementing the prioritiser function is to provide a means of measuring the computation time of an R-R transfer circuit of variable active size (dependent upon input data) that is simple enough to study without the theoretical principles involved becoming obscured.

2.1 The implementation of boolean functions

Let $\mathbf{B} = \{1, 0\}$ be the set of booleans. A boolean function $f : \mathbf{B}^m \to \mathbf{B}^n$ has n coordinate functions $f_1, \ldots, f_n$ where

$$f(\underline{x}) = \langle f_1(\underline{x}), \ldots, f_n(\underline{x}) \rangle$$

with $\underline{x} \in \mathbf{B}^m$.

The function f can be computed by a circuit-of-interest C_f embedded into service circuitry, as shown in Figure 8.1. The input word $\underline{x}$ is supplied by input register I with cells $I_1, \ldots, I_m$. The circuit is then *enabled*, and, after a known period of time, defined to be the *computation time* $\tau_c(\underline{x})$, it writes the word $f(\underline{x})$ to output register O with cells $O_1, \ldots, O_n$. During initialisation of the circuit, all the output cells are reset to logic 0. Note that the registers I and O are *not* part of the circuit-of-interest C_f to

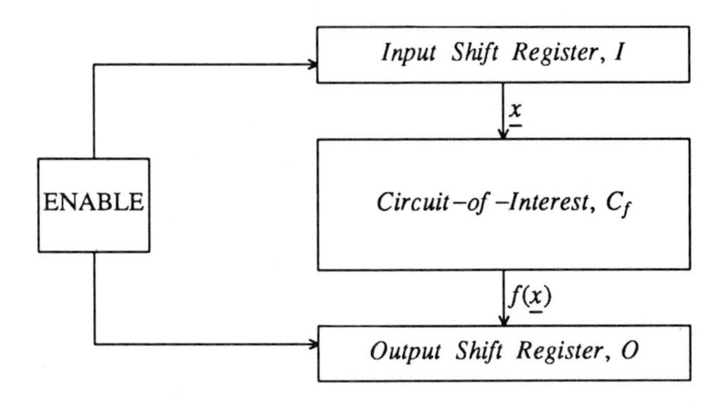

Figure 8.1. R-R architecture for measuring computation time

be analysed: C_f is isolated from the service circuitry by $ENABLE$ gates (the time for signals to cross the $ENABLE$ gates must however be included in estimations of the computation time). We say that C_f with this service circuitry is an implementation of the boolean function f, or, equivalently, that f is a functional specification of C_f with this service circuitry.

2.2 Computational time

In order to proceed with measurements of the computation time for a particular circuit, a suitable definition is required for use in an experimental environment.

2.2.1 Definition of computation time The *computation time* $\tau_c(\underline{x})$ for a circuit C_f computing an R-R transfer function $f : \mathbf{B}^m \rightarrow \mathbf{B}^n$, on an input word $\underline{x} \in \mathbf{B}^m$, is the shortest delay between C_f receiving $\underline{x}$ on its input ports IP_i, $i = 1, \ldots, m$, and producing the correct result $f(\underline{x})$ on its output ports OP_j, $j = 1, \ldots, n$.

In a circuit implementation, the input data word $\underline{x}$ is supplied by input register I (connected to IP), and the result is written to output register O (connected to OP). Thus the input port IP_i receives data stored in input register cell I_i, and output port OP_j writes data to output register cell O_j.

2.2.2 Definition of worst-case computation time The *worst-case computation time* for circuit C_f is the maximum computation time $\tau_c(\underline{x})$, as defined above, required by the circuit for *any* permitted input data word $\underline{x}$ in the *domain* of f.

2.3 Prioritiser function

The *prioritiser* is a boolean function $PRIOR : \mathbf{B}^n \rightarrow \mathbf{B}^n$ defined by

$$PRIOR(\underline{x}) = \langle p_1(\underline{x}), \ldots, p_n(\underline{x}) \rangle$$

where

$$p_i(\underline{x}) = \begin{cases} 1 & \text{if } x_1, \ldots, x_{i-1} = 0 \text{ and } x_i = 1 \\ 0 & \text{otherwise} \end{cases}$$

with $\underline{x} = \langle x_1, \ldots, x_n \rangle$, $\underline{x} \in \mathbf{B}^n$, and $x_i \in \mathbf{B}$, $i = 1, \ldots, n$.

The prioritiser function can be implemented using the R-R transfer architecture described above. In operation, if sufficient $ENABLE$ time has been allowed, then the prioritiser returns a logic 1 to O_i if I_i is the first input register holding a 1, and returns 0 to all other cells in the output register. For example, if $\underline{x} = \langle 0,0,1,1,0,0,1,0 \rangle$ (where $\underline{x} \in \mathbf{B}^8$), then

$$PRIOR(\underline{x}) = \langle 0,0,1,0,0,0,0,0 \rangle$$

2.4 Carry-chain implementations for the prioritiser

There are several ways of implementing the prioritiser circuit as an R-R transfer circuit. We have chosen to use two implementations based upon a carry-chain circuit because

(i) carry-chains are widely used in practical circuits,
(ii) they can be produced using small, simple modular sub-circuits, and
(iii) the *active size* of the circuit is easily controllable using the input data.

In the implementations used, the first cell in the carry chain must receive a *header-signal* c applied to channel cin_1. In addition, both $\underline{x}$ and $\overline{\underline{x}} = NOT(\underline{x})$ are required by the prioritiser in order to avoid skew within the cells p_i: the prioritiser is supplied with the complement data word $\overline{\underline{x}}$ where $\overline{x}_i = NOT(x_i)$, $i = 1, \ldots, n$. Thus the functional description of the prioritiser given above is modified for the implementation by the addition of extra inputs c and $\overline{\underline{x}}$. These are included in the formal specification for the circuit-of-interest given in Section 3.

Figure 8.2 shows the architecture of the circuit implementation to be used in the experiment. The input word $\underline{x}$ is supplied by input register cells I_i and $\overline{\underline{x}}$ is supplied by cells $\overline{I}_i$. Cell C supplies the header-signal to the start of the carry-chain. The output

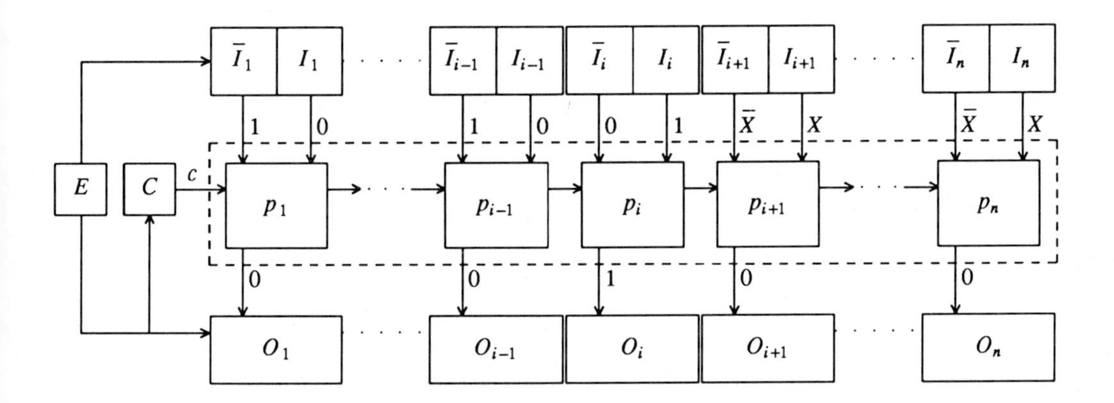

Figure 8.2. Behaviour of prioritiser

word $PRIOR(\underline{x})$ is written to output register cells O_i. Cell E provides the necessary $ENABLE$ signals. The circuit-of-interest consists solely of certain *prioritiser cells* p_i enclosed in the dashed box.

Figure 8.2 also shows the circuit behaviour when a typical input word $\underline{x}$ is supplied, where x_i is the first non-zero bit (X means 'don't care'). The circuit can only receive data from C, I and $\overline{I}$, and write to O when $ENABLE$ is raised, thus allowing the computation time of the circuit to be determined. Each input and output channel is labelled with the appropriate data value (i.e., c, 0, 1, X or $\overline{X}$), assuming that the computation has been completed successfully.

The experimental chip contains two different implementations of the prioritiser circuit. Both implementations have the same architecture as shown above, and only differ in the construction of the basic cells p_i.

Note that the following design rule must be obeyed:

> The output from a pass transistor network must *not* connect to the gate of another pass transistor (see, for example, Ullman [1984], p.18).

This means that inverters must be introduced into the input lines *yin* and $\overline{yin}$, since these will be supplied via the $ENABLE$ pass transistors (see Figures 8.5 and 8.7).

2.4.1 Implementation 1: use of a pass-transistor carry-chain In order to examine the effects on computation time of composing a sequence of *non-level-restoring* circuit cells, we have deliberately broken the Mead & Conway rule-of-thumb (see Mead & Conway [1980], p. 23), which can be given as follows.

> To obtain minimum signal delay within pass-transistor carry-chains, level-restoring buffers should be introduced at regular intervals along the chain. The optimal number of pass-transistors between each buffer and the next, required to produce minimum delay, depends on the capacitive loading on the chain.

In this implementation, the carry-chain consists simply of pass-transistors with *no* signal level-restoring logic included (this should be compared with the second implementation given in Section 2.4.2). Each cell p_i consists of a basic cell called *pcell* (see Figure 8.3), which performs the following boolean function:

$$pcell(yin, \overline{yin}, cin) = \langle yout(yin, \overline{yin}, cin), cout(yin, \overline{yin}, cin)\rangle$$

where

$$yout(yin, \overline{yin}, cin) = \begin{cases} 1 & \text{if } cin = 1 \text{ and } yin = 1, \\ 0 & \text{otherwise,} \end{cases}$$

$$cout(yin, \overline{yin}, cin) = \begin{cases} 1 & \text{if } cin = 1 \text{ and } \overline{yin} = 1, \\ 0 & \text{otherwise.} \end{cases}$$

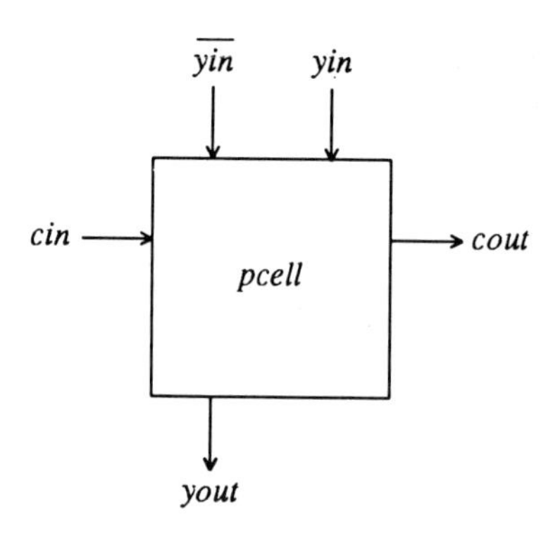

Figure 8.3. Basic pcell

When two cells are concatenated, then $cin_i = cout_{i-1}$ $(i = 2, \ldots, n)$, and $cin_1 = c$. Each cell adds two more pass-transistors to the carry-chain, when the carry signal (a logic 1) is propagated (i.e., $cin = 1$ and $\overline{yin} = 1$). The header node, C, must supply a logic 1 to the start of the chain (cin_1).

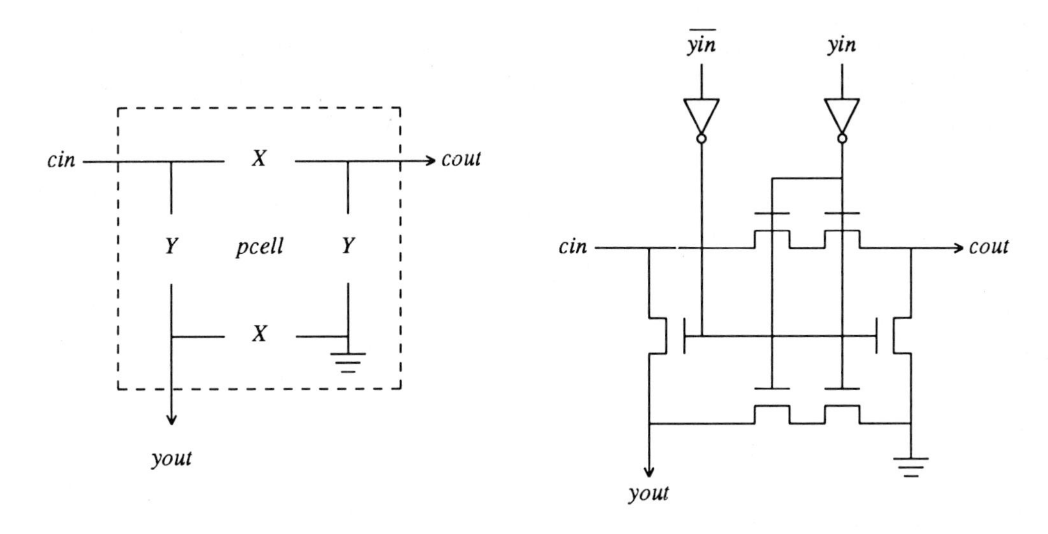

Figure 8.4. Figure 8.5.

In Figure 8.4, X and Y represent switches:
X is closed *iff* $yin = 0$,
Y is closed *iff* $\overline{yin} = 0$.

Figure 8.5 gives the circuit used in this implementation.

2.4.2 Implementation 2: level-restoring carry-chain According to 'good design principles' (see Mead & Conway [1980], p. 23) an nMOS pass-transistor carry-chain should be broken up by the introduction of signal-level-restoring buffers, in order to contain the otherwise rapid increase in computation time. In this second implementation, each cell contains a buffer so that the carry-chain contains only two pass-transistors before the carry signal-level is restored:

In this implementation the *pcell* has the following function:

$$pcell(yin, \overline{yin}, cin) = \langle yout(yin, \overline{yin}, cin), cout(yin, \overline{yin}, cin) \rangle$$

where

$$yout(yin, \overline{yin}, cin) = \begin{cases} 1 & \text{if } cin = 0 \text{ and } yin = 1, \\ 0 & \text{otherwise,} \end{cases}$$

$$cout(yin, \overline{yin}, cin) = \begin{cases} 1 & \text{if } cin = 0 \text{ and } \overline{yin} = 1, \\ 0 & \text{otherwise.} \end{cases}$$

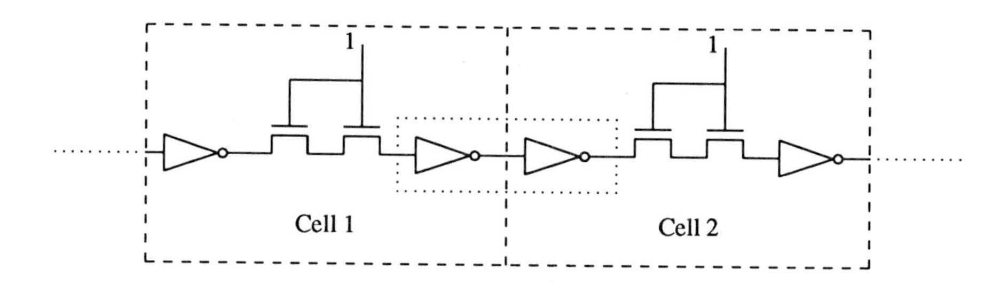

Figure 8.6. Level-restoring carry-chain.

When two cells are composed, the inverters in $cout_{i-1}$ and cin_i join to form the level-restoring buffer, as indicated by the dotted box in Figure 8.6.

Each cell adds two pass transistors and a buffer to the carry-chain if the carry signal (a logic 0) is propagated. The header node C must supply a logic 0 to the start of the chain at input channel cin_1. Figure 8.7 shows the circuit used in the second implementation.

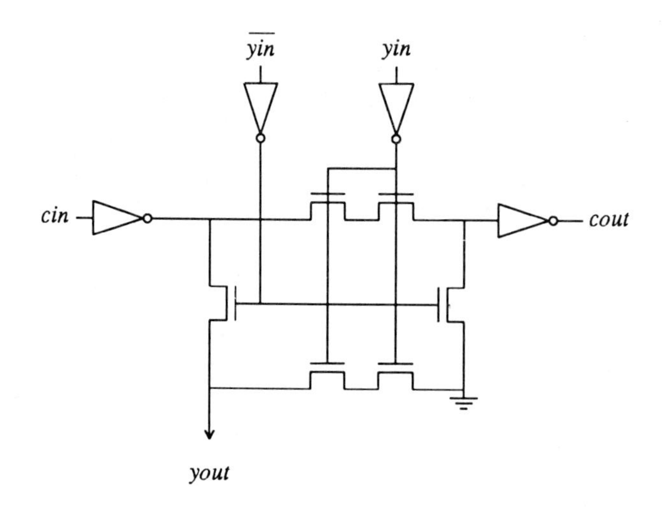

Figure 8.7. Level-restoring pcell

3 FORMAL SPECIFICATION

3.1 Introduction

A major part in the design process of an algorithm involves the formal specification of the algorithm. This requires both a logical specification, written in a formal specification language, and a performance specification, based upon a model of computational complexity, developed for the specification language. It is important to note that consideration of algorithm performance is an integral part of the design process.

The main purpose of a formal specification is to provide a mathematically rigorous presentation of the required function, against which algorithms (and their implementations) can be formally proved correct, or against which they can be tested. In this way the user will know precisely what a device implementing the function should be able to do (see Harman & Tucker [1987]).

In this section, the *PR algorithmic specification language* is used to give formal specifications for both implementations of the prioritiser circuit-of-interest, described in Section 2. Using a complexity model tailored to the language the performance of both prioritiser implementations will be found. The predictions can then be compared with experimental results (see Section 6) in order to test the validity of assumptions made in the complexity model.

3.2 The PR formal specification language

The algorithmic specification language PR allows a collection of functions to be defined, over a *many-sorted algebra* A. The basic operations of the structure, given by the basic operation symbols of A, define the *level of computational abstraction* to be used in the specification process. Basic functions include the basic operations of the algebra A, together with *projection* and *definition by cases.* Complex functions are constructed by the function-building operations of *vectorisation*, *composition*, and *primitive recursion* (see, for example, Thompson [1987] for a full definition of PR).

In general, a specification consists of a collection of communicating functions. In order to control data selection and scheduling between functions, a synchronous algorithm specified in PR can be considered to be embedded into a network of *modules* and *channels.* Let there be k modules $m_1, \ldots, m_k$. Each module m_i has $p(i)$ inputs, $q(i)$ outputs, and contains function $f_i : A^{p(i)} \to A^{q(i)}$ where

$$f_i(\underline{a}) = \langle f_i^1(\underline{a}), \ldots, f_i^{q(i)}(\underline{a}) \rangle,$$
$$\underline{a} \in A^{p(i)}.$$

The network has a global discrete clock $\mathbf{T} = \{0, 1, 2, \ldots\}$ which synchronises the computation.

For a particular module m_i, the function $val_{m_i}(t, \underline{a}, \underline{b})$ gives the value of data on the output channels of m_i at a time t, when the network is initialised with values $\underline{b}$ and is computing on input data $\underline{a}$:

$$val_{m_i}(t, \underline{a}, \underline{b}) = \langle val^1_{m_i}(t, \underline{a}, \underline{b}), \ldots, val^{q(i)}_{m_i}(t, \underline{a}, \underline{b})\rangle.$$

If m_i has data values $x_1(t), \ldots, x_{p(i)}(t)$ on its input channels at time t, then the value on output channel c at time $t+1$ will be given by

$$val^c_{m_i}(t+1, \underline{a}, \underline{b}) = f^c_i(x_1(t), \ldots, x_{p(i)}(t)).$$

Thus $val^c_{m_i}(t, \underline{a}, \underline{b})$ is used to select an item of data from output channel c for input into the next function in sequence (see Eker & Tucker [1987] for further details).

3.3 The formal specification of the prioritiser

The above scheme is used for the formal specification of the prioritiser described in Section 2. The algebraic data type to be used is given first.

3.3.1 Algebraic data type A The underlying *algebraic data type A* is a three-sorted algebra, consisting of the binary data set $\mathbf{B} = \{1, 0\}$ with constants 1 and 0, and operations NOT and AND, together with a discrete clock $\mathbf{T} = \{0, 1, 2, \ldots\}$ with constant 0 and operation $succ(t)$, and the booleans $\mathbf{Bool} = \{tt, ff\}$ with constants tt and ff and operations $\neg$ and $\wedge$. Note that $\mathbf{T}$ and $\mathbf{Bool}$ are always present in PR specifications, and the binary data set $\mathbf{B}$, although isomorphic to $\mathbf{Bool}$, is included to distinguish computational data items from PR constructs. The operations NOT, AND, $succ$, $\neg$ and $\wedge$ have their usual mathematical meanings.

3.3.2 Module network Consider the prioritiser to be represented by the network of modules and channels shown in Figure 8.8 (compare this with the architecture of the circuit implementing the prioritiser function given in Section 2.2).

The labels C, I_i, $\overline{I}_i$, p_i and O_i, $i = 1, \ldots, n$, give the names of modules used in the prioritiser. The labels cin_1, yin_i, $\overline{yin}_i$, $yout_i$, $i = 1, \ldots, n$, and $cout_j$, $j = 2, \ldots, n-1$ give the channel names to be used, which are to be distinguished from data values appearing on the channels.

The specification captures the carry-chain structure of the modular circuit with each module p_i containing an identical function $pcell$ analogous to the basic prioritiser cell

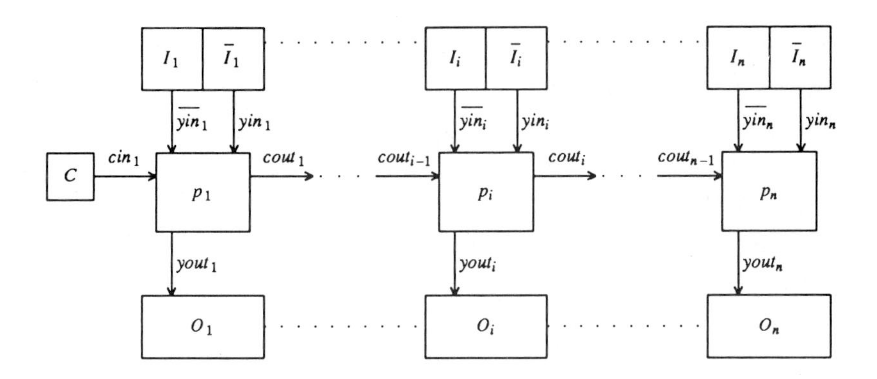

Figure 8.8. Module diagram for prioritiser

and independent of i. i.e., *pcell* has input channels yin, $\overline{yin}$, and cin, and output channels $yout$ and $cout$. The specifications for the two implementations of the prioritiser will differ only in the definitions used for *pcell*, the allowed value for header-node signal c appearing on channel cin_1, and the initial values $\underline{b}$ given to the module outputs (see Section 3.3.4 and Section 3.3.5).

Modules C, I_i and $\overline{I}_i$ are input modules. They are initialised with values given by $\underline{b}$ and receive an input word $\underline{a} = \langle c, \underline{x}, \underline{\overline{x}} \rangle$, (where $\underline{x}, \underline{\overline{x}} \in \mathbf{B}^n$, $c \in \mathbf{B}$, and thus $\underline{a} \in \mathbf{B} \times \mathbf{B}^n \times \mathbf{B}^n$). Modules O_i are output modules from which the results of the computation are read. The value of $\underline{a}$ will be constant for the duration of the computation and so the prioritiser is said to be *statically* specified.

3.3.3 Formal specification In order to simplify notation in the specification, let $\delta_i(t)$ represent the data value present on channel δ_i of module m_i at clock tick t:

$$\delta_i(t) = val^{\delta}_{m_i}(t, \underline{a}, \underline{b}).$$

For example, the data value appearing on channel name $yout_1$ of module name p_1 at time t (see Figure 8.8) will be represented by $yout_1(t)$ where

$$yout_1(t) = val^{yout_1}_{p_1}(t, \underline{a}, \underline{b})$$

The prioritiser specification can now be given. A valid result can only be produced if the correct data configuration is supplied. Thus the header-signal c must be of the correct (implementation dependent) value, and the complement data word $\underline{\overline{x}}$ must be supplied, where $\overline{x}_i = NOT(x_i)$, $i = 1, \ldots, n$:

$$PRIOR : \mathbf{B} \times \mathbf{B}^n \times \mathbf{B}^n \rightarrow \mathbf{B}^n \cup \{u\},$$

$$PRIOR(\underline{a}) = \begin{cases} u & \text{if } x_i \neq NOT(\overline{x}_i),\ i = 1, \ldots, n, \\ & \text{or if } c \neq chead, \\ P(\underline{a}) & \text{otherwise.} \end{cases}$$

('u' means 'undefined')
where the value *chead* is implementation dependent (see Section 3.3.4 and Section 3.3.5).

$$P : \mathbf{B} \times \mathbf{B}^n \times \mathbf{B}^n \rightarrow \mathbf{B}^n$$

$$P(\underline{a}) = \langle yout_1(\lambda(\underline{a})), \ldots, yout_n(\lambda(\underline{a}))\rangle$$

(note that $yout_i(t) = O_i(yout_i(t)) = val_{O_i}(t, \underline{a}, \underline{b})$)

where $\lambda : \mathbf{B} \times \mathbf{B}^n \times \mathbf{B}^n \rightarrow \mathbf{T}$ is a function that represents the number of clock 'ticks' required to complete the computation of P, and

$$\begin{aligned} yout_i(t+1) &= yout(cout_{i-1}(t), yin_i(t), \overline{yin}_i(t)), \\ cout_i(t+1) &= cout(cout_{i-1}(t), yin_i(t), \overline{yin}_i(t)), \\ &\text{for } i = 2, \ldots, n, \\ yout_1(t+1) &= yout(cin_1(t), yin_1(t), \overline{yin}_1(t)), \\ cout_1(t+1) &= cout(cin_1(t), yin_1(t), \overline{yin}_1(t)). \end{aligned}$$

Input modules I_i, $\overline{I}_i$ and C contain the following functions:

$$\begin{aligned} I_i &: In_i(\underline{a}) = x_i \\ \overline{I}_i &: \overline{In}_i(\underline{a}) = \overline{x}_i \\ C &: C(\underline{a}) = c \end{aligned}$$

(thus $In_i(\underline{a}) = proj_i^n(\underline{x})$ and $\overline{In}_i(\underline{a}) = proj_i^n(\underline{\overline{x}})$).

Each module p_i contains the function $pcell : \mathbf{B}^3 \rightarrow \mathbf{B}^2$ consisting of the two coordinate functions $yout, cout : \mathbf{B}^3 \rightarrow \mathbf{B}$. The definitions for *yout* and *cout* will depend upon which implementation of the basic prioritiser cell is being used (see below).

3.3.4 Specification 1: pass-transistor carry-chain The following is a formal specification for the first implementation of the *pcell* function described in Section 2.4.1:

$$pcell(cin, yin, \overline{yin}) = \langle yout(cin, yin, \overline{yin}), cout(cin, yin, \overline{yin})\rangle$$

where

$$yout(cin, yin, \overline{yin}) = AND[cin, NOT(\overline{yin})],$$
$$cout(cin, yin, \overline{yin}) = AND[cin, NOT(yin)].$$

In addition the header signal must have the following value during computation to be valid:

$$chead = cin_1(t) = 1 \quad (t > 0).$$

For Implementation 1 the output channels of each module are initialised to the following values (refer to Figure 8.8):

$$yin_i(0), \overline{yin}_i(0) = 1,$$
$$cin_1(0) = 0,$$
$$yout_i(0) = 0,$$
$$cout_i(0) = 0.$$

3.3.5 Specification 2: level-restoring carry-chain In the following specification for *pcell* the use of level-restoring inverters in the carry-chain (see Section 2.4.2) is reflected by the complementation of *cin* and *cout*:

$$pcell(cin, yin, \overline{yin}) = \langle yout(cin, yin, \overline{yin}), cout(cin, yin, \overline{yin})\rangle$$

where

$$yout(cin, yin, \overline{yin}) = AND[NOT(cin), NOT(\overline{yin})],$$
$$cout(cin, yin, \overline{yin}) = NOT(AND[NOT(cin), NOT(yin)]).$$

The header-signal must have the following value during computation to be valid:

$$chead = cin_1(t) = 0 \quad (t > 0).$$

The outputs of each module are initialised to the following values (again, refer to Figure 8.8):

$$yin_i(0), \overline{yin}_i(0) = 1,$$
$$cin_1(0) = 1,$$
$$yout_i(0) = 0,$$
$$cout_i(0) = 1.$$

3.4 Time complexity for prioritiser specifications

In this section we will use the definition of the *length of computation*, $c_A(f, \underline{a})$ of $f(\underline{a})$, given in Thompson & Tucker [1985], in order to calculate the worst-case computation time for a prioritiser of data size n. Each *atomic* operation is assumed to cost one time unit, with the cumulative cost of function-building operations following intuition, i.e., if a function f is defined by vectorisation, with $f = \langle f_1, \ldots, f_n \rangle$ for some $f_1, \ldots, f_n \in PR$, then the cost is the maximum cost of the f_is:

$$c_A(f, \underline{a}) = 1 + \mathrm{MAX}[c_A(f_1, \underline{a}), \ldots, c_A(f_n, \underline{a})],$$

and if a function f is defined by composition with $f = h \bullet g$ for some $g, h \in PR$ then

$$c_A(f, \underline{a}) = 1 + c_A(g, \underline{a}) + c_A(h, g(\underline{a})).$$

The atomic operations used in the prioritiser specification are

$$AND, \ NOT.$$

3.4.1 Linear composition rule In calculating the computation time for a sequence of functions, the complexity model used in PR uses the linear composition rule, which can be stated as follows:

Given a function $f : A^{w_1} \rightarrow A^{w_{n+1}}$ and functions $f_i : A^{w_i} \rightarrow A^{w_{i+1}}$, $i = 1, \ldots, n$, where f is defined by

$$f(\underline{a}_1) = f_n \bullet f_{n-1} \bullet \ \ldots \ \bullet f_1(\underline{a}_1)$$

then

$$c_A(f, \underline{a}_1) = n - 1 + c_A(f_1, \underline{a}_1) + \sum_{j=2}^{n} c_A(f_j, f_{j-1}(\underline{a}_{j-1})),$$

where

$$\underline{a}_j = f_{j-1}(\underline{a}_{j-1}), \quad j = 2, \ldots, n$$
$$(\underline{a}_i \in A^{w_i}, \quad i = 1, \ldots, n).$$

Of particular interest here is the asymptotic computation time, given by

$$c_A(f, \underline{a}_1) = O(n).$$

3.4.2 Complexity The asymptotic *performance* of the prioritiser can now be established, and is independent of the method of implementation of *pcell*:

$$\begin{aligned} c_A(PRIOR, \underline{a}) &= 1 + \mathrm{MAX}_i(c_A(O_i, \underline{a})) \\ &= 1 + c_A(O_n \bullet yout_n \bullet cout_{n-1} \bullet \ \ldots \ \bullet cout_1 \bullet \overline{yin}_1, \underline{a}). \end{aligned}$$

Since this is the linear composition of a set of functions, then the linear composition rule (Section 3.4.1) can be used to say immediately that the performance for the prioritiser, according to this complexity model, is given by

$$c_A(PRIOR, \underline{a}) = O(n).$$

The scheme assumes that the computation time of a complex function is the linear composition of the computation times of the individual functions from which it is constructed, and does not depend upon the details of implementation of the basic cells. Also note that the asymptotic result would not be affected if it were assumed that the cost of evaluating $\sigma(\underline{a})$ for each atomic function $\sigma : A^u \rightarrow A^v$ computing on $\underline{a} \in A^u$ were given by an arbitrary cost function $p_A(\sigma, \underline{a})$.

The complexity model is typical of those used with formal specification languages and yet, as will be seen in Section 6, it does not predict the correct asymptotic performance even for a simple function such as the prioritiser, when results are compared with those obtained empirically from a VLSI device implementing the function.

4 PERFORMANCE ESTIMATION FOR THE PRIORITISER USING RC NETWORK MODELLING OF NMOS CIRCUITS

4.1 Introduction

In this section, an *RC network modelling* for nMOS circuits, commonly used in electrical engineering applications (e.g., for logic circuit simulators), will be employed to obtain estimates of circuit computation times, and in particular, for both implementations of the prioritiser. These results can then be compared with those obtained using traditional complexity theory methodology, in Section 3.4, and from experiments performed upon the fabricated chips (see Section 6).

A suitable set of *leaf-cell characteristics*, similar to those supplied with nMOS circuit cell-libraries (see, for example, Newkirk & Mathews [1983]) will form the basis of a hierarchical RC network model, together with composition rules for the cell characteristics. When several circuit cells are composed, these rules will be used to calculate the cell characteristics required by the resultant circuit cell. In this way, the computation time for a complete circuit, constructed from a set of sub-circuit cells can be estimated. The work is based largely upon that given in Lin & Mead [1984].

The following problems must be addressed.

(i) How can a complete nMOS circuit be modelled using RC networks?
(ii) How can the delay in the RC model be estimated?
(iii) How can this model be adapted to cater for an hierarchical cellular circuit based upon a set of leaf cells, and specified using computation graphs?

For the purposes of this chapter we will limit ourselves to the simplest of circuit structures, namely those having single inputs and outputs, where the cells are joined serially. The circuits compute register-to-register transfer functions.

4.2 Representation of nMOS circuits by RC networks

At circuit level, the delay in an output port, O_i, receiving the correct result, $proj_i^n(f(\underline{x}))$, is the time taken, after the circuit has received $\underline{x}$, for that node to achieve the voltage which represents the correct logic value (e.g., this will be equivalent to the time taken for an output register to reach its switching *threshold voltage*, having initially held either a logic '0' voltage or a logic '1' voltage).

In order to estimate this signal delay, an nMOS circuit is modelled by a series of *RC networks*. For each network, the signal delay is calculated, and then these individual delays can be combined to estimate the overall (worst-case) delay through the complete nMOS circuit. This will be an estimation of the *computation time* for the circuit, as defined in Section 2.2. The problem is reduced to one of modelling an nMOS circuit by a set of suitable RC networks and then estimating the signal delay through the networks. The structure of the RC networks is determined by the state of the nMOS circuit. The term 'RC network' is used to mean only those networks which model nMOS circuits.

4.3 Definition of RC networks

For the purposes of this chapter, an RC network can be defined recursively.

(i) Basis. Resistor r in series with capacitor c (see Figure 8.10(b)), the output node being the common connection of r and c, and the input node being the free end of r. The free end of c is connected to earth.

(ii) Recursion. For two RC networks, Z_1 and Z_2, which have already been formed, a new RC network, Z, can be formed by connecting the output of Z_1 to the input of Z_2. The input to Z is identified with the input to Z_1, and the output from Z with the output from Z_2 (see Figure 8.9). All RC networks constructed in this way are *ladder networks*. Each node in such a network has a capacitance to ground, with the nodes being connected by resistors.

We begin by detailing RC network modelling, sufficient to estimate the computation time of the two prioritiser circuit implementations described in Section 2.

4.4 Definition of signal delay in an RC network

The signal delay, $\tau_D(n)$, for a node, n, in an RC network, representing an nMOS circuit, and receiving a unit step input voltage, $v(t)$, at time $t = 0$ is given by the *modified Elmore's Delay* (see Lin & Mead [1984]):

$$\tau_D(n) = \int_0^\infty [1 - y_n(t)]dt$$

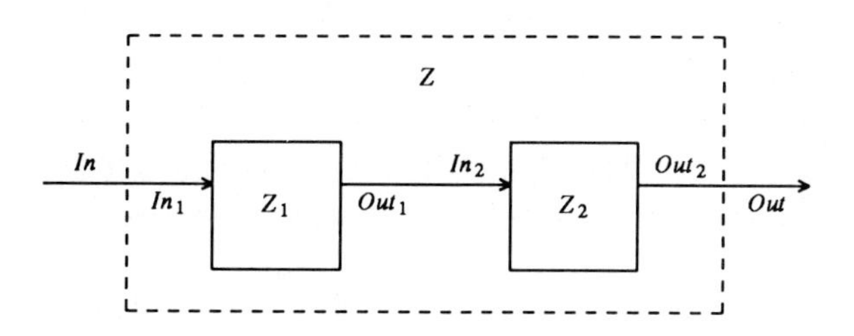

Figure 8.9. Serial composition of two RC networks

where $y_n(t)$ is the *normalised* transient voltage response of node, n (i.e., $y_n(t) \in [0, 1]$). There may be initial charge present on the nodes of the RC network. The delay $\tau_D(n)$ is equal to the area above the curve for $y_n(t)$ but below 1 (see Figure 8.10(a)).

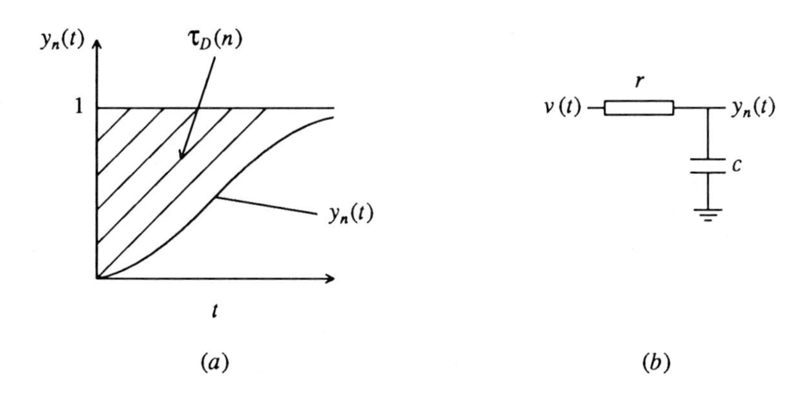

Figure 8.10. (a) Modified Elmore's delay; (b) Simple RC circuit example

As an example, the normalised transient voltage response, $y_n(t)$, for a resistor, r, in series with a capacitor, c (see Figure 8.10(b)) is given by

$$y_n(t) = 1 - \exp(-t/rc).$$

Thus, the Elmore Delay, $\tau_D(n)$, for this circuit will be given by

$$\begin{aligned}\tau_D(n) &= \int_0^\infty [1 - (1 - \exp(-t/rc))]dt \\ &= rc.\end{aligned}$$

This example shows that $\tau_D(n)$ is equal to the *time constant* for certain RC networks.

This definition of delay does not entirely agree with the discussion in Section 4.2; it relates delay to the area enclosed above the voltage response graph, instead of relating it to a specific 'threshold voltage'. It is used because it is easy to calculate (unlike the 'threshold voltage' delay), from a small set of electrical parameters for an RC network. Of particular interest to us is that a value for $\tau_D(n)$ can be calculated hierarchically.

We thus assume that the delay $\tau_D(n)$ is closely related to the notion of 'threshold voltage' delay so that it can be used to obtain an estimate for the computation time for an nMOS circuit.

4.5 Modelling nMOS circuit elements by RC networks

Before describing the methods used to calculate the delay through RC networks, we will consider how to construct a suitable network to represent an nMOS circuit.

In order to model the two different implementations of the prioritiser circuit described in Section 2.4.1 and Section 2.4.2, we need to be able to represent the following nMOS circuit elements:

(1) Pass-transistors;
(2) Signal-level-restoring inverters;
(3) Interconnect (metal, poly and diffusion wires).

As stated above, the RC representation for these basic nMOS circuit elements is in general data-dependent; it also depends upon the way the element is being used. The idea is to associate a capacitance with each node in the nMOS circuit, and then connect the nodes together with resistances, to represent nMOS transistors. The particular resistance value to be used will depend upon the type of transistor, the way it is being used, and the state of its gate; for example, an enhancement-mode transistor with a 'logic 0' voltage on its gate is represented by an open circuit (an infinite resistance), since it is switched off. Circuit nodes are charged by being connected, via resistances, to Vdd, and discharged by being connected to ground. Generally a new set of RC networks will be formed to represent the nMOS circuit every time a node changes state.

Figure 8.11 gives the necessary RC network models for the nMOS circuit elements listed above.

(1) Pass transistor:

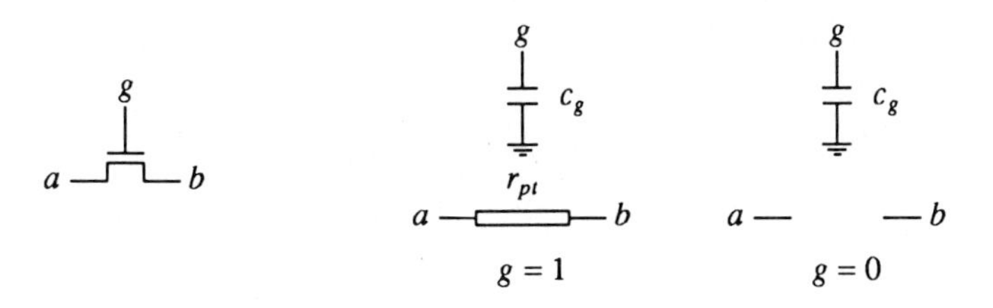

(2) Level-restoring inverter:

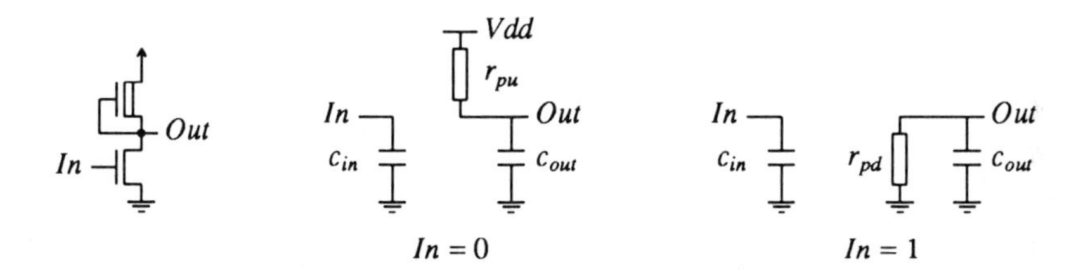

Figure 8.11. nMOS circuit elements and their representation as RC networks

For interconnect, the capacitance of each node is calculated from the circuit geometry. In most cases, interconnect resistance is negligible with respect to transistor resistances, and so can be ignored (for long wires, see Section 4.6.3). In Figure 8.11, notice that when the inverter is switching from logic 1 to logic 0, the (initially uncharged) capacitor, c_{out}, is being *charged up* via resistor, r_{pu}. When the inverter switches from logic 0 to logic 1, then (fully charged) c_{out} is *discharged* through r_{pd}. If a node contains initial charge, then its *effective* capacitance will be altered; if the node is fully charged, for example, then its effective capacitance is zero.

We will now construct simplified data dependent RC networks for each leaf cell used in the two prioritiser implementations. These will be used to estimate the circuit computation times.

4.5.1 Construction of RC network model for Prioritiser-1 For the first implementation of the prioritiser described in Section 2.4.1 no level-restoring logic is included in the pass-transistor carry-chain. Consider the prioritiser being supplied with an input word $\underline{x}$ with $x_i = 1$ and $x_j = 0$, for $j = 1, \ldots, i-1, i+1, \ldots, n$. We assume for simplicity that the pass-transistor gates in the carry-chain have already been set to their required values; this reduces each leaf-cell to one which has a single input and output. For the purpose of timing analysis, three types of leaf cell can be identified, which have distinct (data-dependent) RC network models. The following diagrams show those parts of the nMOS circuit required, together with their appropriate RC models, where the labels $r_1, \ldots, r_3$ and $c_a, \ldots, c_f$ refer to actual resistance and capacitance values which are calculated from circuit geometry. Thus for example r_1 is the resistance of a switched-on pass-transistor, and c_a is the capacitance of circuit node *cin* in prioritiser cells $p_1, \ldots, p_i$.

(i) The *head* cell, contains the header node and initial *ENABLE* pass-transistor:

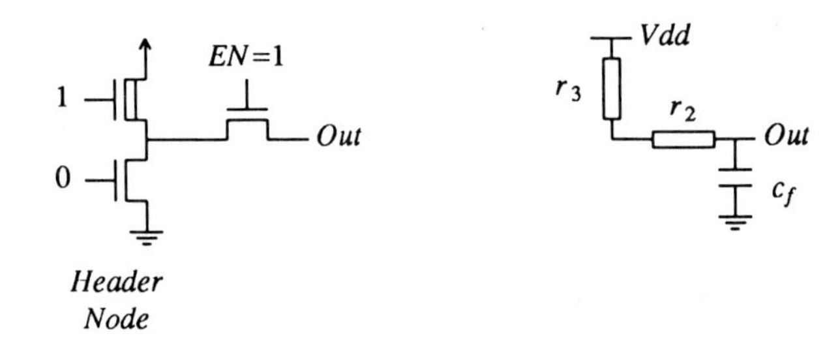

(ii) The *middle* Prioritiser-1 cells p_j, for $j = 1, \ldots, i-1$, where $\overline{yin}_j = 1$ and $yin_j = 0$:

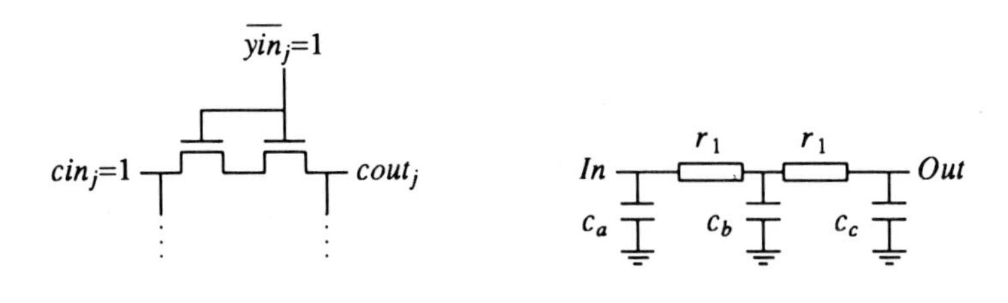

(iii) The *tail* cell, containing prioritiser cell p_i, with $yin_i = 1$ and $\overline{yin_i} = 0$, the output $ENABLE$ transistor and the output register transistor gate:

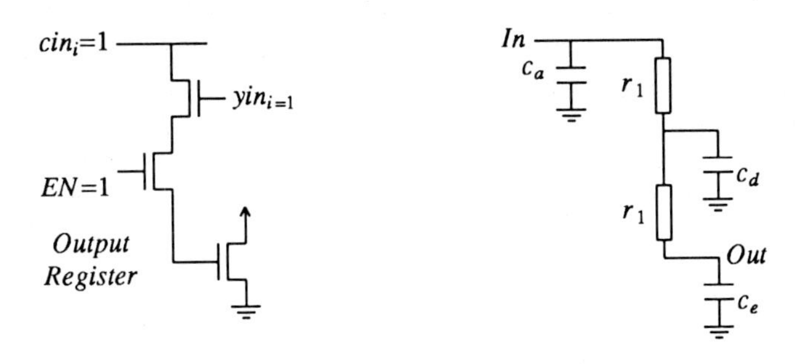

The leaf cells can now be joined together to form the complete prioritiser circuit. Figure 8.12 shows the complete nMOS circuit.

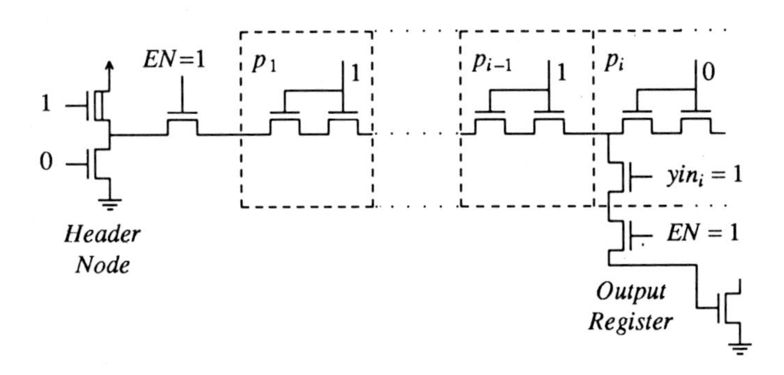

Figure 8.12. Prioritiser circuit: Implementation 1

The carry signal originates in the header node (a superbuffer), and passes through the $ENABLE$ pass transistor into the prioritiser circuit-of-interest. It propagates along the carry-chain through $i-1$ prioritiser cells, $p_1, \ldots, p_{i-1}$, and is finally steered into output register cell O_i by prioritiser cell p_i.

This is represented by the RC network shown in Figure 8.13, where the individual RC networks for each leaf-cell join to form an *RC ladder* network.

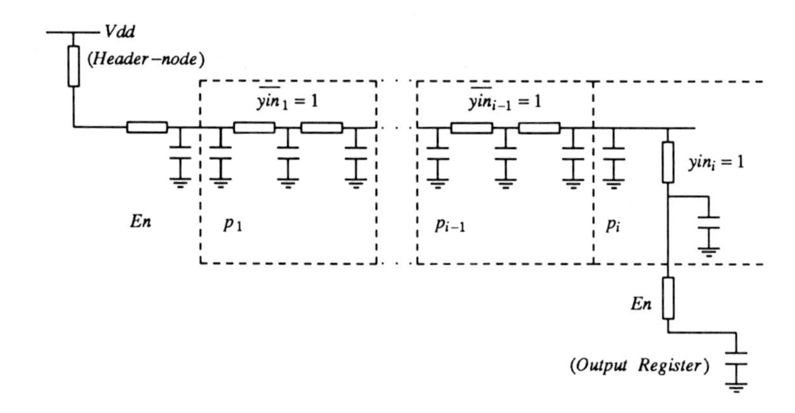

Figure 8.13. RC network representation of prioritiser circuit: Implementation 1

4.5.2 Construction of RC network model for Prioritiser-2 Due to the introduction of level-restoring inverters into the prioritiser cells, the pass-transistor carry-chain is broken into sections containing two pass-transistors. Again, we consider the circuit receiving an input word $\underline{x}$ with $x_i = 1$ and $x_j = 0$, for $j = 1, \ldots, i-1, i+1, \ldots n$, and assume that the pass-transistor gates in the carry-chain have already been set to their required values. The labels $r_1, \ldots, r_5$ and $c_1, \ldots, c_8$ refer to circuit resistance and capacitance values. The following set of diagrams show the leaf cells used in the second implementation, together with their equivalent RC networks, which are data dependent.

(i) The *front* cell, containing, as above, the header node, and initial *ENABLE* pass-transistor:

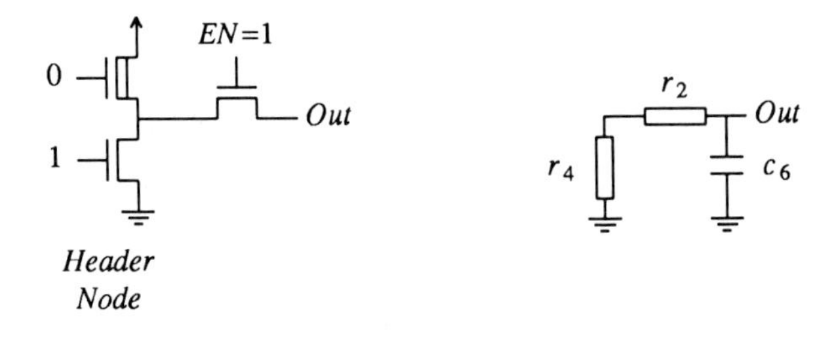

(ii) The *centre* level-restoring Prioritiser-2 cells, p_j, for $j = 1, \ldots, i-1$, with $\overline{yin}_j = 1$ and $yin_j = 0$:

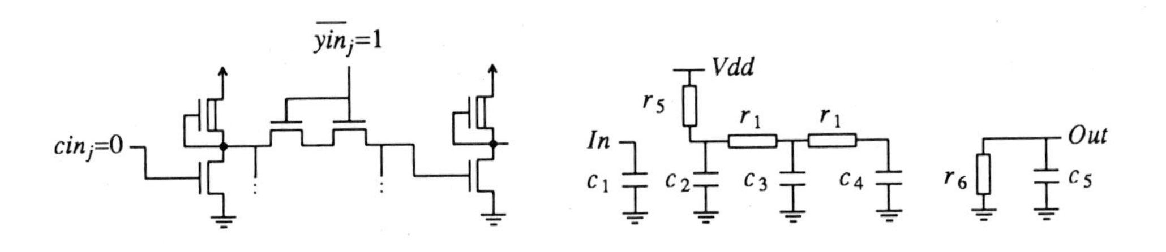

(iii) The *back* cell, containing Prioritiser-2 cell p_i, with $yin_i = 1$ and $\overline{yin}_i = 0$, the output $ENABLE$ transistor and the output register transistor gate:

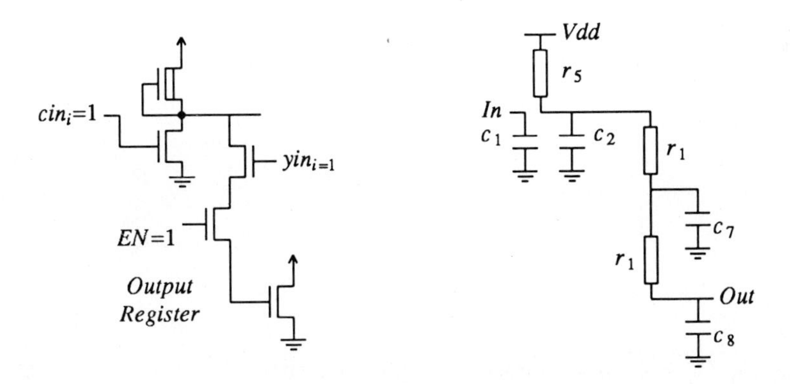

We can now join these leaf cells to form the complete Prioritiser-2 circuit. Figures 8.14 and 8.15 show the complete nMOS circuit together with its RC network model equivalent.

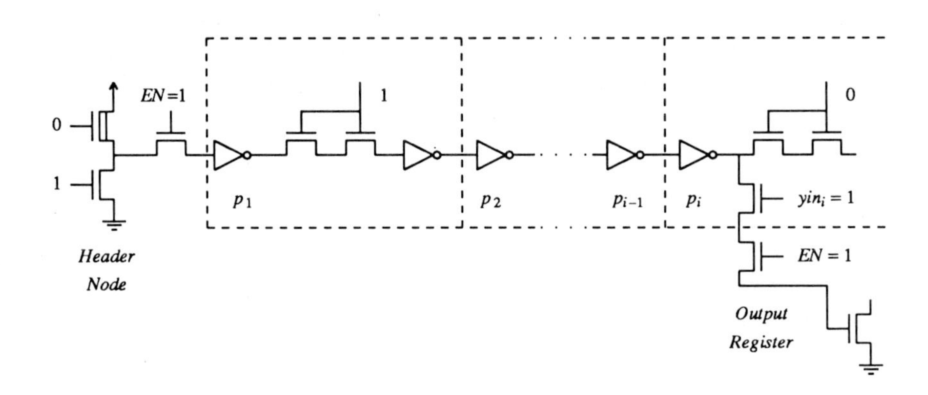

Figure 8.14. Prioritiser circuit: Implementation 2

The RC model consists of a sequence of small independent RC networks, with each network input node being controlled by the previous network output node, similar to the example used in Section 4.6.1.

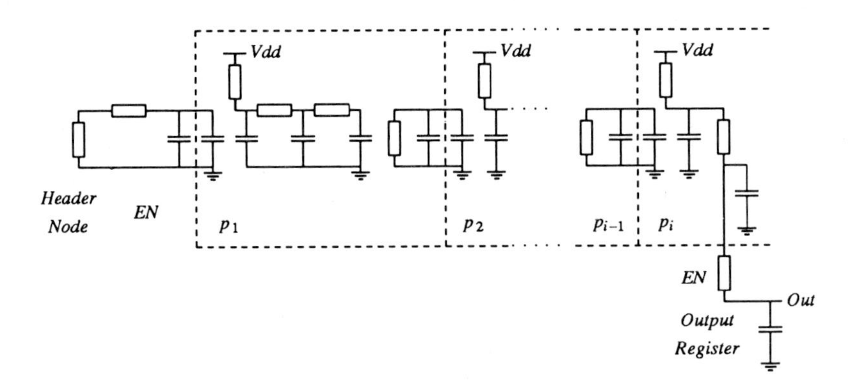

Figure 8.15. RC network representation of prioritiser circuit: Implementation 2

4.6 Calculation of signal delay through RC networks

4.6.1 Signal delay through a sequence of RC networks In general, several RC networks will be used to model a single nMOS circuit, C_f. It is assumed that C_f has a single input port, IN, and output port, OUT, and is modelled by a sequence of independent two-port RC networks, $Z_1, \ldots, Z_k$, where, for example, the output node, N_1 of network Z_1 determines the time at which network Z_2 receives its input. (The output of Z_i typically will represent the gate of a pass transistor modelled in Z_{i+1}, where the transistor is connected either to Vdd or to Ground.)

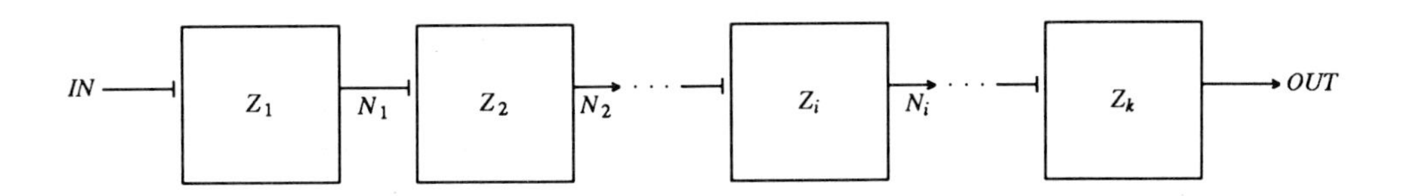

Figure 8.16. Sequence of interacting RC networks

Z_2 cannot operate until Z_1 has completed, i.e., until a time $\tau_D(N_1)$. If the delays through the individual networks are given by $\tau_D(N_i)$, as defined in Section 4.4, then the signal delay in reaching output port OUT through C_f is given by:

$$\tau_{OUT} = \sum_{i=1}^{k} \tau_D(N_i) \tag{1}$$

i.e., the delay through a sequence of *interacting* RC networks modelling an nMOS circuit is simply the *linear composition* of their individual delays.

4.6.2 Calculation of delay through individual RC networks In Section 4.6.1, the delay through a sequence of independent (but *interacting*) RC networks was given as the linear composition of the individual network delays. We will now calculate the delay for the individual RC networks, that are required to model the prioritiser circuit implementations. These are all *RC ladder* networks, as defined in Section 4.3.

For each network, three parameters are required.

R: series input–output resistance.
C: Total network capacitance to ground.
D: Internal network delay, when its output node is open.

For the primitive case these parameters are given by

$$R = r,$$
$$C = c,$$
$$D = rc \quad \text{(the \textit{time constant} of the network).}$$

When two networks Z_1 and Z_2 with parameters R_1, C_1, D_1, and R_2, C_2, D_2 respectively, are connected serially (the output of Z_1 is connected to the input of Z_2 as shown in Figure 8.9), then the new parameters for the resulting network are given by

$$R = R_1 + R_2, \tag{2a}$$
$$C = C_1 + C_2, \tag{2b}$$
$$D = D_1 + D_2 + R_1 \cdot C_2, \tag{2c}$$

Thus it can be shown, for example, that the delay for a uniform n-element RC Ladder, with resistances, R, and total node capacitances, C, is given by

$$D = \sum_{i=1}^{n} i \cdot RC$$
$$= \frac{n(n+1)}{2} R \cdot C.$$

4.6.3 Delay in long wires The delay through a wire with significant resistance can be approximated by the Π-network shown in Figure 8.17, where r is the total resistance of the wire, and c is the total capacitance.

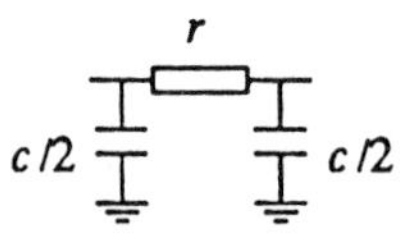

Figure 8.17. Π-network approximation for interconnect

This information can be used in circuit design, in order to select the appropriate size of superbuffer to drive the wire load. For most of the wires used in the prioritiser, the resistances are insignificant compared with the resistance of a transistor and can therefore be ignored (however, see Section 6.9 for examples of its use).

4.6.4 Calculation of delay parameters for circuit leaf-cells The basic nMOS circuit elements which we wish to model using RC networks are circuit *leaf-cells* (see Section 5.4 for definition). In Implementation 1 of the prioritiser, the leaf cells, p_i, are modelled essentially by single RC networks (i.e., for the carry-chain). The leaf-cell parameters can thus be calculated using equation (2).

However, in Implementation 2, there are several independent RC network elements in each prioritiser leaf cell (see Figure 8.15). Both equation (1) and equation (2) must be combined to indicate that there is no continuous input–output resistance through the leaf cell.

Let an additional cell parameter, B, be introduced for each circuit cell, c_i, where

$$B = \begin{cases} 1 & \text{if there } is \text{ an electrical connection between input} \\ & \text{and output for cell } c_i, \\ 0 & \text{if there is } no \text{ electrical conection.} \end{cases}$$

Equation (2) can be suitably modified to incorporate the linear composition rule (equation (1))

$$R = B_2 \cdot R_1 + R_2, \tag{3a}$$
$$C = C_1 + B_1 \cdot C_2, \tag{3b}$$
$$D = D_1 + D_2 + R_1 \cdot C_2, \tag{3c}$$

with an additional equation to compose the B parameters:

$$B = B_1 \cdot B_2. \tag{3d}$$

We can now proceed to calculate the *leaf-cell parameters* for each implementation of the prioritiser, and thus estimate their computation times.

4.6.5 Calculation of delay through RC network model for Prioritiser-1 Consider the prioritiser receiving an input word, $\underline{x} \in \mathbf{B}^n$, where $x_i = 1$ and $x_j = 0$, $j = 1, \ldots, i-1, i+1, \ldots, n$. We begin by calculating the *leaf-cell parameters* for each of the three different types of leaf cell given in Section 4.5.1 (capital letters are used, with suitable subscripts, to refer to the required leaf-cell parameters).

(1) The *head leaf cell* with parameters R_h, C_h, D_h and B_h:

$$R_h = r_2 + r_3,$$
$$C_h = 0,$$
$$D_h = (r_2 + r_3) \cdot c_f,$$
$$B_h = 0.$$

(2) The *middle* Prioritiser-1 cells $p_1, \ldots, p_{i-1}$ with parameters R_m, C_m, D_m, and B_m:

$$R_m = 2r_1,$$
$$C_m = c_a + c_b + c_c,$$
$$D_m = r_1 \cdot (c_b + 2c_c),$$
$$B_m = 1.$$

(3) The *tail cell* with parameters R_t, C_t, D_t and B_t:

$$\begin{aligned} R_t &= 2r_1, \\ C_t &= c_a + c_d + c_e, \\ D_t &= r_1 \cdot (c_d + 2c_e), \\ B_t &= 1. \end{aligned}$$

Note that equation (3) is used to calculate these parameters.

4.6.5.1 Computation time for Prioritiser-1 For an n-bit prioritiser (Implementation 1), $prior1_n$, with input word, $\underline{x}$, the computation time can be estimated, using only leaf-cell parameters, as follows:

$$\begin{aligned} \tau_{prior1_n}(\underline{x}) \approx D_h + (i-1)D_m + D_t + \frac{(i-2)(i-1)}{2} \cdot R_m C_m \\ + (R_h + (i-1)R_m) \cdot C_t + (i-1) \cdot R_h C_m \end{aligned} \tag{4}$$

where $\underline{x} \in \mathbf{B}^n$, $x_i = 1$, and $x_j = 0$, $j = 1, \ldots, i-1, i+1, \ldots, n$.

This estimate is obtained by repeated use of equation (3), in order to combine the leaf-cell parameters.

4.6.5.2 Worst-case computation time for Prioritiser-1 From equation (4) it can be seen that the worst case computation time for an n-bit prioritiser, as defined in Section 2.2, will occur for $i = n$, i.e.,

$$\begin{aligned} \underset{\underline{x}\in\mathbf{B}^n}{\mathrm{MAX}}(\tau_{prior1_n}(\underline{x})) \approx D_h + (n-1)D_m + D_t + \frac{(n-2)(n-1)}{2} \cdot R_m C_m \\ + (R_h + (n-1)R_m) \cdot C_t + (n-1) \cdot R_h C_m. \end{aligned}$$

Thus, using an n-bit prioritiser circuit, the *worst-case* computation time of an i-bit prioritiser, $prior1_i$, can be estimated, for $i = 1, \ldots, n$:

$$\begin{aligned} \underset{\underline{y}\in\mathbf{B}^i}{\mathrm{MAX}}(\tau_{prior1_i}(\underline{y})) \equiv \tau_{prior1_n}(\underline{x}) \approx D_h + (i-1)D_m + D_t + \frac{(i-2)(i-1)}{2} \cdot R_m C_m \\ + (R_h + (i-1)R_m) \cdot C_t + (i-1) \cdot R_h C_m, \end{aligned}$$

with

$$\underline{y} \in \mathbf{B}^i, y_l = 1, y_k = 0, k = 1, \ldots, l-1, l+1, \ldots, i,$$

and

$$\underline{x} \in \mathbf{B}^n, x_i = 1, x_j = 0, j = 1, \ldots, i-1, i+1, \ldots, n.$$

4.6.6 Calculation of delay through RC network model for Prioritiser-2 We first calculate the leaf-cell parameters (identified using capital letters with subscripts) required for each of the three leaf-cell types described in Section 4.5.2, before combining them together to estimate the delay through the complete prioritiser circuit. The prioritiser is supplied with input word $\underline{x}$, with $x_i = 1$, and $x_j = 0$, for $j = 1, \ldots, i-1, i+1, \ldots, n$.

(i) The *front* cell has cell parameters R_f, C_f, D_f and B_f, given by

$$\begin{aligned} R_f &= r_4 + r_2, \\ C_f &= 0, \\ D_f &= (r_2 + r_4) \cdot c_6, \\ B_f &= 0. \end{aligned}$$

(ii) The *centre* Prioritiser-2 cells, $p_1, \ldots, p_{i-1}$, with cell parameters R_c, C_c, D_c, and B_c:

$$\begin{aligned} R_c &= r_6, \\ C_c &= c_1, \\ D_c &= r_5 c_2 + (r_1 + r_5) \cdot c_3 + (2r_1 + r_5) \cdot c_4 + r_6 c_5, \\ B_c &= 0. \end{aligned}$$

(c) The *back cell* with parameters R_b, C_b, D_b, and B_b:

$$\begin{aligned} R_b &= r_5 + 2r_1, \\ C_b &= c_1, \\ D_b &= r_5 c_2 + (r_1 + r_5) \cdot c_7 + (2r_1 + r_5) \cdot c_8, \\ B_b &= 0. \end{aligned}$$

4.6.6.1 Computation time for Prioritiser-2 For an n-bit prioritiser (Implementation 2), $prior2_n$, the computation time can be estimated by

$$\tau_{prior2_n}(\underline{x}) \approx D_f + (i-1) \cdot D_c + D_b + R_f C_c + (i-2) \cdot R_c C_c + R_c C_b. \tag{5}$$

4.6.6.2 Worst-case computation time for Prioritiser-2 For an n-bit prioritiser (Implementation 2), it can be seen from equation (5) that the worst-case computation time will occur, as with Implementation 1, for $i = n$:

$$\underset{\underline{x} \in \mathbf{B}^n}{\mathrm{MAX}}(\tau_{prior2_n}(\underline{x})) \approx D_f + (n-1) \cdot D_c + D_b + R_f C_c + (n-2) \cdot R_c C_c + R_c C_b.$$

Thus the *worst-case* computation time of an i-bit prioritiser, $prior2_i$, can now be estimated, for $i = 1, \ldots, n$:

$$\underset{\underline{y} \in \mathbf{B}^i}{\mathrm{MAX}}(\tau_{prior2_i}(\underline{y})) \equiv \tau_{prior2_n}(\underline{x}) \approx$$
$$D_f + (i-1) \cdot D_c + D_b + R_f C_c + (i-2) \cdot R_c C_c + R_c C_b,$$

with

$$\underline{y} \in \mathbf{B}^i, y_l = 1, y_k = 0, k = 1, \ldots, l-1, l+1, \ldots, i,$$

and

$$\underline{x} \in \mathbf{B}^n, x_i = 1, x_j = 0, j = 1, \ldots, i-1, i+1, \ldots, n.$$

5 RC DELAY MODEL AND COMPUTATION GRAPHS

5.1 Introduction

In the previous section, a set of leaf cell parameters was formed for each different (data-dependent) leaf cell used in a simplified model of the prioritiser. These were then combined, using equation (3) in order to estimate the delay through the complete prioritiser. The composition rules (equation (3)) will now be adapted so that they apply to a circuit modelled by *computation graphs* (see McEvoy [1986]). Each node in a graph will be restricted to single input and output, with serial composition of nodes. Resulting circuit models will therefore consist of linear sequences of sub-circuit modules. This restricted computation graph representation will be sufficient to construct the simplified model of the prioritiser (where the pass-transistor gates along the carry-chain are assumed to have been set to their appropriate values before the header $ENABLE$ gate is switched on at the start of computation). Therefore each leaf cell has a single input and output.

In this section R_v, C_v, D_v, and B_v are used to represent the leaf cell parameters of primitive node v, such as those calculated for the prioritiser in Section 4.6.5 and Section 4.6.6. A set of composition functions is now introduced in order to determine electrical (delay) characteristics for a composition node; these are given by $R(v)$, $C(v)$, $D(v)$ and $B(v)$ where v is a node in a computation graph, $G \in G(\Gamma, n)$. The priorister will be used as an example to illustrate the use of these functions.

5.2 RC network delay composition functions

Let v_{in}, v and v_{out} be any nodes, within a computation graph $G \in G(\Gamma, n)$.

(i) $R(v)$ is the total driving resistance at the output of v, summed from v_{in} to v, and taking into consideration the electrical continuity (given by B) of the intervening nodes.

(ii) $C(v)$ is the total capacitance to be driven at the input to v, summed from v_{out} to v, again taking into consideration the electrical continuity of the intervening nodes.

(iii) $D(v)$ is the total signal delay between v_{in} and the open-circuit output of v.

(iv) $B(v)$ is a flag indicating the electrical continuity between v_{in} and the output of v

5.2.1 Functionality

(i) $D : V \cup \{v_{in}, v_{out}\} \to \mathbf{R}$,
(ii) $R, B : V \cup \{v_{in}\} \to \mathbf{R}$,
(iii) $C : V \cup \{v_{out}\} \to \mathbf{R}$.

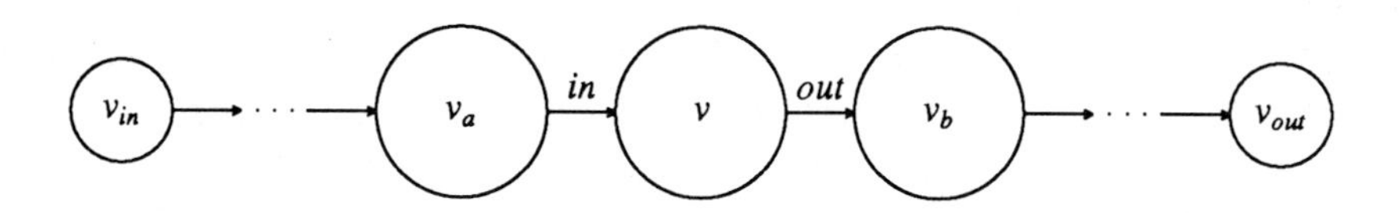

Figure 8.18. Node v embedded in a computation graph

5.2.2 Primitive case If v is a primitive node (i.e., v is a node in a primitive computation graph $G \in G(\Gamma, 0)$), and the *leaf-cell parameters* for v are R_v, C_v, D_v and B_v, then

$$R(v) = R_v, C(v) = C_v, D(v) = D_v, \quad \text{and} \quad B(v) = B_v,$$

i.e., the composition functions are simply equal to the leaf-cell parameters.

5.2.3 Special cases

(i) For the input and output nodes v_{in} and v_{out},

$$R(v_{in}) = 0, D(v_{in}) = 0, B(v_{in}) = 1, \quad \text{and} \quad C(v_{out}) = 0,$$

(ii) If $HEADS_D(v) = \emptyset$,

$$\begin{aligned} R(v) &= R(tl \bullet cout_v(out)), \\ D(v) &= D(tl \bullet cout_v(out)), \\ B(v) &= 0, \\ C(v) &= 0 \quad (\text{ there is no input to } v) \end{aligned}$$

(If $TAILS_D(v) = \emptyset$ then v has no outputs and is thus not computationally useful.)

5.2.4 Composite case with $HEADS_D(v) \neq \emptyset$ If v is a node in computation graph $G \in G(\Gamma, n)$ (see Figure 8.18) then the composition functions are given by

$$R(v) = B(tl \bullet cout_v(out)) \bullet R(v_a) + R(tl \bullet cout_v(out)), \tag{6a}$$

$$D(v) = D(v_a) + D(tl \bullet cout_v(out)) + R(v_a) \bullet C(hd \bullet cin(in)), \tag{6b}$$

$$C(v) = B(tl \bullet cout_v(out)) \bullet C(v_b) + C(hd \bullet cin_v(in)), \tag{6c}$$

$$B(v) = B(v_a) \bullet B(tl \bullet cout_v(out)). \tag{6d}$$

5.3 Prioritiser-1 example

In this section, the Implementation 1 prioritiser from Section 4.5.1 will be represented as a computation graph in order to calculate its delay using the composition functions defined above. Recall that leaf cell parameters R_x, C_x, D_x, and B_x, where $x = h, m, t$, were calculated in Section 4.6.5.

5.3.1 Computation graph representation of prioritiser The following primitive computation graphs $H, M, T \in G(\Gamma, 0)$ represent the leaf cells given in Section 4.5.1:

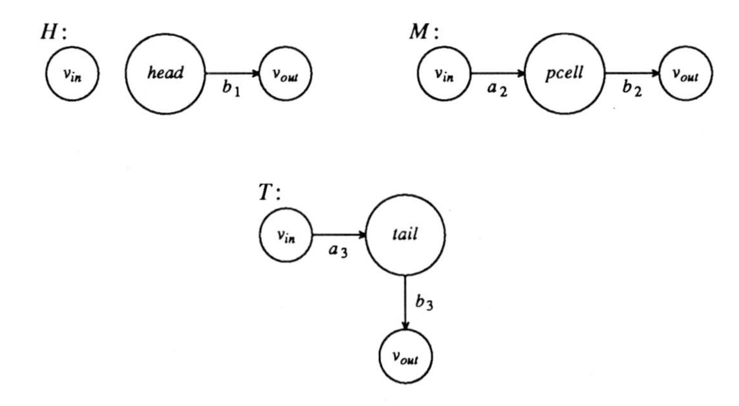

Now the complete prioritiser is represented by computation graph *PRIOR*:

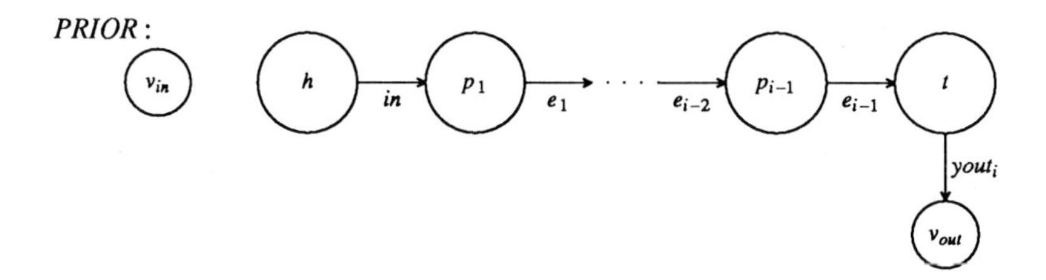

$$gf(h) = H, \quad cout_h(in) = b_1,$$

$$gf(p_j) = M, \quad cout_{p_j}(e_j) = b_2, j = 1, \ldots, i-1,$$

$$cin_{p_1}(in) = a_2, \quad cin_{p_k}(e_{k-1}) = a_2, k = 2, \ldots, i-1,$$

$$gf(t) = T, \quad cout_t(yout_i) = b_3, \quad cin_t(e_{i-1}) = a_3,$$

$$tl \bullet cout_h(in) = head, \quad tl \bullet cout_{p_j}(e_j) = pcell, \quad tl \bullet cout_t(yout_i) = tail.$$

5.3.2 Calculation of delay through computation graph PRIOR For the primitive computation graphs H, M, and T the composition functions are equal to the leaf-cell parameters:

$$R(head) = R_h, \quad C(head) = 0, \quad D(head) = D_h, \quad B(head) = B_h,$$

$$R(pcell) = R_m, \quad C(pcell) = C_m, \quad D(pcell) = D_m, \quad B(pcell) = B_m,$$

$$R(tail) = R_t, \quad C(tail) = C_t, \quad D(tail) = D_t, \quad B(tail) = B_t,$$

The composition functions of Section 5.2 can now be applied to computation graph *PRIOR*; the resulting equation should be identical to equation (4) from Section 4.6.5.1:

$$\begin{aligned} D(t) &= D(p_{i-1}) + D(tl \bullet cout_t(yout_i)) + R(p_{i-1}) \bullet C(hd \bullet (cin_t(e_{i-1})) \\ &= D(p_{i-1}) + D(tail) + R(p_{i-1}) \bullet C(tail) \\ &= D(p_{i-1}) + D(tail) + R(p_{i-1}) \bullet C_t \end{aligned} \tag{7}$$

where

$$D(p_{i-1}) = D(p_{i-2}) + D(pcell) + R(p_{i-2}) \bullet C(pcell) \tag{8}$$

and

$$R(p_{i-1}) = B(tl \bullet cout_{p_{i-1}}(e_{i-1})) \bullet R(p_{i-2}) + R(tl \bullet cout_{p_{i-1}}(e_{i-1})) \tag{9}$$

Equations (8) and (9) are fully expanded down to leaf-cell parameters, and substituted into equation (7), which then becomes

$$\begin{aligned} D(t) =& D_h + (i-1)D_m + D_t + \frac{(i-2)(i-1)}{2} \cdot R_m C_m \\ &+ (R_h + (i-1)R_m) \cdot C_t + (i-1) \cdot R_h C_m. \end{aligned}$$

This is the same as equation (4) as required.

5.4 Leaf cells and leaf-cell characteristics

In this chapter, the following definition of a circuit leaf cell is used.

5.4.1 Definition A *leaf cell* is a circuit module that is represented by a primitive computation graph.

Every primitive node (i.e., a node that is in a primitive computation graph) will be given a set of leaf-cell characteristics by the circuit designer. The set would include physical characteristics such as cell dimensions (to be used by layout tools), functional characteristics (to be used by formal specification languages) and electrical characteristics, required by the RC network delay model. The characteristics of a composite node will be supplied via appropriate composition functions, equivalent to equation (6).

We will now describe how the electrical leaf-cell characteristics can be estimated.

5.4.2 Estimation of leaf-cell characteristics

(i) **Interconnect capacitance and resistance**: can be estimated by examining circuit geometry, with knowledge of sheet capacitances and resistances for the target fabrication process (see for example Pucknell & Eshraghian [1985]).

(ii) **Electrical continuity**: can be found by examining the device-level circuit contained within the leaf cell

(iii) **Transistor resistances**: the device level simulator SPICE is used in order to estimate resistances for the different types and usage of transistor. By varying capacitance, c, and obtaining the signal delay, τ_i $(i = 1, \ldots, 3)$ from SPICE, transistor resistances r_1, r_2 and r_3 in Figure 8.19 can be found from

$$\tau_1 = r_2 \cdot c',$$
$$\tau_2 = (r_1 + r_2) \cdot c',$$
$$\tau_3 = (r_2 + r_3) \cdot c'.$$

c' consists of a fixed capacitance c_0, to account for transistor capacitances, together with the capacitance, c which is varied in the circuit modelled by SPICE.

Once all resistances and capacitances are known for the circuit elements contained in a leaf cell v, then R_v and D_v can be found by using equation (3).

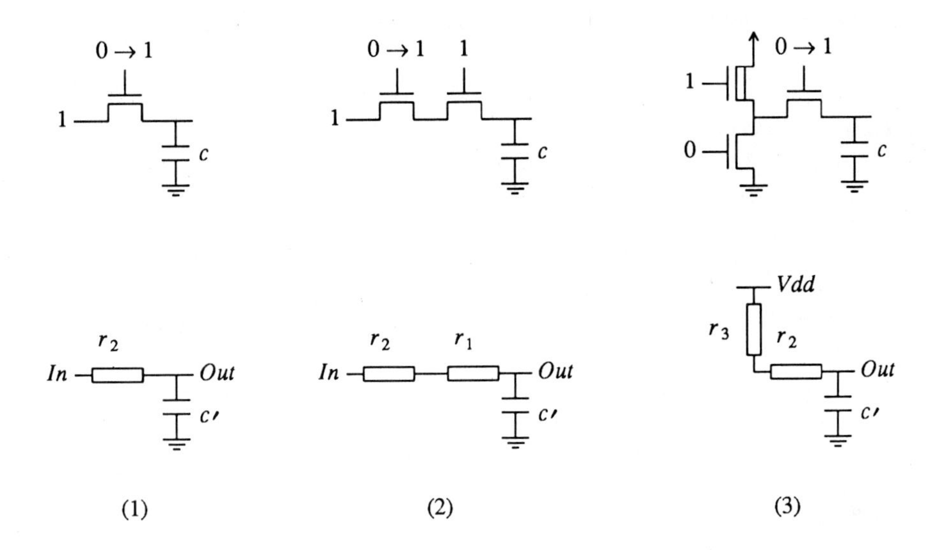

Figure 8.19. Examples of nMOS circuits used to find transistor resistances.

6 IMPLEMENTATION OF THE PRIORITISER EXPERIMENT

6.1 INTRODUCTION

In this section, an outline of the architecture for the prioritiser chip will be given, together with a summary of its main features. The prioritiser experiment will be described, the results being compared with those obtained in Section 4.

6.2 Prioritiser chip architecture

Each chip contains the following three independent prioritiser circuits:

(1) one 50-bit 'simple pass-transistor' carry-chain circuit (Circuit-1) using Implementation 1 pcells described in Section 2.4.1
(2) one 50-bit 'level-restoring' carry-chain circuit (Circuit-2) using Implementation 2 pcells described in Section 2.4.2
(3) one 13-bit 'simple pass-transistor' carry-chain circuit (Circuit-3), where a 'dummy cell' is inserted between each Implementation 1 pcell and the next. A dummy cell simply consists of convoluted diffusion routing linking the carry line (see Section 6.9.2).

Input and output registers are included with each prioritiser circuit-of-interest, together with a *comparator* circuit, to compare input word $\underline{x}$ and output word $PRIOR(\underline{x})$.

The comparator produces a single-bit result for each attempted $ENABLE$ of the prioritiser. Figure 8.20 shows the architecture of a single prioritiser circuit embedded in its service circuitry.

The chip uses a two-phase non-overlapping clock, where data is loaded and unloaded during ϕ_1 and control signals (except for $ENABLE$ which is unclocked) are loaded during ϕ_2. The following signals are required.

> *LOAD-CYCLE (L)*: to load/cycle data in input register,
> *RESET-NOTRESET and RESET-VALUE (RV)*: to initialise prioritiser circuit of interest and output register,
> *ENABLE (E)*: to allow prioritiser computation to take place,
> *UNLOAD* and *CYCLEOUT (U)*: to cycle/unload data in output register,
> *COMPARE (COMP)*: to compare contents of input and output registers.

6.2.1 Restriction on input word Noticc that if the input word $\underline{x}$ is restricted as follows:

$$x_j = 0, j = 1, \ldots, i-1, i+1, \ldots, n,$$
$$x_i = 1,$$

i.e., $\underline{x}$ contains a single logic 1 in position i, then the output word $PRIOR(\underline{x})$, will be the same as $\underline{x}$, if sufficient $ENABLE$ time has been allowed for the implemented circuit to function correctly. By comparing the input and output registers, it is possible to determine easily whether or not the circuit has computed the correct result, without having to inspect the whole output register. In the implemented prioritiser, an on-chip comparator is used in conjunction with the input word restriction.

6.3 Cell descriptions

We give a brief description of the leaf-cells and composite-cells used in the prioritiser chip design:

(i) Three types of pcell to form the prioritiser circuit.

- (a) *simple pass-transistor pcell*: basic pass-transistor network described in Section 2.4.1.
- (b) *level-restoring pcell* (see Section 2.4.2): two level-restoring inverters are inserted in the carry chain, and the pcell pass-transistor ratios

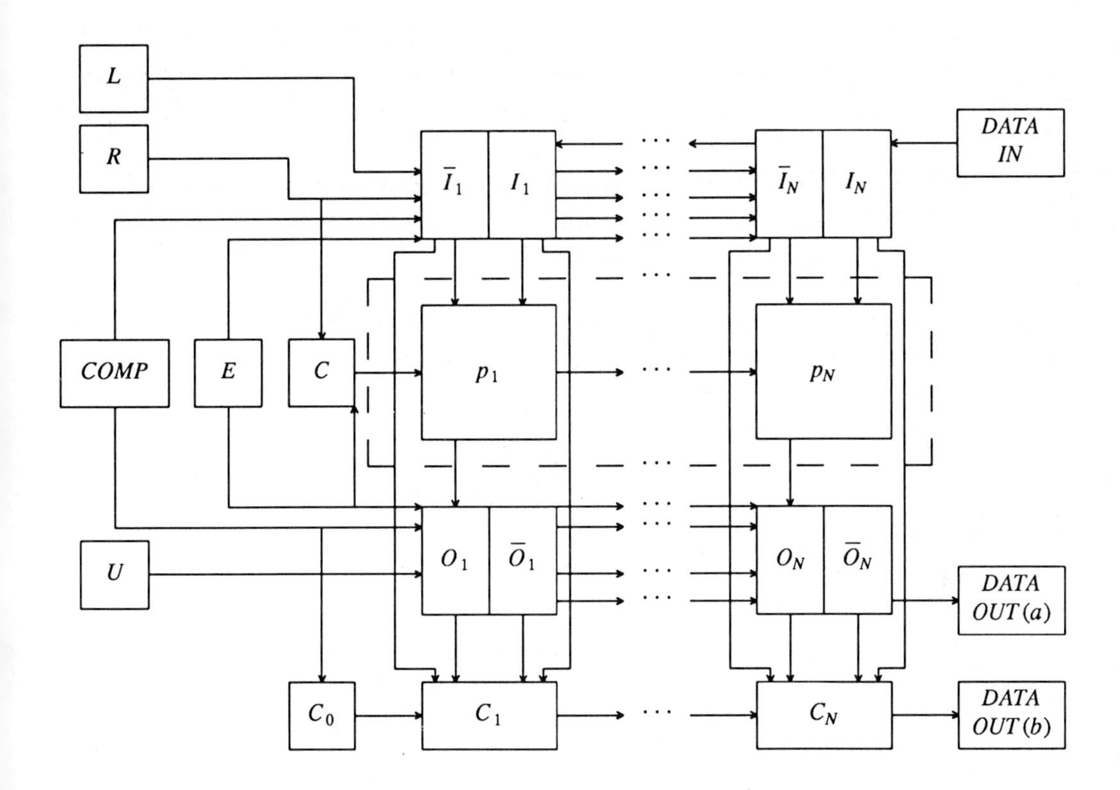

Key to diagram:
p_i - prioritiser pcell (see Section 2.4.1 and Section 2.4.2).
C - Carry-chain Header-node supplying cin_1 signal.
$I_j, \overline{I}_j$ - Input cell holding j input bit (and its complement) and associated control circuitry.
$O_i, \overline{O}_i$ - Output cell holding ith output bit (and its complement) and associated control circuitry.
C_i - Comparator cell, comparing ith input and output bits.
C_0 - Signal node to first comparator cell (i.e., logic 0).
$L, R, COMP, E, U,$ - control cells providing control signals.
$DATA\ OUT(a)$ produces the serial N-bit word from the output shift register.
$DATA\ OUT(b)$ produces the single output bit from comparator circuit.

Note that power and clock lines are not included. The dashed box encloses the *circuit-of-interest*, for which the computation time is to be found for various input words.

Figure 8.20. Architecture of prioritiser

are adjusted (to W:L = 1:3); with the addition of inverters, the pcell is equivalent to a three-input NAND gate and the pull-up:pull-down transistor resistance ratio must be maintained at an overall 4:1 (see Pucknell & Eshraghian [1985]).

(c) *dummy cell*: contains only convoluted diffusion track of varying length to continue the carry line, and allows data to flow from input to output register cells without change.

(ii) *bit-slice*: pcell with input/output shift register cells. A comparator is included, except with the dummy cell where it is unnecessary.

(iii) *word*: 25 bit-slices (a word with 25 basic pcells would form a 25-bit prioritiser, and two words can then be joined to form a 50-bit prioritiser).

Additional cells to service the prioritiser.

(iv) *shift register cell*: standard 2-inverter cell with cycle and shift, together with required control circuitry for enabling, resetting and comparison.
(v) *Comparator Cell*, C_j: output = 0 if $I_j = O_j$ and $C_{j-1} = 0$; output = 1 otherwise (i.e., *negative logic* is used).
(vi) *control-drivers*: superbuffer to drive control lines, one per line, per word.
(vii) *input/output pads*: standard pads to interface with external circuitry and provide protection to chip.
(viii) *interconnect, power/ground lines.*

All geometry is contained within leaf cells which then butt together to form the complete circuit.

6.4 Main features of circuit design

(1) *Timing.* The most important feature on the chip is the *ENABLE* control line, which isolates the circuit-of-interest from its service circuitry; only when this signal is raised can the prioritiser receive data from its input register, I and carry-chain header node, C, and write to its output register, O. The *ENABLE* pass transistor gates are supplied via a superbuffer, with each superbuffer supplying only 25 transistor gates, thus allowing for fast and symmetric rise/fall edges.
(2) *Reset.* It is vital to initialise the chip to a known state between each *ENABLE*. This is achieved using *RESET-NOTRESET* and *RESET-VALUE* control lines, which allow the prioritiser circuit-of-interest (both the carry-line and all pass-transistor gates) and the output register to be initialised to logic 0.

(3) *Comparator circuit.* In order to reduce the amount of data required to be output after each *ENABLE*, the on-chip comparator is included, to compare input and output words, producing a one-bit output. This helps to simplify the automatic test program used to control the DAS 9100 (see Section 6.6). An additional control line, *COMP*, is used to isolate the comparator circuitry (which includes long diffusion/poly wires) from the input and output cells, and thus reduces loading, especially on the input cells, when the comparator is not being used.

6.5 Fabrication

The design was laid out using the KIC design tool, conforming to an extended set of design rules based on those given in Mead & Conway [1980], so that the design would pass on Design Rule Checkers (DRCs) at all potential fabrication plants. The chips were fabricated at the Edinburgh SERC facility, and of twenty samples returned, four function correctly.

6.6 Chip test environment

Figure 8.21 shows the test equipment used for testing both the correct functioning of the prioritiser chips and for determining their computation times.

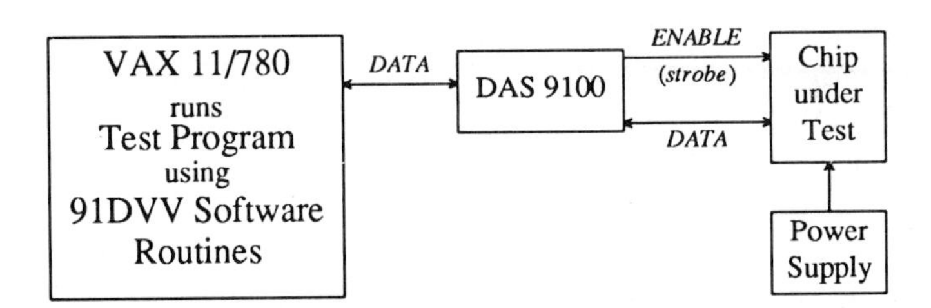

Figure 8.21. Prioritiser chip test environment

A Tektronix DAS 9100 Digital Analysis System forms the basis of the test equipment. It applies test vector sequences to the prioritiser chip and reads the resultant data words off the chip.

The Tektronix 91DVV software package routines on the VAX 11/780 allow the DAS to be controlled remotely, thus permitting full automation of chip testing. An automatic test program, employing these routines to communicate with the DAS, generates the necessary vectors to test the chip, sets a suitable *ENABLE* duration, and obtains results from the DAS.

The chip is mounted in a simple VERO wirewrap board that facilitates connection with the DAS probes. The *ENABLE* signal line is supplied by a DAS *strobe* line. This allows a single pulse of variable duration to be applied (minimum duration of 40 ns with increments of δw =40 ns), and connects directly to the chip pin, to reduce possible signal distortion from unnecessary wiring on the VERO-board.

6.7 Experimental procedure

Due to the availability of the necessary software to control the DAS remotely, all that is required to determine the computation times for each prioritiser is a program of the following form:

```
C PROGRAM on VAX:
  Initialise DAS 9100
  LOOP 1:
    Load DAS with appropriate test vector sequence (see below)
    LOOP 2:
      Set DAS strobe width (i.e., required ENABLE duration)
      Run chip test: DAS applies vector sequence to chip
                     and reads results
      Read COMPARE result from DAS
      Change strobe length as required (see below)
    END OF LOOP 2
    Write results
  END OF LOOP 1
```

Note that *COMPARE* is true if and only if $\underline{x} = PRIOR(\underline{x})$. Due to the input data word restriction (see Section 6.2.1) this will only be so if the prioritiser circuit has had sufficient *ENABLE* time in order to compute the correct result.

The test vector sequence performs the following operations on the chip:

LOAD data word into input register and CYCLE
RESET prioritiser circuit-of-interest and output register
Raise ENABLE control signal for known duration (using DAS strobe line) to allow prioritiser computation.
CYCLE result word in output register
COMPARE input and output words producing one-bit result

LOOP 1 runs for each of the three prioritiser circuits on a chip, over all (N) input words. LOOP 2 continues until *COMPARE* is true for one particular strobe width, w, but is not true for the next smallest strobe width, $w - \delta w$ (i.e., the computation time has been found to within the accuracy of the test equipment). In order to achieve this, the strobe width is changed using a simple bisection algorithm.

6.8 Accuracy of Results

Having obtained measurements for the computation time of the prioritiser it is important to consider how accurate these results are likely to be, and to introduce techniques to improve their accuracy if possible. The following issues have been addressed, the main consideration being of course with the *ENABLE* control signal:

(1) ENABLE signal

(i) **On-chip**: No matter how accurate the *ENABLE* signal is when it arrives at the chip input pad, only by careful circuit design on the chip can the signal remain undistorted. Because of the amount of circuitry involved with the *ENABLE* line, several signal drivers must be included, that have, if possible, small and equal rise and fall times. The important consideration with the *ENABLE* signal is that its width remains unchanged. Thus, the *ENABLE* signal is driven across the chip by a series of buffers, the final buffers driving only a fraction of the necessary *ENABLE* circuitry. These buffers in turn are all driven by a buffer on the input pad.

(ii) **DAS probe-chip pin**: The DAS strobe line is screened and attached directly to the prioritiser chip *ENABLE* pin thus eliminating unnecessary VERO-board wiring and possible consequent signal distortion.

(iii) **Accuracy of DAS strobe signal**: Using an oscilloscope attached directly to the chip pin, the *ENABLE* signal can be examined for accuracy; it exhibits a small overshoot (but this is well-damped and is not sufficient to affect chip logic), and possesses symmetric, fast rising and falling edges. Most important of all is that it is of an accurate duration (to within 5 ns) even for 40 ns pulses.

(2) Other sources of inaccuracy

(i) **Variations between different chips**: Due to variations in fabrication then individual chips in the batch may give different results for the prioritiser computation times. This problem is

(ii) **Initial charge on prioritiser/output register**: Unless the prioritiser circuit-of-interest is initialised to the same state before every run, then results will not be valid; the *RESET* facility ensures that the same known state is adopted by the chip before each *ENABLE*.

(iii) **Order of testing**: It may be possible that results will vary depending on when a particular reading is taken within a sequence of readings; the automatic test-program partly caters for this by performing the tests in reverse order (i.e., testing from bit 50 down to bit 1).

(iv) **Environmental factors**: Factors such as temperature variations are likely to affect readings.

In order to reduce the effect of spurious errors such as (i) or (iv) above, multiple readings are taken for each prioritiser bit, using several different chips, with the final result being the mean over all of these results. The uncertainty is taken to be a combination of standard deviation of results, and an estimate of the possible error made in taking the readings, due to limitations in the accuracy of *ENABLE* duration.

6.9 Practical refinements in calculated prioritiser delay

On the fabricated chip, two 50-bit prioritisers are formed (Circuit-1 and Circuit-2) by joining together two 25-bit prioritiser words (see Section 6.3 (iv)) with long wires. The delay due to the wire capacitance is significant especially for Circuit-1 and must therefore be included in the delay calculation. In addition, Circuit-3 contains dummy cells consisting solely of long diffusion wires, in order to investigate the effects on delay of interconnect. A further calculation therefore must be performed in order to find Circuit-3 delay. This uses the wire model given in Section 4.6.3

6.9.1 Delay due to link wire for Circuit-1 and Circuit-2 The delay, D_w, due to the link wire joining the two 25-bit words in Circuit-1 and Circuit-2 must be included in the delay calculation. If the wire capacitance is C_w, then this delay is given by

$$D_{w|i>25} = (R_h + 25R_m) \cdot C_w.$$

This delay must be included in equation (4) and equation (5) when $x_1 = \ldots = x_{i-1} = 0$ and $x_i = 1$, where $i > 25$. Since the wire is run mainly in metal, its resistance is negligible and can thus be ignored.

6.9.2 Calculation of delay for Circuit-3 The second Implementation 1 prioritiser included on the fabricated chip contains alternating prioritiser cells, p_i, $i = 1, \ldots, 13$, and *dummy* cells, dp_j, $j = 1, \ldots, 12$. The input word, $\underline{z}$ to this prioritiser has $z_i = 1$, and $z_k = 0$, $k = 1, \ldots, i-1, i+1, \ldots, 13$. Each dummy cell wire is modelled by a Π-network as described in Section 4.6.3, with resistance, RP_j, capacitance CP_j and internal delay DP_j, where

$$DP_j = \frac{RP_j \cdot CP_j}{2}.$$

In a similar way to the calculation in Section 4.6.5.1, the total computation time for this implementation of the prioritiser can be given:

$$\begin{aligned}
\underset{\underline{y}\in\mathbf{B}^i}{\mathrm{MAX}}(\tau_{prior1_i}(\underline{y})) &\equiv \tau_{prior1_n}(\underline{z}) \\
&\approx D_h + (i-1)D_m + D_t + \frac{(i-2)(i-1)}{2} \cdot R_m C_m \\
&+ (R_h + (i-1)R_m) \cdot C_t + (i-1) \cdot R_h C_m \\
&+ \sum_{k=1}^{i-1} (DP_k + (R_m + R_h) \cdot CP_k + RP_k \cdot C_t) \\
&+ \sum_{k=1}^{i-2} \left(k \cdot R_m \cdot CP_{k+1} + (C_m + CP_{k+1}) \cdot \sum_{l=1}^{k} RP_l \right) \qquad (10)
\end{aligned}$$

where

$$\underline{y} \in \mathbf{B}^i, y_l = 1, y_k = 0, k = 1, \ldots, l-1, l+1, \ldots, i$$

and

$$\underline{z} \in \mathbf{B}^{13}, z_i = 1, z_j = 0, j = 1, \ldots, i-1, i+1, \ldots, 13$$

6.10 Results

Each reading gives the maximum *ENABLE* time, E_{max} on a 40 ns interval (the minimum strobe increment). Thus the mean *ENABLE* time is $E_{mean} = E_{max} - 20$ ns, with a 20 ns uncertainty. Figures 8.22 – 8.24 summarise the experimental and calculated results for each implementation, and each prioritiser size i. The plotted experimental time τ_{expt} is the average result over all *ENABLE* times E_{mean}, recorded from the chips tested. The calculated computation time τ_{calc} is the result of using the appropriate value of i in equation (4) (Circuit-1), equation (5) (Circuit-2) or equation (10) (Circuit-3).

The results are also tabulated in the appendix. Column *Percent-1* gives the standard deviation* of the set of readings E_{mean} as a percentage of τ_{expt} for each prioritiser size i. Column *Percent-2* gives the percentage difference between τ_{expt} and τ_{calc}.

---

* The standard deviation, σ, of a set of readings, a_i, $i = 1, \ldots, n$, is given by

$$\sigma = \left(\frac{\sum_{i=1}^{n} (a_i - \bar{a})^2}{n} \right)^{1/2}$$

where $\bar{a}$ is the average of readings $a_1, \ldots, a_n$.

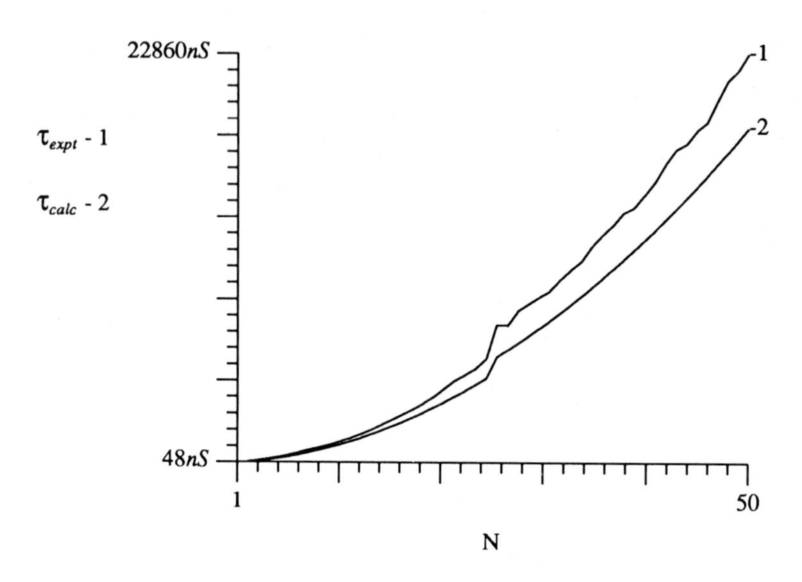

Figure 8.22. Circuit-1

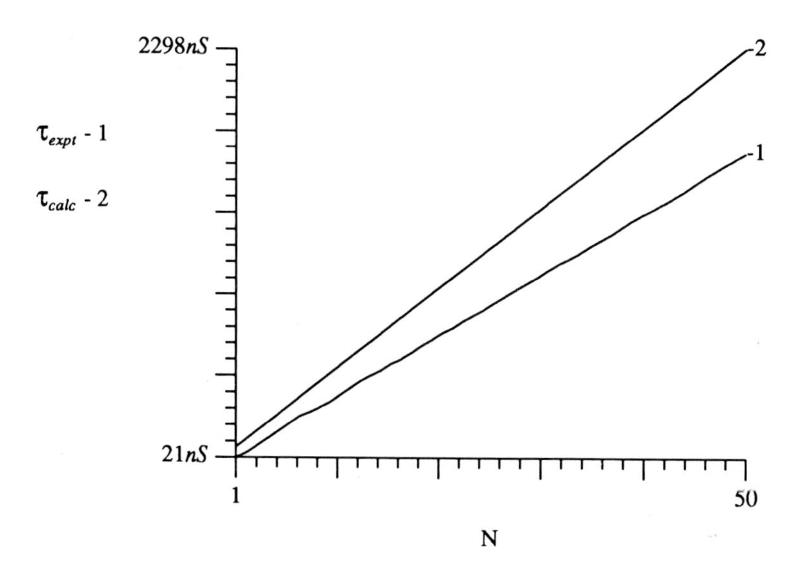

Figure 8.23. Circuit-2

6.10.1 Comments on results A comparison of experimental and calculated results shows that, for all but the smallest bit-values, the RC network model estimates the correct computation time to within at least 40% of the average experimental computation time measured from a set of fabricated circuits. In particular, the RC network model has provided good results when long wires have been used, especially in Circuit-3. The shape of each graph indicates that the *qualitative* result for the delay model estimates are correct. The *quantitative* accuracy of the calculated results depends entirely upon the accurate determination of device resistances and node capacitances in each leaf cell. Note that even within the experimental readings there is an appreciable spread in computation time readings, largely due to variations between individual chips tested. A more fabrication-dependent calibration scheme would therefore be desirable, instead of relying on general resistance and capacitance figures for a particular feature size.

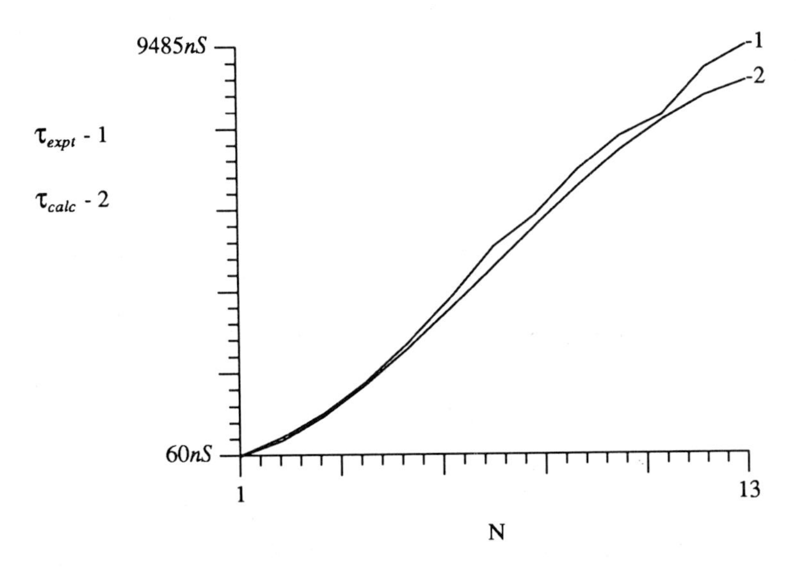

Figure 8.24. Circuit-3

The results show, by the non-linearity of Figure 8.22 and Figure 8.24, that a time complexity model based solely upon linear composition, such as that used in Section 3.4, would not be able to capture the correct asymptotic behaviour of circuits, where the interaction between sub-circuit cells (due to drive capability and loading) significantly changes the computation time of the complete circuit. However, the RC network model obtains the correct asymptotic behaviour and a reasonable estimate of measured computation time for the prioritiser circuit.

APPENDIX

A1 Tables of results from prioritiser chips

Table 8.1. Results for prioritiser Circuit-1

Bit	τ_{expt}	Percent-1	τ_{calc}	Percent-2
1	48	58.3	35	−27.1
2	99	36.4	78	−21.2
3	178	20.2	134	−24.7
4	260	15.0	203	−21.9
5	361	13.9	286	−20.8
6	486	13.6	382	−21.4
7	614	11.1	491	−20.0
8	772	11.1	614	−20.5
9	905	9.3	751	−17.0
10	1055	11.1	900	−14.7
11	1242	11.3	1063	−14.4
12	1422	11.5	1240	−12.8
13	1652	10.5	1430	−13.4
14	1900	11.2	1633	−14.1
15	2164	10.1	1850	−14.5
16	2441	11.0	2080	−14.8
17	2732	10.2	2323	−15.0
18	3019	10.2	2580	−14.5
19	3360	12.8	2851	−15.1
20	3730	12.9	3134	−16.0
21	4194	11.1	3431	−18.2
22	4638	7.6	3742	−19.3
23	4962	6.7	4066	−18.1
24	5306	8.4	4403	−17.0
25	5834	9.9	4754	−18.5
26	7758	10.9	5951	−23.3
27	7708	10.9	6328	−17.9
28	8526	11.7	6719	−21.2
29	8911	7.2	7124	−20.1
30	9286	8.7	7541	−18.8
31	9628	8.8	7972	−17.2
32	10257	11.9	8417	−17.9
33	10796	10.5	8875	−17.8

Table 8.1. (continued) Results for prioritiser Circuit-1

Bit	τ_{expt}	Percent-1	τ_{calc}	Percent-2
34	11264	10.7	9346	−17.0
35	12090	9.1	9831	−18.7
36	12720	9.2	10329	−18.8
37	13268	9.3	10840	−18.3
38	13963	11.1	11365	−18.6
39	14276	10.5	11904	−16.6
40	14985	10.5	12455	−16.9
41	15733	11.7	13020	−17.2
42	16717	8.9	13599	−18.7
43	17495	11.2	14191	−18.9
44	17815	10.3	14796	−16.9
45	18556	10.3	15415	−16.9
46	19024	10.0	16047	−15.6
47	20222	11.1	16692	−17.5
48	21341	10.0	17351	−18.7
49	21928	8.7	18024	−17.8
50	22860	12.7	18709	−18.2

Table 8.2. Results for prioritiser Circuit-2

Bit	τ_{expt}	Percent-1	τ_{calc}	Percent-2
1	21	100.0	79	276.2
2	48	58.3	124	158.3
3	86	31.4	169	96.5
4	126	21.4	214	69.8
5	166	16.3	260	56.6
6	206	13.1	305	48.1
7	246	11.0	350	42.3
8	270	11.1	396	46.7
9	297	10.8	441	48.5
10	326	8.6	486	49.1
11	366	7.4	532	45.4
12	406	6.7	577	42.1
13	446	6.1	622	39.5

Table 8.2. (continued) Results for prioritiser Circuit-2

Bit	τ_{expt}	Percent-1	τ_{calc}	Percent-2
14	477	5.9	667	39.8
15	506	6.3	713	40.9
16	544	5.9	758	39.3
17	570	5.3	803	40.9
18	605	4.5	849	40.3
19	645	4.2	894	38.6
20	679	4.1	939	38.3
21	716	3.9	985	37.6
22	745	4.3	1030	38.3
23	784	4.1	1075	37.1
24	815	4.0	1120	37.4
25	845	3.4	1166	38.0
26	884	3.3	1211	37.0
27	920	3.0	1256	36.5
28	956	3.0	1302	36.2
29	989	3.3	1347	36.2
30	1024	3.5	1392	35.9
31	1064	3.0	1438	35.2
32	1100	2.9	1483	34.8
33	1128	2.8	1528	35.5
34	1160	2.5	1573	35.6
35	1198	2.4	1619	35.1
36	1232	2.4	1664	35.1
37	1265	2.5	1709	35.1
38	1304	2.5	1755	34.6
39	1342	2.4	1800	34.1
40	1376	2.4	1845	34.1
41	1404	2.3	1891	34.7
42	1437	2.1	1936	34.7
43	1471	2.2	1981	34.7
44	1505	2.1	2026	34.6
45	1545	2.1	2072	34.1
46	1582	2.1	2117	33.8
47	1615	2.2	2162	33.9
48	1652	2.1	2208	33.7
49	1684	2.0	2253	33.8
50	1716	1.9	2298	33.9

Table 8.3. Results for prioritiser Circuit-3

Bit	τ_{expt}	Percent-1	τ_{calc}	Percent-2
1	60	33.3	35	−41.7
2	477	12.6	392	−17.8
3	1023	10.7	952	−6.9
4	1746	11.1	1678	−3.9
5	2640	10.8	2513	−4.8
6	3672	14.5	3430	−6.6
7	4856	7.6	4388	−9.6
8	5593	10.8	5334	−4.6
9	6626	13.5	6241	−5.8
10	7405	12.9	7072	−4.5
11	7893	10.7	7767	−1.6
12	8950	7.2	8313	−7.1
13	9485	10.1	8667	−8.6

A2 Computation graphs

A *computation graph*, $G(\Gamma) \in \cup\{G(\Gamma, k) \mid k \in N\}$, is an extension of the signal flow graph, allowing a functional description of a circuit to be presented in a hierarchical way; $G(\Gamma, k)$ is given by:

$$G(\Gamma, k) = \langle D, gf, cn, st, d\rangle$$

where the communication graph $D = \langle V \cup \{v_{in}, v_{out}\}, E, tl, hd\rangle$. The node-set V represents the *processors* in a circuit, and the edge-set E represents the *communication channels*, which transfer data between the nodes. Each node either represents a *primitive* function symbol $\sigma \in \Sigma$, or can be substituted by a computation graph $G(\Gamma, k-1)$ from a lower level. $\Gamma = \langle S, \sigma, tp\rangle$ is the *signature* of the graph, $gf : V \rightarrow G(\Gamma, k-1)$ is the *graph function* to substitute nodes at level k by computation graphs from lower levels, $st : E \rightarrow S$ is the *sort function*, cn is the *connection function* to associate each channel incident on a node with an input or output channel from the computation graph at a lower level, and $d : E \rightarrow N$ is a delay function. $cn = \langle cin, cout\rangle$ is a pair of families of bijections, $\langle cin_v \mid v \in V\rangle$ and $\langle cout_v \mid v \in V\rangle$, where $cin_v : HEADS_D(v) \rightarrow gf(v)_{in}$ and $cout_v : TAILS_D(v) \rightarrow gf(v)_{out}$.

For a node $v \in V$, the mapping $HEADS(v) = \{e \in E \mid hd(e) = v\}$ will be the set of inputs to that node, and $TAILS(v) = \{e \in E \mid tl(e) = v\}$ will be the set of outputs. $TAILS(v_{in})$ and $HEADS(v_{out})$ will be the sets of circuit inputs and outputs respectively, with $HEADS(v_{in}) = \emptyset$ and $TAILS(v_{out}) = \emptyset$. (See McEvoy [1986] for further details.)

A3 Values of resistances and capacitances used in prioritiser delay calculations

The following list gives the values of device resistances and node capacitances used in Section 4.6.5, Section 4.6.6 and Section 6.9.2.

(i) Resistances:

$$r_1 = 19500\Omega, \quad r_2 = 17700\Omega, \quad r_3 = 2000\Omega,$$
$$r_4 = 500\Omega, \quad r_5 = 56800\Omega, \quad r_6 = 7100\Omega.$$

(ii) Capacitances:

$$c_a = 140 \text{ fF}, \quad c_b = 23 \text{ fF}, \quad c_c = 182 \text{ fF},$$
$$c_d = 110 \text{ fF}, \quad c_e = 470 \text{ fF}, \quad c_f = 53 \text{ fF},$$
$$c_1 = 37 \text{ fF}, \quad c_2 = 391 \text{ fF}, \quad c_3 = 30 \text{ fF},$$
$$c_4 = 206 \text{ fF}, \quad c_5 = 64 \text{ fF}, \quad c_6 = 53 \text{ fF},$$
$$c_7 = 161 \text{ fF}, \quad c_8 = 448 \text{ fF}.$$

The link-wire used to join the two 25-bit words together in Word1 and Word3 has a capacitance, $C_w = 833$ fF.

For the dummy cells used in Word5 Table 8.4 provides the necessary data.

Table 8.4.

Cell	C(fF)	$R(\Omega)$	Delay (ns)
1	4937	7520	19
2	4570	6960	16
3	4190	6380	13
4	3717	5660	11
5	3310	5040	9
6	2902	4420	7
7	2442	3920	5
8	2022	3080	3
9	1602	2440	2
10	1130	1720	1
11	710	1080	0
12	276	420	0

In order to calculate these figures from the circuit geometry, the constants shown in Tables 8.5 and 8.6 were used, taken from Pucknell & Eshraghian [1985]. A unit square (□) has area $2\lambda \times 2\lambda$, and a unit edge has length 2λ, where $\lambda = 3\ \mu$m.

Table 8.5. Area

Layers	Cap (fF/μm^2)	Cap (fF/ □)	Cap (relative)
Gate-Channel	0.4	14.40	1
Diffusion-Substrate	0.1	3.60	0.25
Polysilicon-Substrate	0.04	1.44	0.1
Metal-Substrate	0.03	1.08	0.075
Metal-Polysilicon	0.03	1.08	0.075

Table 8.6. Periphery

Layer	Cap (fF/μm)	Cap (fF/edge)	Cap (relative)
Diffusion	0.8	4.8	0.33

For resistance, diffusion has a value of 20Ω/□.

8 REFERENCES

Baudet [1983]
G. M. Baudet (1983) 'Design and complexity of VLSI algorithms', in *Foundations of Computer Science IV, Part 1*, ed. J. W. deBakker & J. van Leeuwan, Mathematisch Centrum, Amsterdam.

Bilardi *et al.* [1981]
G. Bilardi, M. Pracchi and F. P. Preparata (1981) 'A critique and an appraisal of VLSI models of computation', in *CMU Conference on VLSI Systems and Computations*, ed. H. T. Kung, Sproule & Steele, Computer Science Press.

Brent & Kung [1981]
R. P. Brent and H. T. Kung (1981) 'The area–time complexity of binary multiplication', *JACM* **28** No. 3, 521–34.

Chazelle & Monier [1985]
B. Chazelle and L. Monier (1985) 'A model of computation for VLSI with related complexity results', *JACM* **32** No. 3, 573–88.

Eker & Tucker [1987]
S. M. Eker and J. V. Tucker (1987) *Specification, Derivation and Verification of Concurrent Line Drawing Algorithms and Architectures*, School of Computer Studies, University of Leeds.

Harman & Tucker [1987]
N. A. Harman and J. V. Tucker (1987) *The Formal Specification of a Digital Correlator I: Abstract User Specification*, Report 9.87, School of Computer Studies, University of Leeds.

Kühnel & Schmeck [1986]
L. Kühnel and H. Schmeck (1986) 'A closer look at VLSI multiplication', in *International Workshop on Systolic Arrays*, University of Oxford.

Lin & Mead [1984]
T.-M. Lin and C. Mead (1984) 'Signal delay in general RC networks', *IEEE Trans. on Computer-Aided Design* **CAD-3** No. 4, 331–49.

Lin & Mead [1986]
T.-M. Lin and C. Mead (1986) 'A hierarchical timing simulation model', *IEEE Trans. on Computer-Aided Design* **CAD-5** No. 1, 188–97.

Martin & Tucker [1987]
A. R. Martin and J. V. Tucker (1987) *The Concurrent Assignment Representation Of Synchronous Systems*, Report 8.87, School of Computer Studies, University of Leeds.

McEvoy [1986]
K. McEvoy (1986) *A Formal Model for the Hierarchical Design of Synchronous and Systolic Algorithms*, Report 7.86, School of Computer Studies, University of Leeds.

Mead & Conway [1980]
C. Mead and L. Conway (1980) *Introduction to VLSI Systems*, Addison-Wesley.

Newkirk & Mathews [1983]
J. Newkirk and R. Mathews (1983) *The VLSI Designer's Library*, Addison-Wesley.

Ousterhout [1983]
J. K. Ousterhout (1983) 'Crystal: a timing analyser for nMOS VLSI circuits', in *Third Caltech Conference on VLSI*, ed. R. Bryant, Springer-Verlag.

Pucknell & Eshraghian [1985]
D. A. Pucknell and K. Eshraghian (1985) *Basic VLSI Design*, Prentice-Hall.

Thompson [1980]
C. D. Thompson (1980) 'A Complexity Theory For VLSI', Thesis, Department of Computer Science, Carnegie-Mellon University, Pittsburgh.

Thompson [1987]
B. C. Thompson (1987) 'A Mathematical Theory Of Synchronous Concurrent Algorithms', PhD Thesis, School of Computer Studies, University of Leeds.

Thompson & Tucker [1985]
B. C. Thompson and J. V. Tucker (1985) 'Theoretical Considerations in Algorithm Design', in *Fundamental Algorithms for Computer Graphics*, ed. R. A. Earnshaw, Springer-Verlag.

Ullman [1984]
J. D. Ullman (1984) *Computational Aspects of VLSI*, Computer Science Press.

9 Superpolynomial lower bounds on monotone network complexity

PAUL E. DUNNE

ABSTRACT
Combinational networks are a widely studied model for investigating the computational complexity of Boolean functions relevant both to sequential computation and parallel models such as VLSI circuits. Recently a number of important results proving non-trivial lower bounds on a particular type of restricted network have appeared. After giving a general introduction to Boolean complexity theory and its history this chapter presents a detailed technical account of the two main techniques developed for proving such bounds.

1 INTRODUCTION

An important aim of Complexity Theory is to develop techniques for establishing non-trivial lower bounds on the quantity of particular resources required to solve specific problems. Natural resources, or *complexity measures*, of interest are Time and Space, these being formally modelled by the number of moves made (resp. number of tape cells scanned) by a Turing machine. 'Problems' are viewed as functions, $f : D \rightarrow R$; D is the domain of inputs and R the range of output values. D and R are represented as words over a finite alphabet Σ and since any such alphabet can be encoded as a set of binary strings it is sufficiently general to consider D to be the set of Boolean valued n-tuples $\{0,1\}^n$ and R to be $\{0,1\}$. Functions of the form $f : \{0,1\}^n \rightarrow \{0,1\}$, are called n-input single output Boolean functions. B_n denotes the set of all such functions and $\mathbf{X_n} = \langle x_1, x_2, \ldots, x_n \rangle$ is a variable over $\{0,1\}^n$. The number of arguments, n, provides a precise concept of the *size* of a problem. Complexity analyses are normally concerned with time and space needs expressed as a function of the problem size. In this way, a Turing machine computing any function whose result depends on all its n inputs must make at least n moves. By a *non-trivial* lower bound on time we mean one which is $\omega(n)$,* that is superlinear in n.

* If $p(n)$ and $q(n)$ are functions from $\mathbf{N}$ to $\mathbf{R}^+$ then $p(n) = \omega(q(n))$ if $\lim_{n \rightarrow \infty} q(n)/p(n) = 0$.

The Turing machine is only one of a number of models of computation. In this chapter we are concerned with an alternative, Boolean networks, for which the main complexity measures are *Size* (the number of gates) and *Depth* (the length of the longest path). This model is of relevance to sequential computation since any function computable in T moves by a deterministic Turing machine can be realised by a Boolean network of size $O(T\log_2 T)$, (Fischer & Pippenger[1979]). It also provides a basic model of *parallel* computation: size is an indication of the amount of hardware used; depth a measure of parallel time. Efficient simulations of apparently more general parallel models by Boolean networks have been exhibited by several authors. For example Savage [1981] relates the Thompson, Brent and Kung model of VLSI chips to a certain class of Boolean network.

Definition 1.1 Let $\Omega \subseteq B_2$. A *Boolean Ω-network,* T, is a directed acyclic graph containing 2 disjoint sets of nodes; I is the set of nodes with in-degree 0 (the *inputs* of T); G the set of nodes with in-degree 2 (the *gates* of T). Each $x_i \in \mathbf{X_n}$ is associated with exactly one input node e_i of T, any remaining input nodes are labelled with constant functions. Each gate, g of T, is associated with some function $h \in \Omega$, denoted by $op(g) = h$ or g is an h-gate. Ω is the *basis* of T. If $\Omega = B_2$ T is called a *combinational network.* With each node, v, of an Ω-network T a Boolean function $res(v)(\mathbf{X_n})$ is associated as follows.

$$res(v)(\mathbf{X_n}) = \begin{cases} 0 & \text{if } v \text{ is an input node labelled } 0, \\ 1 & \text{if } v \text{ is an input node labelled } 1, \\ x_i & \text{if } v \text{ is the input node } e_i \text{ of } T, \\ op(v)(res(v_L), res(v_R)) & \text{otherwise,} \end{cases}$$

v_L and v_R being the nodes of T which supply the Left (resp. Right) input * of v, that is the nodes such that $\langle v_L, v\rangle$ and $\langle v_R, v\rangle$ are distinct directed edges of T.

T computes $f(\mathbf{X_n}) \in B_n$ if and only if there is some node t of T for which $res(t)(\mathbf{X_n}) \equiv f(\mathbf{X_n})$. •

We employ the following notation

$$\mathbf{C}(T) \triangleq |\{v : v \text{ is a gate node of } T\}|;$$
$$\mathbf{C}(f) \triangleq \min\{\mathbf{C}(T) : T \text{ computes } f\}.$$

$\mathbf{C}(f)$ is called the *combinational complexity* of f.

The study of the properties of different types of Boolean network is almost fifty years old and can be traced back to the work of Shannon [1938] in the U.S, Shestakov [1938]

* 'left' and 'right' are distinguished by viewing the gate as having one output wire; the first input wire encountered on rotating this clockwise is taken as the 'left' input.

in the Soviet Union and Nakasima [1936] in Japan. The first important result on the combinational complexity of Boolean functions was derived by Shannon [1949].

A predicate, Π, is said to hold for *almost all* $f \in B_n$ if the number, $P(n)$ of functions not satisfying the predicate is $o(|B_n|)$.

Fact 1.1 For almost all $f \in B_n$ $\mathbf{C}(f) \geq 2^n/n$ □

Lupanov [1958] demonstrated that this is the best possible by giving a construction which asymptotically achieves this bound for any Boolean function.

Although Shannon's theorem establishes the *existence* of functions with exponential combinational complexity, it is not of any use in identifying specific explicitly defined instances of hard functions. At present even modest superlinear lower bounds are beyond the best methods known and the largest lower bound proved is only $3n$ (Blum [1984]). Since little progress has been made towards non-trivial lower bounds on combinational complexity a number of restricted forms of Boolean network have been introduced in the hope that these may be easier to analyse. In relation to combinational networks there are two aims of such models: to find proof techniques applicable to the restricted case which may be adapted to work for more general forms; or to show that the simplified model may efficiently simulate combinational networks. Among the many different restricted types of network which have been studied are formulae, in which gate nodes are constrained to have out-degree at most 1; planar networks, in which the underlying graph must be planar; and monotone networks in which the basis is $\Omega = \{\wedge, \vee\}$. Formulae are of interest since bounds on formula depth provide identical bounds on network depth. Planar networks are used by Savage (1981) to compare Boolean networks and VLSI circuits. It can be shown that the size of a minimal planar network realising any $f \in B_n$ is at most $\mathbf{C}(f)^2$ (Lipton & Tarjan [1979]).

Definition 1.2 Let f and g be n-input Boolean functions. A partial order $\leq$ is defined by saying $f \leq g$ if and only if for all $\alpha \in \{0,1\}^n$, $f(\alpha) = 1 \Rightarrow g(\alpha) = 1$. $f \in B_n$ is a *monotone Boolean function* if and only if for $1 \leq i \leq n$, $f_i^0 \leq f_i^1$, where f_i^j is the $(n-1)$-input Boolean function given by,

$$f_i^j(x_1, \ldots, x_{(n-1)}) = f(x_1, \ldots, x_{(i-1)}, j, x_{(i+1)}, \ldots, x_{(n-1)})$$

M_n denotes the class of all n-input monotone Boolean functions. It is easy to show that $f \in M_n$ if and only if f can be computed by a monotone Boolean network. •

Monotone Boolean networks have been the most widely examined restricted model, but it is only recently that sound theoretical justifications have been found for it.

These come from two groups of results, one demonstrating that non-trivial lower bounds on combinational complexity can be derived from appropriate bounds on monotone network size, the other consisting of general techniques which can be applied to prove exponential lower bounds on monotone network size. The latter is the main topic of this chapter.

The significance of these results can be appreciated by considering earlier work on monotone complexity. The first indications that this was a tractable model came in the mid 1970s with the appearance of the first superlinear lower bounds on the size of monotone networks computing *sets* of functions. Among these were the precise characterisation of optimal networks computing $\{\wedge, \vee\}$-Boolean matrix product (Paterson [1975]) and an almost quadratic lower bound of Wegener [1982]. However, even within this restricted model, the best lower bound obtained for any specific single output function, by 1985, was merely $4n$ (Tiekenheinrich [1984]). Thus reasoning about the monotone complexity of functions in M_n had seemed to be almost as difficult as deriving non-trivial bounds on combinational complexity, this impression being heightened by the work of Berkowitz ([1982]. Berkowitz exhibited a connection between the monotone complexity of a special class of functions and combinational complexity. Let

$$\mathbf{C^m}(S) \triangleq |\{v : v \text{ is a gate in the monotone network } S\}|,$$
$$\mathbf{C^m}(f) \triangleq \min\{\mathbf{C^m}(S) : S \text{ computes } f\}.$$

$T^n_k \in M_n$ is the function which takes the value 1 if and only if at least k of its arguments are 1. The *k-slice* of $f \in B_n$, is the function $f_k \in M_n$ defined by,

$$f(\mathbf{X_n}) \wedge T^n_k(\mathbf{X_n}) \vee T^n_{k+1}(\mathbf{X_n}).$$

So f_k is the function which is 1 if more than k inputs are 1; 0 if fewer than k inputs are 1; and which agrees with f on all inputs containing exactly k 1's.

Berkowitz [1982] proved the following.

Fact 1.2 For all $f \in B_n$,

(i) $\mathbf{C}(f) \leq \sum_{k=1}^{n} \mathbf{C}(f_k) + O(n)$,
(ii) $\mathbf{C}(f_k) \leq \mathbf{C}(f) + O(n)$,
(iii) $\mathbf{C^m}(f_k) = O(\mathbf{C}(f_k) + n(\log_2 n)^2)$. □

(i)–(iii) establish that f has 'large' combinational complexity if and only if some k-slice of f has 'large' *monotone* complexity. In particular lower bounds of $h(n) = \omega(n(\log_2 n)^2)$ on the monotone complexity of f_k imply $\mathbf{C}(f) = \Omega(h(n))$.

So by the end of 1984 it was known that monotone networks were a suitable model with which to study issues in combinational complexity and that, for certain sets of functions, non-trivial results about monotone complexity could be proved. No super-linear bound had been obtained for any single output function. Then, in 1985, the Soviet mathematician Razborov proved *superpolynomial* lower bounds, i.e., $\omega(n^k)$ for all fixed k, on the monotone complexity of some graph-theoretic predicates. The papers Razborov [1985a,b] introduce several innovative concepts and techniques which seem to be more generally applicable. Subsequently there have been two further significant developments. Andreev [1985], obtained the first exponential lower bounds on monotone complexity using a different approach from Razborov. Finally Alon and Boppana working from the outline proof of Razborov [1985a], demonstrated that certain combinatorial results could be improved and thereby showed that exponential bounds could be obtained using Razborov's methods.

In this chapter we give a detailed technical description of both techniques; that of Razborov as enhanced by Alon & Boppana [1986], and that of Andreev. In the remainder of this section we give some basic definitions and notation. In Section 2 the Razborov, Alon and Boppana results are presented. Section 3 describes the conceptually simpler approach employed in Andreev [1985]. We conclude with a discussion and comparison of the two methods.

Definitions A *monom* is a monotone Boolean function of the form

$$m(\mathbf{X_n}) = x_{i_1} \wedge x_{i_2} \wedge \ldots \wedge x_{i_r}.$$

Here $\{x_{i_1}, \ldots, x_{i_r}\} \subseteq \mathbf{X_n}$. The set of variables defining a monom, m, is denoted by $var(m)$. It will sometimes be convenient to regard a monom as the set of variables defining it, as well as a function. In this way we may write $m_1 \cap m_2$ and $m_1 \cup m_2$ instead of $var(m_1) \cap var(m_2)$ and $var(m_1) \cup var(m_2)$. A monom, m, is said to be an *implicant* of $f(\mathbf{X_n}) \in M_n$ if $m \leq f$. m is a *prime implicant* of $f(\mathbf{X_n})$ if $m \leq f$ and for all monoms p such that $var(p) \subset var(m)$ p is not an implicant of f. $\mathbf{I}(f)$ denotes the set of implicants of $f \in M_n$, $\mathbf{PI}(f)$ its set of prime implicants. It is well known that any monotone Boolean function may be expressed as the disjunction of its prime implicants.

In the next section we will be interested in the following monotone Boolean functions.

$\mathbf{X_n^U} = \{x_{ij} : 1 \leq i < j \leq n\}$ denotes a set of $n(n-1)/2$ Boolean variables. $G(\mathbf{X_n^U})$ is a function from assignments, α, to $\mathbf{X_n^U}$ onto n-vertex undirected graphs in which $G(\alpha)$ contains an edge $\{i, j\}$ if and only if $x_{ij} = 1$ in the assignment α to $\mathbf{X_n^U}$. For $1 \leq k \leq n$ the function k-*clique* $(\mathbf{X_n^U})$ takes the value 1 if and only if $G(\mathbf{X_n^U})$ contains a k-clique, i.e., a set $\{v_1, \ldots, v_k\} \subseteq \{1, \ldots, n\}$ of vertices such that for all $1 \leq i \neq j \leq k$, $\{v_i, v_j\}$ is an edge of $G(\mathbf{X_n^U})(\alpha)$.

$\mathbf{X_{n,n}} = \{x_{i,j} : 1 \leq i, j \leq n\}$ denotes a set of n^2 Boolean variables. $B(\mathbf{X_{n,n}})$ is a mapping from assignments to $\mathbf{X_{n,n}}$ onto $2n$-vertex bipartite graphs. $B(\alpha)$ is a bipartite graph over two disjoint sets V and W of vertices ($|V| = |W| = n$) in which there is an edge between $v_i \in V$ and $w_j \in W$ if and only if $x_{i,j} = 1$ under α. $PM(\mathbf{X_{n,n}})$ is the monotone Boolean function which takes the value 1 if and only if the bipartite graph $B(\mathbf{X_{n,n}})(\alpha)$ contains a *perfect matching*, i.e., there is a $2n$-vertex subgraph (factor) of $B(\mathbf{X_{n,n}})$ in which every vertex is the endpoint of exactly one edge. $PM(\mathbf{X_{n,n}})$ may be expressed as

$$PM(\mathbf{X_{n,n}}) = \bigvee_{\sigma \in S_n} \bigwedge_{i=1}^{n} x_{i,\sigma(i)}$$

Notation For any finite set F, 2^F denotes the set of all subsets of F and $P_s(F)$ the set of all subsets of F having cardinality $\leq s$. We use $2^{\mathbf{X_n}}$ to denote the set $\mathbf{I}(1)$ and $\emptyset$ to denote $\mathbf{I}(0)$.

2 THE LATTICE METHOD OF RAZBOROV, ALON AND BOPPANA

This technique is presented in Razborov [1985a,b] and was improved by Alon and Boppana [1986]. Our description below combines results from all of these papers. The method divides naturally into three stages, of which only the last is heavily dependent on the monotone function considered. At the core of the approach is a novel interpretation of monotone Boolean networks as a particular type of combinatorial structure called a regular lattice.

Definition 2.1 A *regular lattice*, $\mathbf{M}$, is a lattice whose elements are subsets of $2^{\mathbf{X_n}}$ and which satisfies

(R1) $\{\mathbf{I}(x_1), \ldots, \mathbf{I}(x_n), \mathbf{I}(0), \mathbf{I}(1)\} \subseteq \mathbf{M}$,
(R2) Elements of $\mathbf{M}$ are ordered by set containment $\subseteq$.

$\sqcap$ and $\sqcup$ denote the usual lattice operations *meet* and *join*. Given A, B elements of a regular lattice $\mathbf{M}$ these operations satisfy

$$A \cup B \subseteq A \sqcup B, \qquad A \sqcap B \subseteq A \cap B. \qquad \bullet$$

In general both these containments will be strict. This motivates the ideas of *surplus* and *deficiency*. For A, B as before these are given respectively by

$$\begin{aligned} \delta_+(A, B) &= (A \cap B) - (A \sqcap B), \\ \delta_-(A, B) &= (A \sqcup B) - (A \cup B). \end{aligned} \tag{1}$$

Finally in order to model monotone networks by regular lattices some concept of the complexity of a function with respect to any such lattice is needed.

Definition 2.2 Let $f(\mathbf{X_n}) \in M_n$ and $\mathbf{M}$ be any regular lattice. The *distance* of f in $\mathbf{M}$, denoted $\rho(f, \mathbf{M})$, is the least t such that

$\exists t$ pairs of lattice elements $\langle A_i, B_i \rangle$ and an element D of $\mathbf{M}$ for which

$$D \subseteq \mathbf{I}(f) \cup \bigcup_{i=1}^{t} \delta_{-}(A_i, B_i), \tag{2}$$

$$\mathbf{I}(f) \subseteq D \cup \bigcup_{i=1}^{t} \delta_{+}(A_i, B_i). \tag{3}$$

•

Using these ideas the three sections of the technique consist of the following.

(S1) Showing that for all $f \in M_n$ and all regular lattices $\mathbf{M}$, $\mathbf{C^m}(f) \geq \rho(f, \mathbf{M})$.
(S2) Constructing a particular class of regular lattices $CLOSED(f)$.
(S3) Proving a lower bound on the monotone complexity of some specific functions, f, by obtaining a lower bound on the distance of f in $CLOSED(f)$. Here the method relies on the structure of the given f in order to derive large bounds.

2.1. Distance in regular lattices and monotone complexity

Lemma 2.1 $\forall f(\mathbf{X_n}) \in M_n$, $\forall$ *regular lattices* $\mathbf{M}$,

$$\mathbf{C^m}(f) \geq \rho(f, \mathbf{M}).$$

Proof Let S be any monotone network realising f with $\mathbf{C^m}(S) = t$. Number the gates of S in topological order. Let l_i, r_i denote the functions computed by the left (resp. right) input nodes for the gate numbered i.

With each node i of S we associate an element $\lambda(v)$ of $\mathbf{M}$ as follows:

$$\lambda(v) = \begin{cases} \mathbf{I}(x_j) & \text{if } v \text{ is the input } e_j, \\ \mathbf{I}(0) & \text{if } v \text{ is an input labelled with } 0, \\ \mathbf{I}(1) & \text{if } v \text{ is an input labelled with } 1, \\ \lambda(v_L) \sqcap \lambda(v_R) & \text{if } op(v) = \wedge, \\ \lambda(v_L) \sqcup \lambda(v_R) & \text{if } op(v) = \vee. \end{cases}$$

We claim that choosing the t pairs $\langle A_i, B_i \rangle$ by $A_i = \lambda(i_L)$, $B_i = \lambda(i_R)$, where i is the gate numbered i by the topological ordering of S, (thus $1 \leq i \leq t$), and choosing

$D = \lambda(t)$ satisfies relations (2) and (3) above. This is easily shown by induction on $t \geq 0$. If $t = 0$ then f is either a constant function or the variable x_i. In any case D is defined as $\mathbf{I}(f)$ and so the inductive base is trivial.

Now suppose the assertion holds for all values $\leq t-1$. We shall prove it holds for t also. Since $t \geq 1$, S contains at least one gate. Consider the output gate of S, which must be labelled t in the topological ordering. By definition we have that $f = l_t \; op(t) \; r_t$. By the inductive hypothesis, since both l_t and r_t are computed using at most $t-1$ gates in S, we have

$$A_t \subseteq \mathbf{I}(l_t) \cup \bigcup_{i=1}^{t-1} \delta_-(A_i, B_i), \tag{4}$$

$$B_t \subseteq \mathbf{I}(r_t) \cup \bigcup_{i=1}^{t-1} \delta_-(A_i, B_i), \tag{5}$$

$$\mathbf{I}(l_t) \subseteq A_t \cup \bigcup_{i=1}^{t-1} \delta_+(A_i, B_i), \tag{6}$$

$$\mathbf{I}(r_t) \subseteq B_t \cup \bigcup_{i=1}^{t-1} \delta_+(A_i, B_i). \tag{7}$$

First suppose that $op(t) = \vee$. In this case, $D = \lambda(t) = A_t \sqcup B_t$, and so, using (4) and (5),

$$\begin{aligned} D &= A_t \cup B_t \cup \delta_-(A_t, B_t) \\ &\subseteq \mathbf{I}(l_t) \cup \mathbf{I}(r_t) \cup \bigcup_{i=1}^{t} \delta_-(A_i, B_i) \\ &= \mathbf{I}(f) \cup \bigcup_{i=1}^{t} \delta_-(A_i, B_i) \end{aligned}$$

so (2) holds in this case. Similarly using (6) and (7);

$$\begin{aligned} \mathbf{I}(f) &= \mathbf{I}(l_t) \cup \mathbf{I}(r_t) \\ &\subseteq A_t \cup B_t \cup \bigcup_{i=1}^{t-1} \delta_+(A_i, B_i) \\ &\subseteq (A_t \sqcup B_t) \cup \bigcup_{i=1}^{t-1} \delta_+(A_i, B_i) \\ &\subseteq D \cup \bigcup_{i=1}^{t} \delta_+(A_i, B_i) \end{aligned}$$

so that (3) holds. The case $op(t) = \wedge$ can be shown by a similar argument. This completes the inductive argument and hence, $\mathbf{C^m}(f) \geq \rho(f, \mathbf{M})$ □

2.2 The class of regular lattices $CLOSED(f)$

Lemma 2.1 demonstrates that in order to prove lower bounds on monotone complexity it is sufficient to prove lower bounds on distance in regular lattices. There are infinitely many possible choices of regular lattice. A lattice with too few elements e.g., containing only $\mathbf{I}(x_i)$, $\emptyset$ and $2^{\mathbf{X_n}}$, would have insufficient structure to allow large bounds to be easily derived. On the other hand, using a lattice with too many elements e.g., containing every subset of $2^{\mathbf{X_n}}$, it would only be possible to derive trivial lower bounds on distance; in the cited example every function has distance 0 only.

Razborov defined a class of regular lattices based on the prime implicant structure of any given monotone function and a novel closure relation. Unfortunately no motivation for the chosen representation is given and the reader should note that much of the development below is non-intuitive.

Let $f(\mathbf{X_n}) \in M_n$ and recall that $\mathbf{PI}(f)$ is the set of prime implicants of f. Below $r \geq 2$ and $s \geq 1$ are natural numbers.

Definition 2.3 For $f \in M_n$, $\mathbf{U_s}(f) \subseteq P_s(\mathbf{X_n})$ is the set,

$$\mathbf{U_s}(f) = \{m : |m| \leq s \text{ and } \exists p \in \mathbf{PI}(f) \text{ such that } p \leq m\}$$

(here we are regarding m both as a function and as a set of variables).

Informally $\mathbf{U_s}$ is the set of shortenings of prime implicants of $f(\mathbf{X_n})$ such that these shortenings contain at most s variables.

The relation $\vdash$ (read 'yields') is a subset of $\mathbf{U_s}(f)^r \times \mathbf{U_s}(f)$ defined by saying

$$\langle E_1, \ldots, E_r \rangle \vdash E_0$$

(where $E_i \in \mathbf{U_s}(f)$, $0 \leq i \leq r$) if and only if

$$\bigcup_{1 \leq i < j \leq r} E_i \cap E_j \subseteq E_0. \tag{8}$$

If $U_1 \subseteq \mathbf{U_s}$ and $E \in \mathbf{U_s}$ we write $U_1 \vdash E$ if there are sets $E_1, \ldots, E_r$ (not necessarily distinct) in U_1 such that $\langle E_1, \ldots, E_r \rangle \vdash E$. A subset U_1 of $\mathbf{U_s}$ is said to be *closed* if $\forall E \in \mathbf{U_s}(f)(U_1 \vdash E \Rightarrow E \in U_1)$. We denote by $\overline{U_1}$ the smallest cardinality closed subset of $\mathbf{U_s}(f)$ which contains U_1, thus if U_1 is closed then $\overline{U_1} = U_1$. This is called the *closure* of U_1.

Finally, for any $E \in \mathbf{U_s}(f)$ the *cover* of E, denoted $\lceil E \rceil$ is the set

$$\lceil E \rceil = \{F : F \supseteq E\}.$$

For a subset U_1 of $\mathbf{U_s}(f)$, this is naturally extended by defining, the set $\lceil U_1 \rceil$ as $\bigcup_{E \in U_1} \lceil E \rceil$. •

Definition 2.4 Let $f(\mathbf{X_n}) \in M_n$. The lattice $CLOSED(f)$ is the lattice which contains exactly the elements

$$\{\lceil U_1 \rceil : U_1 \subseteq \mathbf{U_s}(f) \text{ and } U_1 \text{ is closed}\}.$$

$CLOSED(f)$ is the lattice used to derive the lower bounds established below.*

Lemma 2.2 For all $f \in M_n$, $CLOSED(f)$ is a regular lattice, whose $\sqcap$ and $\sqcup$ operations satisfy for all $\lceil A \rceil, \lceil B \rceil \in CLOSED(f)$,

$$\lceil A \rceil \sqcap \lceil B \rceil = \lceil A \cap B \rceil,$$
$$\lceil A \rceil \sqcup \lceil B \rceil = \lceil \overline{A \cup B} \rceil.$$

Proof It is clear that $CLOSED(f)$ contains $\mathbf{I}(x_i) = \lceil \overline{\{x_i\}} \rceil$ for all $1 \leq i \leq n$ and also $\mathbf{I}(0) = \lceil \emptyset \rceil$ and $\mathbf{I}(1) = \lceil \mathbf{U_s}(f) \rceil$. For the second part of the lemma consider $SUB(f) = \{C : C \text{ is a closed subset of } \mathbf{U_s}(f)\}$. Obviously $SUB(f)$ under the ordering $\subseteq$ forms a lattice with operations $inf(A, B) \equiv A \cap B$ and $sup(A, B) \equiv \overline{(A \cup B)}$. The cover operation $\lceil .. \rceil : SUB(f) \to CLOSED(f)$ is an order preserving lattice homomorphism, $A \subseteq B \Rightarrow \lceil A \rceil \subseteq \lceil B \rceil$. The lemma will follow if $\lceil .. \rceil$ is actually an *isomorphic mapping.* Thus if for any closed subsets A, B of $\mathbf{U_s}(f)$ we have $\lceil A \rceil \subseteq \lceil B \rceil \Rightarrow A \subseteq B$. So suppose A, B are two closed subsets of $\mathbf{U_s}(f)$ with $\lceil A \rceil \subseteq \lceil B \rceil$. Let $E \in A$. Since $E \in A \subseteq \lceil A \rceil \subseteq \lceil B \rceil$ it follows from the definition of $\lceil .. \rceil$ that there is some $F \in B$ such that $F \subseteq E$, we therefore have from (8) that $F, F, \ldots, F \vdash E$ and so $E \in B$ since this is a closed subset of $\mathbf{U_s}$. So we have established that $A \subseteq B$ proving the lemma. □

$C \subseteq 2^{\mathbf{X_n}}$ (i.e., set of monoms or of variable sets) is said to be *independent* if for each distinct $A, B \in C$, $A \not\subseteq B$ and $B \not\subseteq A$.

The lower bound proofs require upper bounds on two measures to be established.

(UPB1) For any closed set $W \subseteq \mathbf{U_s}(f)$ the value of $|base_k(W)|$, where

$$base_k(W) = \{E \in W : |E| \leq k \text{ and } \forall F \in W \; F \subseteq E \Rightarrow F = E\}.$$

* For the lower bound on k-clique this is not strictly true. Instead of using $\mathbf{U_s}$ as the basis for defining 'closure', $\vdash$ and 'covers' a particular subset of $\mathbf{U_s}$ is used. The combinatorial results proved subsequently all hold for the lattice structure that arises. This will be clear when the actual form used is defined later.

So $base_k(W)$ consists of the minimal (w.r.t. $\subseteq$) sets in W of cardinality at most k.

(UPB2) For any set $W \subseteq \mathbf{U_s}(f)$ the value of $|\overline{W} - W|$, i.e., the number of sets added to W to render it closed.

The main contribution made by Alon and Boppana was in improving the upper bounds on these quantities originally obtained in Razborov [1985a,b]. In practice only the improvement of (UPB1) led to larger lower bounds, although in principle that made to (UPB2) could also yield better results. At present no examples where this is so are known.

We tackle the problem posed by (UPB1) in two stages: first it is shown that for any independent subset, W, of $P_k(\mathbf{X_n})$ for which there does not exist any $(r+1)$-tuple $\langle E_0, E_1, \ldots, E_r\rangle \in W^{r+1}$ such that $\bigcup_{1\leq i<j\leq r} E_i \cap E_j \subset E_0$ contains at most $(r-1)^k$ members. Note that the containment is *strict*. Using this it is easy to derive an upper bound on $I_k = \max\{|base_k(W)| : W \subseteq \mathbf{U_s}(f) \text{ and } W \text{ is closed}\}$.

Following Alon and Boppana, we say that an independent subset W of $P_k(\mathbf{X_n})$ is *r-stable* if there does not exist any $(r+1)$-tuple in W^{r+1} with the property described in the preceding paragraph.*

Lemma 2.3 (Alon & Boppana [1986], improving Razborov [1985a]) *Let* $W \subseteq P_k(\mathbf{X_n})$ *be independent. If* W *is r-stable then* $|W| \leq (r-1)^k$.

Proof By induction on $r \geq 2$. For the inductive base, $r = 2$ let $W \subseteq P_k(\mathbf{X_n})$ be independent and 2-stable. If $|W| \geq 2$ then W contains sets W_1 and W_2. But $W_1 \cap W_2 \subset W_1 \in W$ which contradicts the assumption of 2-stability. Thus $|W| = 1$ proving the inductive base.

For the inductive hypothesis assume the lemma holds for all values $\leq r-1$ and let $W \subseteq P_k(\mathbf{X_n})$ be independent and r-stable. We shall show that $|W| \leq (r-1)^k$. Choose any $V \in W$ and for each $C \subseteq V$ define a set $W_C \subseteq P_{k-|C|}(\mathbf{X_n})$ by

$$W_C = \{F - C : F \in W \text{ and } F \cap V = C\}.$$

Clearly W_C is independent. W_C is also $(r-1)$-stable. To see this suppose the contrary. Then there is some r-tuple $\langle E_0, E_1, \ldots, E_{r-1}\rangle \in W_C^r$ such that

$$\bigcup_{1\leq i<j\leq r-1} E_i \cap E_j \subset E_0.$$

* Alon and Boppana [1986] actually uses the term 'has Property $P(r,k)$' to describe such a set.

But this implies that

$$\bigcup_{1 \leq i < j \leq r-1} (E_i \cup C) \cap (E_j \cup C) \subset E_0 \cup C$$

and since $E_i \in W_C$, by definition we have that $\bigcup_{1 \leq i \leq r-1}(E_i \cup C) \cap V = C$. So if we choose $F_i = E_i \cup C$, for each $0 \leq i \leq r-1$ and set $F_r = V$ then $\langle F_0, F_1, \ldots, F_r \rangle \in W^{r+1}$ and, from the previous argument,

$$\bigcup_{1 \leq i < j \leq r} F_i \cap F_j \subset F_0,$$

contradicting the r-stability of W. It follows that W_C is $(r-1)$-stable.

By the inductive hypothesis, since $W_C \subseteq P_{k-|C|}(\mathbf{X_n})$, this gives $|W_C| \leq (r-2)^{k-|C|}$. Now,

$$\begin{aligned} |W| = \sum_{C \subseteq V} |W_C| &\leq \sum_{C \subseteq V} (r-2)^{k-|C|} \\ &= \sum_{i=0}^{|V|} \binom{|V|}{i} (r-2)^{|V|-i} \\ &\leq \sum_{i=0}^{k} \binom{k}{i} (r-2)^{k-i} = (r-1)^k \end{aligned}$$

by the Binomial theorem. The completes the induction, proving the lemma. □

Corollary 2.1 $I_k \leq (r-1)^k$.

Proof (Razborov [1985b]) Let W be any closed subset of $\mathbf{U_s}(f)$. It is sufficient to show that $|base_k(W)| \leq (r-1)^k$. Clearly $base_k(W) \subseteq P_k(\mathbf{X_n})$ and is independent. Suppose $base_k(W)$ is not r-stable. As before we can find $\langle E_0, E_1, \ldots, E_r \rangle \in base_k(W)^{r+1} \subseteq W^{r+1}$ such that

$$\bigcup_{1 \leq i < j \leq r} E_i \cap E_j = E \subset E_0.$$

Hence $base_k(W) \vdash E \Rightarrow W \vdash E \Rightarrow E \in W$ by closure. But $E \in W$ contradicts $E_0 \in base_k(W)$ since $E \subset E_0$. This contradiction shows that $base_k(W)$ is r-stable, thus from Lemma 2.3 $|base_k(W)| \leq (r-1)^k$ as claimed. □

We now turn to the problem posed in (UPB2), that of bounding the number of sets needed to produce $\overline{W}$ from W. In fact all that is needed for subsequent development in the lower bound proofs is an upper bound on the number of iterations of the following algorithm to produce $\overline{W}$.

Input : A, B *where* A, $B \subseteq \mathbf{U_s}(f)$ *and closed*
Output : $\overline{W} = \overline{(A \cup B)}$
Method : (*Alon & Boppana*[1986])
  $W_0 := A \cup B$
  $i := 0$
  while $W_i \neq \overline{W}$ **do**
    $V_{i+1} := E \in base_s(\{F \notin W_i : W_i \vdash F\}$
    $W_{i+1} := W_i \cup (\lceil V_{i+1} \rceil \cap P_s(\mathbf{X_n}))$
    $i := i + 1$
  od

The Closure Algorithm

Let p be the maximal number of iterations of this algorithm, and $\langle V_1, V_2, \ldots, V_p \rangle$ the sequence of (minimal) sets whose cover up to sets of cardinality s is added at each stage. We wish to derive an upper bound on p. Now since $\mathbf{U_s}(f)$ is finite and closed it is obvious that $p \leq |\mathbf{U_s}(f)|$ and this is the measure used in (Razborov [1985a,b]). As we remarked previously even this crude estimate is adequate to derive the exponential bounds obtained in Alon & Boppana [1986]. However the improved bound derived by Alon and Boppana may yet be of value for deriving further results.

Lemma 2.4 *p, the number of iterations of the Closure Algorithm, is* $\leq 2r^s$.

Proof Let $S = \langle V_1, \ldots, V_p \rangle$ be the sequence of minimal sets added by the closure algorithm. This sequence has the following property.

> Each V_i has cardinality $\leq s$ and there do not exist $i_1 \leq i_2 \leq \ldots \leq i_r < i_{r+1}$ for which $\langle V_{i_1}, \ldots, V_{i_r} \rangle \vdash U \subset V_{i_{r+1}}$ We say that any sequence of distinct sets $\langle C_1, \ldots, C_q \rangle$ which satisfy this have *Property T(r,s)*.

Now suppose S is as above but that S does not have property $T(r, s)$. We claim that then $V_{i_{r+1}}$ would not be the set added at the i_{r+1} iteration. This is because $V_{i_{r+1}}$ is supposed to be a minimal set which is not in $W_{i_{r+1}-1}$ but such that $W_{i_{r+1}-1} \vdash V_{i_{r+1}}$. Now if

$$\langle V_{i_1}, V_{i_2}, \ldots, V_{i_r} \rangle \vdash U \subset V_{i_{r+1}}$$

i.e., $T(r, s)$ does not hold, then $V_{i_{r+1}}$ could only be an appropriate set to consider if $U \in W_{i_{r+1}-1}$. This can be true only if $U \in A \cup B$ or $U \in \lceil V_j \rceil \cap P_s(\mathbf{X_n})$ for some $j < i_{r+1}$. In the former case we have, without loss of generality, $U \in A$ so $\lceil U \rceil \cap P_s(\mathbf{X_n}) \subseteq A$ (since A is closed) hence, because $U \subset V_{i_{r+1}} \in P_s(\mathbf{X_n})$, $V_{i_{r+1}} \in A$. This contradicts $V_{i_{r+1}}$ occurring in the sequence of sets S. The latter case is even easier to dismiss for then $\lceil U \rceil \cap P_s(\mathbf{X_n}) \subseteq \lceil V_j \rceil \cap P_s(\mathbf{X_n})$ thus by the previous argument $V_{i_r+1} \in \lceil V_j \rceil \cap P_s(\mathbf{X_n})$ and is again unsuitable.*

So $\langle V_1, \ldots, V_p \rangle$ has property $T(r,s)$. We claim that for all $r \geq 1$, $s \geq 0$ any sequence of distinct sets $\langle C_1, \ldots, C_q \rangle$ with property $T(r,s)$ must have $q \leq 2r^s$. Clearly this proves the lemma.

This claim is established by induction on $r \geq 1$. For the inductive base let $Q = \langle C_1, \ldots, C_q \rangle$ have property $T(1,s)$. For $r = 1$, the relation $\vdash$ satisfies $Q \vdash \emptyset$. Now suppose that $q \geq 3$. Since the C_i are distinct, at least one of C_2, C_3 must be non-empty. Without loss of generality, assume it is C_2. Now we have a contradiction since $C_1 \vdash \emptyset \subset C_2$ and Q does not have property $T(1,s)$. It follows that $q \leq 2$ proving the inductive base.

Assume the claim holds for all values $\leq r-1$ and let Q have property $T(r,s)$. We must show that $q \leq 2r^s$. Put $D = C_1$ and for each $V \subseteq D$ define the sequence Q_V as the sequence of sets $C_i - V$ such that $C_i \cap D = V$, these appearing in the same order as in Q. It is easy to show that Q_V has Property $T(r-1, s-|V|)$ by using methods similar to Lemma 2.3. By the inductive hypothesis, $|Q_V| \leq 2(r-1)^{s-|V|}$ and so,

$$\begin{aligned} q = |Q| &= 2\sum_{i=0}^{|D|} \binom{|D|}{i} (r-1)^{|D|-i} \\ &\leq 2\sum_{i=0}^{s} \binom{s}{i} (r-1)^{s-i} = 2r^s. \end{aligned}$$

This completes the proof by induction. □

---

* The reader familiar with Alon and Boppana [1986] may wonder why we have (i) defined the Closure algorithm for sets of the form $A \cup B$ for closed A, B instead of arbitrary subsets of $\mathbf{U}_s(f)$ as is done in their paper, and (ii) given a detailed exposition that S actually has property $T(r,s)$, when this is just stated in the paper. The Closure algorithm, of course, does work for arbitrary subsets, however property $T(r,s)$ does not always hold. Consider forming the closure of $C = \{x_1x_2\}$ when $r = 2, s \geq 5, \mathbf{U}_s(f) = P_s(\mathbf{X_n})$. Using the Closure algorithm, $C \vdash x_1x_2x_j \forall 3 \leq j \leq n$. Each set $x_1x_2x_j$ is not contained in C and is a minimal such set. It is easy to see that choosing $V_i = x_1x_2x_{2+i}$ is a valid choice of sequence for the closure algorithm to make. But this sequence does not have property $T(2,s)$. $V_1, V_2 \vdash x_1x_2 \subset V_3$. In fact in this case the number of iterations, p, is exactly $n - 2 \geq 2r^s = 2^{s+1}$ whenever $s < \lfloor \log_2 n \rfloor - 1$. Our presentation, which is sufficient for the purpose intended, avoids this problem, showing that for any such U which arises, $U \not\in A \cup B$ by using the closure of A and B.

2.3 Lower bounds on distance in $CLOSED(f)$

In this section we show how the combinatorial results proved in Section 2.2 can be used to produce a general inequality for lower bounds on monotone complexity. Section 2.4 below will then give some specific applications.

The technique used is a probabilistic counting argument, in the style of Erdős & Spencer [1974]. Subsequently the following notation will be used.

(1) $M_+(f)$ is a randomly chosen prime implicant of $f(\mathbf{X_n})$. Each such prime implicant is selected independently with probability $|\mathbf{PI}(f)|^{-1}$.
(2) $M_-(f)$ is a randomly chosen monom, (i.e., subset of $\mathbf{X_n}$).

The exact details of how M_- is defined depend on the function considered.

$$Ext(f,k) = \max_{\{m \in 2^{\mathbf{X_n}} : |m| = k\}} |\{p \in \mathbf{PI}(f) : p \leq m\}|.$$

Let $A \cup B = C_0, C_1, \ldots, C_p = \overline{A \cup B}$, where A and B are closed subsets of $\mathbf{U_s}(f)$, be the sequence of successive sets generated by the closure algorithm. Let $\langle E_1, \ldots, E_p \rangle$ be the minimal sets added at each iteration. Thus $C_i \vdash E_{i+1}$ and $E_{i+1} \not\subseteq C_i$, for each $0 \leq i \leq p-1$.

$$Gap(f) = \max_{0 \leq i \leq p-1} Prob[M_-(f) \in (\lceil E_{i+1} \rceil - \lceil C_i \rceil)].$$

The reason for these random variables is to produce upper bounds on

$$EXCESS(f) = \max_{A,B \in CLOSED(f)} Prob[M_+ \in \delta_+(A,B)], \tag{9}$$

$$DEFICIT(f) = \max_{A,B \in CLOSED(f)} Prob[M_- \in \delta_-(A,B)]. \tag{10}$$

Theorem 2.1 *Let* $t = \rho(f, CLOSED(f))$; *then*

$$t \geq \frac{1 - Prob[M_+(f) \in D]}{EXCESS(f)},$$

$$t \geq \frac{Prob[M_-(f) \in D] - Prob[M_-(f) \leq f]}{DEFICIT(f)}.$$

Proof The first inequality follows from relation (3) in Definition 2.2, using the fact that $M_+(f)$, as a prime implicant of f, occurs in $\mathbf{I}(f)$ with probability 1. The second inequality follows from relation (2) in Definition 2.2. □

Lemma 2.5 $EXCESS(f) \leq (r-1)^{2s} Ext(f, s+1)/|\mathbf{PI}(f)|$.

Proof Let $A = \lceil V \rceil$ and $B = \lceil W \rceil$ be elements of $CLOSED(f)$ where $V \subseteq \mathbf{U}_s(f)$, $W \subseteq \mathbf{U}_s(f)$ are closed. We have, from Lemma 2.2,

$$\begin{aligned}
\delta_+(A,B) &= (A \cap B) - (A \sqcap B) \\
&= (\lceil V \rceil \cap \lceil W \rceil) - (\lceil V \cap W \rceil) \\
&= (\lceil base_s(V) \rceil \cap \lceil base_s(W) \rceil) - (\lceil V \cap W \rceil) \\
&= (\bigcup_{E \in base_s(V)} \lceil E \rceil \cap \bigcup_{F \in base_s(W)} \lceil F \rceil) - (\lceil V \cap W \rceil) \\
&= \bigcup_E \bigcup_F (\lceil E \rceil \cap \lceil F \rceil) - (\lceil V \cap W \rceil) \\
&= \bigcup_E \bigcup_F (\lceil E \cup F \rceil - \lceil V \cap W \rceil)
\end{aligned}$$

Consider any $E \in base_s(V)$ and any $F \in base_s(W)$. We can distinguish three possible cases.

Case 1 $E \cup F \not\subseteq var(p)\ \forall p \in \mathbf{PI}(f)$.

In this case $Prob[M_+(f) \in \lceil E \cup F \rceil] = 0$.

Case 2 $E \cup F \in \mathbf{U}_s(f)$.

Obviously $E \in V$, thus since V is closed $E \cup F \in V$ also. In the same way $E \cup F \in W$. So $E \cup F \in V \cap W$ and therefore the set $\lceil E \cup F \rceil - \lceil V \cap W \rceil$ is empty.

Case 3 $E \cup F \subseteq var(p)$ for some $p \in \mathbf{PI}(f)$, but $E \cup F \notin \mathbf{U}_s(f)$.

This can only be so if $|E \cup F| \geq s + 1$, by the definition of $\mathbf{U}_s(f)$. We now have

$$\begin{aligned}
Prob[M_+(f) \in \lceil E \cup F \rceil] &= Prob[E \cup F \subseteq M_+(f)] \\
&\leq \frac{Ext(f, s+1)}{|\mathbf{PI}(f)|}
\end{aligned}$$

So, in every case, $Prob[M_+(f) \in \lceil E \cup F \rceil]$ is at most $Ext(f, s+1)/|\mathbf{PI}(f)|$. From the preceding analysis and Corollary 2.1,

$$\begin{aligned}
Prob[M_+(f) \in \delta_+(A,B)] &\leq \frac{|base_s(V)||base_s(W)|Ext(f, s+1)}{|\mathbf{PI}(f)|} \\
&\leq \frac{(r-1)^{2s} Ext(f, s+1)}{|\mathbf{PI}(f)|}.
\end{aligned}$$

Since A, B were chosen arbitrarily the upper bound on $EXCESS(f)$ follows. $\square$

To produce an upper bound on $DEFICIT(f)$ we use the result of Lemma 2.4.

Lemma 2.6 $DEFICIT(f) \leq 2r^s Gap(f)$.

Proof As in the proof of Lemma 2.5, let $A = \lceil V \rceil$, $B = \lceil W \rceil$ be elements of $CLOSED(f)$ where V, W are closed subsets of $\mathbf{U}_s(f)$. Let $C_0 = V \cup W$ and $\langle C_1, \ldots, C_p \rangle$ be the sequence of sets created by the closure algorithm i.e., the W_j sets in the description of this algorithm above. Finally let $\langle E_1, \ldots, E_p \rangle$ be the sequence of minimal sets used in the closure algorithm. Recall that $C_i \vdash E_{i+1}$, $E_{i+1} \not\subseteq C_i$ and that the sequence $\langle E_1, \ldots, E_p \rangle$ has property $T(r, s)$ and hence $p \leq 2r^s$. Applying Lemma 2.2 we have

$$\begin{aligned} \delta_-(A,B) &= \lceil \overline{(V \cup W)} \rceil - \lceil V \cup W \rceil \\ &\subseteq \bigcup_{i=0}^{p-1} (\lceil C_{i+1} \rceil - \lceil C_i \rceil) \\ &= \bigcup_{i=0}^{p-1} (\lceil E_{i+1} \rceil - \lceil C_i \rceil). \end{aligned}$$

Note that the second inequality uses the fact that

$$\bigcup_{i=0}^{k-1} (\lceil C_{i+1} \rceil - \lceil C_0 \rceil) \subseteq \bigcup_{i=0}^{k-1} (\lceil C_{i+1} \rceil - \lceil C_i \rceil),$$

this being easily established by induction on $1 \leq k \leq p$.

Now since $p \leq 2r^s$, $C_i \vdash E_{i+1}$, $E_{i+1} \not\subseteq C_i$ it is immediate from the last inequality that

$$Prob[M_-(f) \in \delta_-(A,B)] \leq 2r^s \max_{0 \leq i \leq p-1} Prob[M_-(f) \in (\lceil E_{i+1} \rceil - \lceil C_i \rceil)].$$

Since A, B were arbitrary this establishes the upper bound on $DEFICIT(f)$ stated.

□

Theorem 2.2 *Let* $t = \rho(f, CLOSED(f))$.

$$t \geq \min\left\{ \frac{|\,\mathbf{PI}(f)\,|}{(r-1)^{2s} Ext(f, s+1)}, \frac{Prob[M_-(f) \in D] - Prob[M_-(f) \leq f]}{2r^s Gap(f)} \right\}, \quad (11)$$

$$t \geq \min\left\{ \frac{|\,\mathbf{PI}(f)\,|\,(1 - Prob[M_+(f) \in D])}{(r-1)^{2s} Ext(f, s+1)}, \frac{1 - Prob[M_-(f) \leq f]}{2r^s Gap(f)} \right\}. \quad (12)$$

Proof (11) follows from Theorem 2.1, Lemma 2.5 and Lemma 2.6 by considering the two cases $D = \emptyset$ and $D \neq \emptyset$. (12) follows in the same way by considering the two cases $D \neq \mathbf{U}_s(f)$ and $D = \mathbf{U}_s(f)$.

□

2.4 Lower bounds for specific monotone functions

We conclude this section by deriving non-trivial lower bounds on the monotone complexity of *k-clique* $(\mathbf{X}_{\mathbf{n}}^{\mathbf{U}})$ and $PM(\mathbf{X}_{\mathbf{n},\mathbf{n}})$. The first will be exponential for suitable choices of k.

We noted earlier that the actual lattice structure employed for the bound on k-clique is slightly different from the family $CLOSED$ defined above. The underlying set $\mathbf{U}_s(f)$ is not the set of monoms containing at most s variables, which are shortenings of prime implicants, i.e graphs with at most s edges which are subgraphs of k-cliques. Instead we take $\mathbf{U}_s(f)$ to be the set of all monoms corresponding to *cliques* with at most s *vertices.* In this way $\lceil E \rceil$, where E is a clique of size $\leq s$, is the set of graphs which contain E as a subgraph. In the same style we amend the definition of Ext, for k-cliques, to be the number of k-cliques a clique of size $s+1$ could be extended to. Since $\mathbf{X}_{\mathbf{n}}^{\mathbf{U}}$ is a set of edges, each clique in $\mathbf{U}_s(f)$ has at least two vertices. It is not difficult to verify that the combinatorial analyses of the preceding sections all hold for the new lattice defined. For $PM(\mathbf{X}_{\mathbf{n},\mathbf{n}})$ no such amendments are needed, and $CLOSED(PM)$ is exactly as defined above. We use $CLOSED(k)$ to denote the amended lattice for the k-clique function. The following is obvious and needs no proof.

Fact 2.1

(i) $|\mathbf{PI}(PM(\mathbf{X}_{\mathbf{n},\mathbf{n}}))| = n!$

(ii) $|\mathbf{PI}(k\text{-}clique(\mathbf{X}_{\mathbf{n}}^{\mathbf{U}}))| = \binom{n}{k}$. □

To start we need upper bounds on $Ext(PM, s+1)$ and $Ext(k\text{-}clique,\ s+1)$.

Lemma 2.7

$$Ext(k\text{-}clique, s+1) = \binom{n-s-1}{k-s-1} \qquad (13)$$

$$Ext(PM, s+1) \leq (n-s-1)! \qquad (14)$$

Proof (13) is immediate from the modified definition of $Ext(k\text{-}clique,\ s+1)$. For (14) a bipartite graph containing $s+1$ edges can only be extended to a perfect matching if each vertex has degree at most one, i.e if the graph is a perfect matching on two sets of $s+1$ vertices. It follows that the number of perfect matchings consistent with this is just the number of perfect matchings over two sets of $n-s-1$ vertices. The upper bound now follows from Fact(2.1). □

The problem of bounding $Gap(PM)$ and $Gap(k\text{-}clique)$ is more difficult. We first consider *k-clique* $(\mathbf{X}_{\mathbf{n}}^{\mathbf{U}})$. Define $M_{-}(k)$ to be the following random n-vertex graph.

Select a random colouring of $\{1, 2, \ldots, n\}$ with g colours $\{1', 2', \ldots, g'\}$, each colouring appearing independently with probability g^{-n}. For a given colouring, $\chi \in ([1 \ldots n] \to [1 \ldots g])$, $G(\chi)$ is the graph in which there is an edge between i and j if and only if $\chi(i) \neq \chi(j)$.

Lemma 2.8

$$Gap(k\text{-}clique) \leq \left(1 - \frac{\Pi_{i=0}^{s-1}(g-i)}{g^s}\right)^r.$$

Proof Given the definition of *Gap*, we have to show that $Prob[M_-(k) \in (\lceil E_{i+1} \rceil - \lceil C_i \rceil)]$ is bounded above by the expression in the Lemma statement. Now $M_-(k)$ is a complete g-partite graph and $\lceil E_{i+1} \rceil$ a set of graphs containing the clique E_{i+1} as a subgraph. $M_-(k)$ contains the same clique if and only if the vertices of E_{i+1} are all coloured differently by χ, the random g-colouring which generates $M_-(k)$. Now suppose that there is some set $F \in C_i$ such that χ colours the vertices of F using different colours. In the same way $M_-(k) \in \lceil F \rceil \subseteq \lceil C_i \rceil$. A subset W of $\{1, \ldots, n\}$ is said to be *properly coloured (PC) by* χ if each vertex in W is coloured differently by χ. It follows that $M_-(k) \in (\lceil E_{i+1} \rceil - \lceil C_i \rceil)$ if and only if E_{i+1} is PC by χ but no set in C_i is PC by χ. So to prove the lemma it is sufficient obtain an upper bound for

$$Prob[E_{i+1} \text{ is } PC \text{ by } \chi \text{ and no set in } C_i \text{ is } PC \text{ by } \chi].$$

From the definition of $\vdash$ we can find $V_1, \ldots, V_r$ in C_i such that $\bigcup_{1 \leq j < l \leq r} V_j \cap V_l \subseteq E_{i+1}$.

It follows that

$$Prob[E_{i+1} \text{ is } PC \text{ and no set in } C_i \text{ is } PC]$$

is no more than

$$Prob[E_{i+1} \text{ is } PC \text{ and } V_j \text{ is not } PC\ \forall\, 1 \leq j \leq r]$$

which is

$$\begin{aligned} &\leq Prob[V_j \text{ is not } PC \mid E_{i+1} \text{ is } PC] \\ &\leq \Pi_{j=1}^{r} Prob[V_j \text{ is not } PC \mid E_{i+1} \text{ is } PC]. \end{aligned}$$

The last inequality holds by virtue of the fact that the sets $V_j - E_{i+1}$ are disjoint (by definition of $\vdash$) and hence the events $\langle V_j \text{ is not } PC \mid E_{i+1} \text{ is } PC\rangle$ are mutually independent. Now let $p_j = |V_j \cap E_{i+1}|$ and $q_j = |V_j - E_{i+1}|$ so that $p_j + q_j = |E_{i+1}| \leq s$.

$$\begin{aligned} Prob[V_j \text{ is not } PC \mid E_{i+1} \text{ is } PC] &= 1 - Prob[V_j \text{ is } PC \mid E_{i+1} \text{ is } PC] \\ &= 1 - \frac{\Pi_{l=p_j}^{|E_{i+1}|-1}(g-l)}{g^{q_j}} \\ &\leq 1 - \frac{\Pi_{l=p_j}^{s-1}(g-l)}{g^{s-p_j}} \\ &\leq 1 - \frac{\Pi_{l=0}^{s}(g-l)}{g^s}. \end{aligned}$$

This proves the lemma. □

Lemma 2.9 *Let $g = k - 1$. If $D \in \mathbf{U}_s(k)$ and $D \neq \emptyset$ then*

$$Prob[M_-(k) \leq k\text{-}clique] = 0,$$
$$Prob[M_-(k) \in D] \geq \frac{\Pi_{i=0}^{s-1}(k-1-i)}{(k-1)^s}.$$

Proof The first inequality is obvious. For the second since $D \neq \emptyset$, D contains at least one set E, say. Thus

$$\begin{aligned} Prob[M_-(k) \in D] &\geq Prob[M_-(k) \in \lceil E \rceil] \\ &= Prob[E \subseteq M_-(k)] \\ &= Prob[E \text{ is } PC \text{ by } \chi \text{ s. t. } G(\chi) = M_-(k)] \end{aligned}$$

and this proves the second inequality. □

Theorem 2.3 *For $3 \leq k \leq \frac{1}{4}(n/\log n)^{2/3}$,*

$$\mathbf{C}^{\mathbf{m}}(k\text{-}clique) = \Omega\left(\left(\frac{n}{16k^{3/2}\text{log}n}\right)^{\sqrt{k}}\right).$$

Proof Fix $s = \lceil\sqrt{k}\rceil$, $r = \lceil 4\sqrt{s \log n}\rceil + 1$, $g = k - 1$. This gives from Theorem 2.2 (11), using $t = \rho(CLOSED(k), k\text{-}clique\ )$,

$$\begin{aligned} t &\geq \frac{\binom{n}{k}}{(r-1)^{2s}\binom{n-s-1}{k-s-1}} \\ &= \frac{n!(k-s-1)!}{(n-s-1)!k!(r-1)^{2s}} \\ &\geq \frac{n^n k^{k-s}}{n^{n-s}k^k(r-1)^{2s}} \\ &\geq \left(\frac{n}{k(r-1)^2}\right)^s \\ &\geq \left(\frac{n}{16k^{3/2}\text{log}n}\right)^{\sqrt{k}}. \end{aligned}$$

We leave as an exercise the problem of showing that t exceeds this quantity in the case $D \neq \emptyset$ in (11) of Theorem 2.2. □

For the function $PM(\mathbf{X_{n,n}})$, the random monom ($\equiv$ bipartite graph) M_- is constructed by the method below.

Let V, W be the disjoint sets of n vertices in the bipartite graph $B(\mathbf{X_{n,n}})$. Select a random labelling, h, of the vertices $V \cup W$ with 0 and 1. Each labelling is chosen with probability 2^{-2n}. $M_-(PM)$ is the random bipartite graph formed by choosing such a labelling of $V \cup W$ and adding edges $\{\langle v_i, w_j \rangle : h(v_i) = h(w_j)\}$, where $1 \leq i, j \leq n$.

In order to bound $Gap(PM)$ with this choice of M_-, Razborov [1985b] proves some combinatorial results on properties of $\mathbf{U_s}(PM)$. It should be noted that in graph-theoretic terms, an element of $\mathbf{U_s}(PM)$ corresponds to a matching containing at most s edges. A *matching* is a (bipartite) graph in which every vertex is the endpoint of at most one edge. This interpretation is convenient for developing an upper bound on $Gap(PM)$.

Lemma 2.10 (Razborov [1985b]) *Let* $\mathbf{B} = \{B_1, B_2, \ldots, B_r\} \subseteq \mathbf{U_s}(PM)$ *be a set of* r *non-empty matchings such that* $B_i \cap B_j = \emptyset$ *whenever* $i \neq j$. *There is a subset* $\{T_1, T_2, \ldots, T_p\}$ *of* $\mathbf{B}$ *such that* $p \geq \sqrt{r}/s$ *and for which the bipartite graph with edges* $\bigcup_{i=1}^{p} T_i$ *contains no cycles, i.e., is a* forest.

Proof Let $\{T_1, T_2, \ldots, T_p\}$ be a maximal size subset of $\mathbf{B}$ for which $\bigcup_{i=1}^{p} T_i$ is a forest. It suffices to prove that $p \geq \sqrt{r}/s$. Suppose the contrary and put $E_0 = \bigcup_{i=1}^{p} T_i$. Since $|T_i| \leq s$, by the assumption we have $|E_0| < \sqrt{r}$. Now consider the subsets V_0 of V, W_0 of W, being those vertices in V, respectively W, which occur in at least one edge of E_0. Clearly $|V_0| < \sqrt{r}$ and $|W_0| < \sqrt{r}$ hence $|V_0 \times W_0| < r = |\mathbf{B}|$. It follows from the edge disjointness of matching in $\mathbf{B}$ that we can find some matching $B_j \in \mathbf{B}$ such that $B_j \cap (V_0 \times W_0) = \emptyset$. By definition $E_0 \subseteq V_0 \times W_0$ hence $B_j \cap E_0 = \emptyset$. But E_0 is a forest and B_j a matching and so from the preceding argument the graph with edges $E_0 \cup B_j$ is also a forest. This contradicts the choice of $\{T_1, \ldots, T_p\}$ as being maximal and therefore we must have $p \geq \sqrt{r}/s$.

Lemma 2.11 Let T *be a forest over* $V \cup W$ *which contains exactly* p *edges* $\{\langle i_k, j_k \rangle : 1 \leq k \leq p\}$. *The events* $\{\langle i_k, j_k \rangle$ is an edge of $M_-(PM)\}$ *(for each* k*) occur independently with probability* $\frac{1}{2}$.

Proof It is sufficient to show that for any subset K of $\{1, 2, \ldots, p\}$ the probability of the event

$$\forall k \in K \ \langle i_k, j_k \rangle \in M_-(PM), \quad \forall k \notin K \ \langle i_k, j_k \rangle \notin M_-(PM)$$

is exactly 2^{-p}.

Let $\chi_K : \{1, \ldots, n\} \rightarrow \{0, 1\}$ be the predicate for which $\chi_K(k) = 1$ iff $k \in K$. Now recalling that $M_-(PM)$ arises from a random labelling $h : V \cup W \rightarrow \{0, 1\}$ it is clear that the probability of this event is just the number of labellings, h, which are solutions to the following system of p linear equations over $GF(2)$, divided by 2^{-2n}, i.e., the total number of distinct labellings.

$$\{h(v_{i_k}) \oplus h(w_{j_k}) \equiv \chi_K(k) \oplus 1\}_{1 \leq k \leq p}.$$

So it suffices to show that this system has 2^{2n-p} distinct solutions.

Consider the forest T with p edges $\{\langle i_k, j_k \rangle : 1 \leq k \leq p\}$. Let $\beta \geq 1$ be the number of connected components (i.e., trees) in this T, each component containing at least one edge. Then T contains exactly $p + \beta$ vertices. For each tree there are exactly 2 ways of labelling the vertices to satisfy the system of equations above, one being the logical complement of the other. That there are exactly two such consistent labellings can be proved by an easy induction on the number of edges in a single component. Now since each component may be labelled independently of the others it follows that there are 2^β labellings of the vertices in the forest T which satisfy the system. This leaves $2n - p - \beta$ vertices unlabelled (those not the endpoint of any edge in T) and any labelling of these will be valid. Thus the system of linear equations over $GF(2)$ has exactly $2^\beta . 2^{2n-p-\beta} = 2^{2n-p}$ distinct solutions as required. □

Corollary 2.2 Let D be a non-empty closed subset of $\mathbf{U}_s(PM)$*. Then* $Prob[M_- \in \lceil D \rceil] \geq 2^{-s}$.

Proof Since D is non-empty it contains at least one matching, E say. Note that E is obviously a forest. We therefore have

$$\begin{aligned} Prob[M_- \in \lceil D \rceil] &\geq Prob[M_- \lceil E \rceil] \\ &= Prob[E \subseteq M_-] = 2^{|E|} \geq 2^{-s}. \end{aligned}$$

□

Lemma 2.12 $Gap(PM) \leq (1 - 2^{-s})^{\sqrt{r}/s}$.

Proof From the definition of $Gap(f)$ it is sufficient to show that if $C \subseteq \mathbf{U}_s(PM)$ and $C \vdash E$ then $Prob[M_-(PM) \in (\lceil E \rceil - \lceil C \rceil)]$ is at most $(1 - 2^{-s})^{\sqrt{r}/s}$.

So suppose that $C \subseteq \mathbf{U}_s(PM)$ and we have $\langle E_1, \ldots, E_r \rangle \in \{C\}^r$ for which

$$\langle E_1, \ldots, E_r \rangle \vdash E.$$

Consider the set (of matchings) $\{F_i : F_i = E_i - E\}$. From (8), the definition of $\vdash$, we have $F_i \cap F_j = \emptyset$ whenever $i \neq j$. In addition if any F_i is empty then $E_0 \subseteq E_i$ and hence $\lceil E_0 \rceil \subseteq \lceil C \rceil$ for which the upper bound on $Gap(PM)$ claimed follows trivially. So it may be assumed that each F_i $(1 \leq i \leq r)$ is non-empty. Now the conditions of Lemma 2.10 hold for the set $\{F_1, \ldots, F_r\}$, thus we can find a subset $T = \{T_1, \ldots, T_p\}$ of this such that $p \geq \sqrt{r}/s$ and for which $\bigcup_{j=1}^{p} T_j$ is a forest.

We now have

$$\begin{aligned} Prob[M_- \in (\lceil E \rceil - \lceil C \rceil)] &\leq Prob[M_- \in \lceil E \rceil \ \& \ \forall i \ M_- \notin \lceil E_i \rceil] \\ &= Prob[E \subseteq M_- \ \& \ \forall i \ F_i \not\subseteq M_-] \\ &\leq Prob[\forall i \ F_i \not\subseteq M_-] \\ &\leq Prob[\forall T_j \in T \quad T_j \not\subseteq M_-]. \end{aligned}$$

From Lemma 2.11 the events $T_j \subseteq M_-$ (for each $1 \leq j \leq p$) are independent and occur with probability $2^{T_j} \geq 2^{-s}$. Therefore $Prob[\forall T_j \in T \quad T_j \not\subseteq M_-]$ is equal to $\Pi_{j=1}^{p} Prob[T_j \not\subseteq M_-]$ and hence is no more than $(1 - 2^{-s})^{\sqrt{r}/s}$. This establishes the upper bound on $Gap(PM)$. $\square$

Lemma 2.13 $Prob[M_- \leq PM(\mathbf{X_{n,n}})] \leq 1/\sqrt{n}$.

Proof (Sketch) M_- contains a perfect matching if and only if the number of vertices of V labelled with 1 by a random labelling h equals the number of vertices of W labelled 1 by the same random labelling. The number of ways in which this can happen is asymptotically bounded above by $\binom{n}{n/2}$. Since each labelling occurs with probability 2^{-2n} this yields the upper bound. $\square$

Theorem 2.4 *For any* $\epsilon > 0$ *and* n *sufficiently large,*

$$\mathbf{C^m}(PM(\mathbf{X_{n,n}})) \geq n^{(1/16-\epsilon)\log_2 n}.$$

Proof Fix $s = \lfloor \log_2 n/8 \rfloor$ and $r = \lfloor n^{1/4}(\log_2 n)^8 \rfloor$ and let

$$t = \rho(CLOSED(PM), PM).$$

Using relation (11) of Theorem 2.2 and relation (14) of Lemma 2.7 gives

$$\begin{aligned} t &\geq \frac{n!}{(r-1)^{2s}(n-s-1)!} \\ &\geq \left[\frac{n}{(r-1)^2}\right]^s \\ &\geq n^{(1/16-\epsilon)\log_2 n}. \end{aligned}$$

Now consider the second part of relation (11) in Theorem 2.2, i.e., D is non-empty. Using Corollary 2.2, Lemma 2.11, Lemma 2.12 and Lemma 2.13 and the chosen values of r and s shows that in this case t would be at least $\exp(\log_2^3 n - o(\log_2^3 n))$ and hence t is asymptotically greater than the first case, $D = \emptyset$. This proves the theorem. □

An important consequence of Theorem 2.4 concerns the power of negation in computing Boolean functions.

Corollary 2.3 The basis $\{\wedge, \vee, \neg\}$ *is superpolynomially more powerful than the basis* $\{\wedge, \vee\}$.

Proof $PM(\mathbf{X_{n,n}})$ can be computed using polynomial size networks over any logically complete basis, e.g., by combining the algorithm of Hopcroft & Karp [1973] with the methods of Fischer & Pippenger [1979]. Theorem 2.4 shows that polynomial size monotone networks do not exist for this function. □

It remains an open problem as to whether negation is exponentially powerful which could be established by obtaining an exponential lower bound on the monotone complexity of $PM(\mathbf{X_{n,n}})$, i.e., a bound growing as 2^{n^ϵ} for some $\epsilon > 0$.

3 THE ANDREEV LOWER BOUND METHOD

The techniques applied in Andreev [1985] are developed from the classical inductive gate elimination method. In this approach a lower bound on the monotone complexity of a family of functions $[f_i]$, for $i = 1, 2, \ldots, n, \ldots$, where $f_n \in M_n$ is obtained as follows. For the inductive base it is shown (directly) that $\mathbf{C^m}(f_1) \geq h(1)$. Now if under the assumption that $\mathbf{C^m}(f_i) \geq h(i)$ for all $i \leq n-1$, it is proved that in any optimal monotone network realising f_n there is some assignment, α to a subset $\mathbf{Y}$ of $\mathbf{X_n}$ which allows at least $g(n)$ gates to be eliminated and is such that $f_n^\alpha(\mathbf{X_n} - \mathbf{Y}) = f_{n-|\mathbf{Y}|}$ then, provided that $h(n-|\mathbf{Y}|)+g(n) \geq h(n)$, a lower bound of $h(n)$ on $\mathbf{C^m}(f_n)$, for all n, follows inductively. This method has been extensively used in almost all existing lower bound proofs for network complexity. The first detailed presentation appears to be that of Red'kin [1973].

The template described above turns out to be too crude to yield even modest linear lower bounds e.g., $\geq (2+\epsilon)n$ for any $\epsilon > 0$ although there are exceptional cases such as Tiekenheinrich [1984]. Consequently the method is usually supplemented with some complex and sophisticated arguments to deal with specific cases where the inductive step breaks down, typical of such proofs are the $2.5n$ lower bounds of Stockmeyer [1977] and Paul [1977]; and the $3n$ lower bound of Blum [1984]; these referring to combinational complexity. Dunne [1985] describes similar methods for monotone networks.

Wegener [1980] introduced an important technique for enhancing the basic inductive technique: the concept of providing certain functions 'for free' as additional inputs. In this way the inputs to a network are the variables $\mathbf{X_n}$ and additionally some finite set of functions $G(n) = \langle g_1(\mathbf{X_n}), \ldots, g_p(n)(\mathbf{X_n})\rangle$. If $\mathbf{C^m_{G(n)}}(f)$ denotes the size of an optimal monotone network with inputs $\mathbf{X_n} \cup G(n)$ then clearly $\mathbf{C^m}(f) \geq \mathbf{C^m_{G(n)}}(f)$. Wegener [1980] was able to derive non-trivial bounds on some sets of Boolean functions by making use of the additional properties of such networks. The method of providing functions for free is also utilised by Andreev [1985] in order to obtain exponential bounds of $2^{n^{1/8}-o(1)}$ on the monotone complexity of a specific function in M_n.

Below E^n denotes the set $\{0,1\}^n$, $||\alpha||$ the number of 1's in $\alpha \in E^n$ and D_f the (minimal) *DNF* of $f \in M_n$. The *size* of f (denoted by $|f|$) is the number of prime implicants of f; the *rank* of f (Rf) is the length of the longest prime implicant of f. If f is a constant function then $|f| = Rf = 0$. Given f_1 and f_2 in M_n, we say that $f_1 \subseteq f_2$ if and only if $\mathbf{PI}(f_1) \subseteq \mathbf{PI}(f_2)$. A function, f, is called *(u,r)-regular* if it can be expressed in the form

$$f = x_{i_1} \wedge x_{i_2} \wedge \ldots x_{i_u} \wedge f_1.$$

Here $i_1, \ldots, i_u$ are distinct, f_1 does not depend on $\{x_{i_1}, \ldots, x_{i_u}\}$, $|f_1| = r$ and each dependent variable of f_1 occurs in D_{f_1} exactly once. f is called *r-regular* if it is (u,r)-regular for some $u \geq 0$.

M_n^t denotes those functions $f \in M_n$ such that every prime implicant of f has length t. Define

$$\pi_t(f) = \begin{cases} \min_{g \in M_n^t : f \leq g} |g| & \text{if such a } g \text{ exists,} \\ 1 + \binom{n}{t} & \text{otherwise.} \end{cases}$$

$$l(r,s) = r^s s!,$$

$$R^n_{r,s} = \{f : f \in M_n, Rf \leq s, |f| \leq l(r,s)\}.$$

An (n,r,s)-*scheme* is a monotone network with functions from $R^n_{r,s}$ given free as extra inputs. $L^n_{r,s}$ is the least number of $\wedge, \vee$ gates needed to realise $f \in M_n$ by a (n,r,s)-scheme. It may be assumed that $r \geq 2$ and $s \geq 1$. Clearly, $L^n_{r,s}(f) = 0 \Leftrightarrow f \in R^n_{r,s}$.

Let $0 < p < 1$. For f_1 and f_2 in M_n we define a measure $\rho_p(f_1, f_2)$ as follows:

$$\rho_p(f_1, f_2) = \sum_{\alpha \in E^n : f_1(\alpha) \neq f_2(\alpha)} p^{n-||\alpha||}(1-p)^{||\alpha||}.$$

$\rho_p(f, g)$ may be interpreted in the following way. Consider constructing a random member, β, of E^n by setting x_i to 0 with probability p and to 1 with probability

$1-p$, the events $\{x_i = e : e \in \{0,1\}\}$ for $1 \leq i \leq n$ being independent. In this way $\rho_p(f,g)$ is just the probability that $f(\beta) \neq g(\beta)$. It is easy to see that

$$\rho_p(F(\mathbf{X_n}, f_1), F(\mathbf{X_n}, f_2)) \leq \rho_p(f_1, f_2) \tag{15}$$

Lemma 3.1 *Let* $\{i_1, \ldots, i_u\} \subseteq \{1, \ldots, n\}$ *and suppose that*

$$g(\mathbf{X_n}) = \bigwedge_{j=1}^{u} x_{i_j} \wedge g_1(\mathbf{X_n} - \{x_{i_j} : 1 \leq j \leq u\})$$

where g_1 *is* $(0,r)$*-regular.*

If $Rg \leq s$ *then* $\rho_p(g, x_{i_1} \ldots x_{i_u}) \leq (sp)^r$.

Proof Without loss of generality suppose that $g = x_1 x_2 \ldots x_u \wedge g_1$, where g_1 is $(0,r)$-regular. By definition every variable on which g_1 essentially depends occurs in D_{g_1} exactly once and $|g_1| = r$. Additionally since $Rg \leq s$ it is obvious that $Rg_1 \leq s$ also. Hence

$$\begin{aligned}
\rho_p(1, g_1) &= \prod_{m \in \mathbf{PI}(g_1)} \rho_p(1, m) \\
&= \prod_{m \in \mathbf{PI}(g_1)} \sum_{x \in var(m)} p \\
&\leq (sp)^r
\end{aligned}$$

Thus from (15) $\rho_p(g, x_1 x_2 \ldots x_u) \leq (sp)^r$, by using $F = f_i \wedge x_1 \wedge \ldots \wedge x_u$. □

Lemma 3.2 *If* $f \in M_n$ *and* $Rf \leq s$, *then there exists some function* $\hat{f}$ *in* $R^n_{r,s}$ *such that* $f \leq \hat{f}$ *and* $\rho_p(\hat{f}, f) \leq |f|(up)^r$.

Proof First of all suppose that if $f \in M_n$, $Rf \leq s$ and $|f| \geq l(r,s)$ then there exists an r-regular g such that $g \subseteq f$. Using the result of Lemma 3.1, we can construct a sequence of functions $f \equiv f_0, f_1, \ldots, f_t \equiv \hat{f}$ which for $0 \leq i < t$ satisfy

(i) $|f_i| \geq l(r,s)$,
(ii) $\rho_p(f_i, f_{i+1}) \leq (sp)^r$.

This sequence can be constructed by the following procedure.

$i := 0$; $f_0 := f$
while $|f_i| \geq l(r,s)$ **do begin**
Find an r-regular g such that $g \subseteq f_i$,
{thus $g \equiv m \wedge g_1$, say}
$i := i+1$
$f_i := f_{i-1} \vee m$
od

Note that this procedure terminates because $|f_{i+1}| < |f_i|$. Obviously (i) is satisfied. Also each f_i is of the form $f_i = h_i \vee m_i \wedge g_i$ where $m_i \wedge g_i$ is r-regular. In this way $f_{i+1} = h_i \vee m_i$. Thus,

$$\begin{aligned} \rho_p(f_i, f_{i+1}) &= \rho_p(h_i \vee m_i \wedge g_i, h_i \vee m_i) \\ &\leq \rho_p(m_i \wedge g_i, m_i) \quad (\text{ by } (15)) \\ &\leq (sp)^r \quad \text{by Lemma 3.1.} \end{aligned}$$

Since

$$\rho_p(f, f_i) \leq \rho_p(f, f_{i-1}) + \rho_p(f_{i-1}, f_i)$$

the final function obtained is the $\hat{f}$ of the Lemma statement.

So it suffices to prove that the supposition stated at the start of the proof does in fact hold. We prove this by induction on s.

If $s = 1$ then f itself is $|f|$-regular i.e., trivially g exists. Now suppose the result holds for $s \leq t-1$. Let $s = t$ and f_1 be a $(0, |f_1|)$-regular function and the maximal possible such that $f_1 \subseteq f$. If $|f_1| \geq r$ then trivially g exists. Therefore suppose that $|f_1| \leq r-1$. Wthout loss of generality let $x_1, \ldots, x_k$ be the variables which f_1 depends upon. It is clear that $k \leq s(r-1)$. By the maximality of f_1 every prime implicant of f contains at least one of $x1$, ..., x_k. Without loss of generality suppose that x_1 occurs in the largest number of prime implicants, and let f_2 be the disjunction over all these prime implicants. Then $f_2 = x_1 \wedge f_3$ where f_3 does not depend on x_1. Clearly $Rf_3 = Rf_2 - 1 \leq t-1$. Also $|f_3| = |f_2| \geq |f|/k \geq r^{t-1}(t-1)! = l(r, t-1)$.

So by the inductive hypothesis, there exists an r-regular g_3 such that $g_3 \subseteq f_3$. So if $g = x_1 \wedge g_3$ then g is r-regular and $g \subseteq f_2 \subseteq f$. □

Lemma 3.3 *If $f \in M_n$ such that $L^n_{r,s}(f) > 0$ then there exists some $g \in M_n$ such that*

(i) $\rho_p(1, g) \geq \rho_p(1, f) - (sp)^r l(r,s)^2$,
(ii) $\pi_{s+1}(g) \geq \pi_{s+1}(f) - l(r,s)^2$,
(iii) $L^n_{r,s}(g) \leq L^n_{r,s}(f) - 1$.

Proof Let S be an optimal (n, r, s)-scheme realising f. Consider any gate both of whose inputs are inputs h_1, h_2 in $R^n_{r,s}$ of S. Both are non-constant by the assumption of optimality. The output of this gate is some function $h_1 * h_2$, where $* = \wedge$ or $\vee$. It is clear that in either case

$$|h_1 * h_2| \leq \max\{|h_1|.|h_2|, |h_1| + |h_2|\} \leq l(r,s)^2. \tag{16}$$

Let h_3 and h_4 be functions whose minimal DNF contains each prime implicant of $D_{h_1 * h_2}$ of length more than s, respectively not more than s. Let h be an arbitrary function in M_n. Consider a network $S(h)$ which is obtained from S by removing the output of this gate and replacing it with the function h. Let $G(h)$ denote the function computed by $S(h)$. It is easy to see that

$$G(h_4) \leq f \leq G(h_4) \vee h_3. \tag{17}$$

Therefore $\rho_p(1, G(h_4)) \geq \rho_p(1, f)$. From Lemma 3.2 we can find a function $h_5 \in R^n_{r,s}$ such that

$$h_4 \leq h_5, \quad \rho_p(h_4, h_5) \leq (sp)^r \mid h_4 \mid . \tag{18}$$

Set $g = G(h_5)$, from (15) and (16) it follows that

$$\rho_p(g, G(h_4)) \leq (sp)^r l(r,s)^2.$$

Using the triangle inequality we have

$$\begin{aligned} \rho_p(1, g) &\geq \rho_p(1, G(h_4)) - \rho_p(g, G(h_4)) \\ &\geq \rho_p(1, f) - (sp)^r l(r,s)^2. \end{aligned} \tag{19}$$

Now, from (17) and (18),

$$f \leq g \vee h_3, \quad \pi_{s+1}(f) \leq \pi_{s+1}(g) + \pi_{s+1}(h_3).$$

Applying (17) and the fact that each prime implicant in D_{h_3} contains at least $s+1$ variables it follows that $\pi_{s+1}(h_3) \leq |h_3| \leq l(r,s)^2$ and so

$$\pi_{s+1}(g) \geq \pi_{s+1}(f) - \pi_{s+1}(h_3) \geq \pi_{s+1}(f) - l(r,s)^2. \tag{20}$$

Clearly,

$$L^n_{r,s}(g) \leq L^n_{r,s}(f) - 1 \tag{21}$$

(19), (20) and (21) prove the result. □

Theorem 3.1 If $f \in M_n$ then

$$L^n_{r,s}(f) \geq \frac{1}{l(r,s)^2} \min\left\{\pi_{s+1}(f), \frac{\rho_p(1,f) - sp}{(sp)^r}\right\}.$$

Proof If $f \in R^n_{r,s}$ then the assertion holds since both sides of this inequality are ≤ 0. If $L^n_{r,s}(f) > 0$ then from Lemma 3.3 there exists a sequence of functions $g_1, g_2, \ldots, g_t$ in M_n such that

$$\begin{aligned} \rho_p(1, g_i) &\geq \rho_p(1, f) - il(r, s)^2(sp)^r, \\ \pi_{s+1}(g_i) &\geq \pi_{s+1}(f) - il(r, s)^2, \quad i = 1, 2, \ldots, t, \\ L^n_{r,s}(f) &> L^n_{r,s}(g_1) > \ldots > L^n_{r,s}(g_t) = 0. \end{aligned}$$

Clearly g_t is a member of $R^n_{r,s}$. If $g_t \equiv 0$ then $\pi_{s+1}(g_s) = 0$ and consequently

$$t = \frac{\pi_{s+1}(f)}{l(r, s)^2}. \tag{22}$$

If $g_t \neq 0$ then it follows that $Rg_t \leq s$ and then $\rho_p(1, g_t) \leq sp$. Thus $\rho_p(1, f) - tl(r, s)^2(sp)^r \leq sp$. So

$$t \geq \frac{\rho_p(1, f) - sp}{l(r, s)^2(sp)^r}. \tag{23}$$

(22) and (23) give the result. □

Let $T = [m_{ij}]$, where $1 \leq i \leq m$, $1 \leq j \leq n$, be an $m \times n$ Boolean matrix without any zero rows. Define the function f_T by,

$$f_T(x_1, \ldots, x_n) = \bigvee_{i=1}^{m} \bigwedge_{j | m_{ij} = 1} x_j.$$

Corollary 3.1 If every row of the Boolean (m, n)-matrix T contains at least $t \geq s+1$ 1's and T does not have a $(k, s+1)$-submatrix of 1's then

$$L^n_{r,s}(f_T) \geq \frac{1}{l(r, s)^2} \min \left\{ \frac{m}{k-1}, \frac{1 - me^{-pt} - sp}{(sp)^r} \right\}.$$

This follows since

$$\pi_{s+1}(f_T) \geq \frac{m}{k-1}$$

and

$$\rho_p(1, f_T) \geq 1 - m(1-p)^s \geq 1 - me^{-pt}.$$ □

Let $GF(q)$ be the Galois field of order q and let $n = q^2$. Also let the pairs (α_j, β_j) range over the set $GF(q) \times GF(q)$. F_i, where $i = 1, 2, \ldots, m = q^{s+1}$, is an enumeration of all polynomials over $GF(q)$ whose degree does not exceed s. $T_{n,s}$ denotes the (m, n)-Boolean matrix such that

$$m_{ij} = 1 \Leftrightarrow \alpha_j = F_i(\beta_j).$$

It is easy to see that for $i_1 \neq i_2$ the system of equations

$$y = F_{i_1}(x), \quad y = F_{i_2}(x)$$

has no more than s common solutions; consequently the matrix $T_{n,s}$ does not have a $(2, s+1)$-submatrix of 1's. In addition each row of $T_{n,s}$ contains exactly q 1's. Set $f_{n,s} = F_{T_{n,s}}$ and from Corollary 3.1 we have

Corollary 3.2

$$L^n_{r,s}(f_{n,s}) \geq \frac{1}{l(r,s)^2} \min \left\{ n^{(s+1)/2}, \frac{1 - n^{(s+1)/2} e^{-p\sqrt{n}} - sp}{(sp)^r} \right\}.$$

□

If we fix

$$s \leq \frac{0.5n^{1/8}}{\log_e n - 2},$$
$$p = \frac{(s+1)\log_e n + 2}{2\sqrt{n}},$$
$$r = \lfloor (s+1)\log_e n \rfloor,$$

this gives the following.

Corollary 3.3

$$L^n_{r,s}(f) \geq \left(\frac{\sqrt{n}}{4s^4 \log_e^2 n} \right)^{s+1}.$$

□

This evaluates to

$$\exp\left(\frac{n^{1/8} \log_e \log_e n}{\log_e n} \right).$$

4 CONCLUSION

Andreev [1985] and the work of Razborov [1985a,b] and Alon & Boppana [1986] offer two approaches to proving non-trivial lower bounds on monotone network size. In this final section we consider in what ways the two basic methods are similar.

Both techniques are, in a broad sense, inductive arguments based on Wegener's concept of providing functions for free as additional inputs. However this is only explicit in Andreev's proof. In the Lattice method the inductive argument occurs in the general lemma relating monotone network size to the distance metric in regular lattices, i.e., Lemma 2.1, but is not otherwise applied in the subsequent combinatorial analyses. The elements of the lattice $CLOSED(f)$ (and by extension $CLOSED(k)$ for k-clique) correspond to particular *monotone* Boolean functions. Thus we can define a mapping $REPR : CLOSED(f) \to M_n$ as follows; for $A \in CLOSED(f)$, $\mathbf{I}(REPR(A)) = A$ so that $\mathbf{PI}(REPR(A)) = base_s(A)$. Note that the cover operation $\lceil .. \rceil$ ensures that $REPR(A)$ is in fact a monotone function. The set of functions $\{REPR(A) : A \in CLOSED(f)\}$ form a subset of the set $R^n_{r,s}$ used by Andreev. This is a proper subset since $|\mathbf{PI}(REPR(A))| \leq (r-1)^s$ while for $f \in R^n_{r,s}$ we have only $|\mathbf{PI}(f)| \leq l(r,s) = r^s s!$ Additionally $L^n_{r,s}$ includes all functions with maximal prime implicant length of s, so that the novel closure relation used by Razborov does not seem to have any analogue in Andreev's proof.

In fact it turns out that we may improve Andreev's lower bound inequality by recasting his proof in terms of the functions arising from $REPR$ instead of the set $R^n_{r,s}$ and by appealing to the improved combinatorial analyses of Alon & Boppana [1986]. Below we describe how this is accomplished.

We shall call a monotone Boolean network, S, an *(n,r,s,d)-scheme*, where $d \in M_n$, if S has as inputs exactly the set of monotone functions

$$R^{n,d}_{r,s} = \{REPR(\lceil A \rceil) : \lceil A \rceil \in CLOSED(d)\}.$$

Since $CLOSED(d)$ is a regular lattice this provides the normal network inputs $\mathbf{X_n}$. $L^{n,d}_{r,s}(f)$ denotes the number of gates in a minimal (n,r,s,d)-scheme realising f. It should be noted that d is not required to equal f. The quantities $\pi_t(f)$ and ρ_p retain their meanings of Section 3. The main result to be reproved is Lemma 3.3, which can be sharpened by using Lemma 2.2, Corollary 2.1, Lemma 2.4 and Lemma 2.6. We can dispense entirely with Lemma 3.1 and Lemma 3.2 of Andreev [1985] in the course of this proof.

Lemma 4.1 If $f \in M_n$ such that $L^{n,d}_{r,s}(f) > 0$ then there exists some $g \in M_n$ for which

(i) $\rho_p(1,g) \geq \rho_p(1,f) - 2r^s(sp)^r$,
(ii) $\pi_{s+1}(g) \geq \pi_{s+1}(f) - (r-1)^{2s}$,
(iii) $L^{n,d}_{r,s}(g) \leq L^{n,d}_{r,s}(f) - 1$.

Proof The proof parallels that of Lemma 3.3. Let S be an optimal (n,r,s,d)-scheme realising f. Consider any gate of S whose inputs are functions h_1, $h_2 \in R^{n,d}_{r,s}$. By definition

$$h_1 = REPR(\lceil A \rceil), \quad h_2 = REPR(\lceil B \rceil),$$

for some closed subsets, A and B, of $\mathbf{U}_s(d)$. The output of the selected gate is some function $h_1 * h_2$ where $* = \wedge$ or $\vee$. Thus, from Corollary 2.1.

$$|h_1 * h_2| \leq \max\{|h_1|.|h_2|, |h_1| + |h_2|\} \leq (r-1)^{2s}. \tag{24}$$

Let h_3, resp. h_4, be functions whose minimal DNF consists of all prime implicants of $h_1 * h_2$ having length more than, resp. at most, s variables. For any $h \in M_n$, $S(h)$ denotes the (n, r, s, d)-scheme obtained by replacing the gate computing $h_1 * h_2$ in S, by a node computing h. $G(h)$ denotes the function computed by $S(h)$. It is obvious that

$$G(h_4) \leq f \leq G(h_4) \vee h_3 \tag{25}$$

and so $\rho_p(1, G(h_4)) \geq \rho_p(1, f)$.

At this point the proof diverges from Lemma 3.3. We claim the there is some function $h_5 = REPR(\lceil C \rceil) \in R^{n,d}_{r,s}$ for which

$$h_4 \leq h_5, \quad \rho_p(h_4, h_5) \leq \begin{cases} 0 & \text{if } * = \wedge, \\ 2r^s(sp)^r & \text{if } * = \vee. \end{cases} \tag{26}$$

First consider $* = \wedge$. In this case choosing h_5 to be $REPR(\lceil A \rceil \sqcap \lceil B \rceil)$ satisfies (26). To show this it is sufficient to prove that $h_4 = h_5$. Now, using Lemma 2.2,

$$\mathbf{PI}(h_4) = \mathbf{PI}(REPR(\lceil A \rceil) \wedge REPR(\mathbf{PI}(\lceil B \rceil)) \cap P_s(\mathbf{X_n}),$$
$$\mathbf{PI}(h_5) = base_s(\lceil A \rceil \sqcap \lceil B \rceil) = base_s(\lceil A \cap B \rceil).$$

Suppose $m \in \mathbf{PI}(h_4)$. Then $|var(m)| \leq s$ and there are monoms $m_1 \in base_s(\lceil A \rceil)$, $m_2 \in base_s(\lceil B \rceil)$ for which $m = m_1 \wedge m_2$. Since $|var(m)| \leq s$, $var(m_i) \subseteq var(m)$ and A, B are closed subsets of $\mathbf{U}_s(d)$ it follows that $m \in A$, $m \in B$ thus $m \in \lceil A \cap B \rceil$ and so $m \leq h_5$. On the other hand suppose that $m \in \mathbf{PI}(h_5)$. Then $m \in \lceil A \cap B \rceil$ and $|var(m)| \leq s$. This implies the existence of some $m_3 \in A \cap B$ for which $var(m_3) \subseteq var(m)$ and hence $m \in A$, $m \in B$ by closure. It follows that $m \leq REPR(\lceil A \rceil) \wedge REPR(\lceil B \rceil)$ and since $|var(m)| \leq s$ we have therefore $m \leq h_4$. It has been proved that $h_4 \leq h_5$ and $h_5 \leq h_4$ hence $h_4 = h_5$. This completes the argument for the case $* = \wedge$.

If $* = \vee$ then h_5 is chosen to be $REPR(\lceil A \rceil \sqcup \lceil B \rceil) = REPR(\lceil \overline{A \cup B} \rceil)$. Since $* = \vee$,

$$\mathbf{PI}(h_4) = \mathbf{PI}(REPR(\lceil A \rceil) \vee REPR(\lceil B \rceil)) \cap P_s(\mathbf{X_n}).$$

Obviously $\mathbf{PI}(h_4) \subseteq \mathbf{PI}(REPR(\lceil \overline{A \cup B} \rceil))$ so $h_4 \leq h_5$. It remains to show that $\rho_p(h_4, h_5) \leq 2r^s(sp)^r$.

Let $M_-(f)$ be a random monom in which x_i occurs with probability $1-p$ and does not occur with probability p. By observing that $\mathbf{I}(h_4) = \lceil A \rceil \cup \lceil B \rceil$ it is clear that

$$\begin{aligned}\rho_p(h_4, h_5) &= Prob[M_-(f) \in \delta_-(\lceil A \rceil, \lceil B \rceil)] \\ &\leq 2r^s Gap(f)\end{aligned}$$

from Lemma 2.4 and Lemma 2.6.

We can produce an upper bound on $Gap(f)$ with the chosen $M_-(f)$, by adapting the techniques of Lemma 2.12. So it is sufficient to show that if $C \subseteq \mathbf{U}_s(d)$ and $C \vdash E$ then $Prob[M_-(f) \in (\lceil E \rceil - \lceil C \rceil)] \leq (sp)^r$. Let $\langle E_1, \ldots, E_r \rangle \in \{C\}^r$ which yields E and consider the set of monoms $\{F_i : F_i = E_i - E\}$ which are pairwise disjoint and can be assumed to be non-empty, cf. the proof of Lemma 2.12. We now have

$$\begin{aligned}Prob[M_- \in (\lceil E \rceil - \lceil C \rceil)] &\leq Prob[M_- \in \lceil E \rceil \ \&\ \forall i\ M_- \notin \lceil E_i \rceil] \\ &= Prob[E \subseteq M_- \ \&\ \forall i\ F_i \not\subseteq M_-] \\ &\leq Prob[\forall i\ F_i \not\subseteq M_-] \\ &\leq (\max_{1 \leq i \leq r}\{Prob[F_i \not\subseteq M_-]\})^r \\ &\leq (\sum_{x \in F} Prob[x \notin M_-])^r \\ &\qquad\qquad \text{(where } F \text{ is the maximising } F_i\text{)} \\ &\leq (sp)^r.\end{aligned}$$

The last line follows from the fact that the events $\{x_i \in M_-\}$ are independent.

The completes the proof of the claim made earlier.

The remainder of the proof is identical to that of Lemma 3.3 but making use of the fact that $|\, h_1 * h_2 \,| \leq (r-1)^{2s}$. The details are left to the reader. □

The following is immediate from this result.

Theorem 4.1 *If* $f \in M_n$ *then*

$$L^{n,f}_{r,s}(f) \geq \left\{ \frac{\pi_{s+1}(f)}{(r-1)^{2s}}, \frac{\rho_p(1,\ f)\ -\ sp}{2r^s\ (sp)^r} \right\}.$$

□

With this expression, the explicit lower bound obtained by Andreev [1985] is improved to $2^{n^{1/4} - O(1)}$. This is the same as that acheived in Alon & Boppana [1986] for the same function, using the Lattice method directly.

REFERENCES

Alon & Boppana [1986]
N. Alon and R. Boppana (1986) 'The monotone circuit complexity of Boolean functions', *Combinatorica*, to appear.

Adreev [1985]
A. E. Andreev (1985) 'A method of proving lower bounds on the complexity of monotone Boolean functions', *Doklady Akademii Nauk SSSR*, **282**, 1033–7 [In Russian] Transl. *Sov. Math. Doklady*, *31*, 530–4.

Berkowitz[1982]
S. Berkowitz (1982) 'On some relationships between monotone and non-monotone circuit complexity', Ph.D Dissertation, Univ. of Toronto.

Blum [1984]
N. Blum (1984) 'A Boolean function requiring $3n$ network size', *Theoret. Comput. Sci.*, **28**, 337–45.

Brent & Kung [1981]
R. Brent and H. T. Kung (1981) 'The area–time complexity of binary multiplication', *Jnl. of the ACM*, **28**, 521–34.

Dunne [1985]
P. E. Dunne (1985) 'A $2.5n$ lower bound on the monotone network complexity of T_3^n' *Acta Informatica*, **22**, 229–40.

Erdős & Spencer [1974]
P. Erdős and J. Spencer (1974) *Probabilistic Methods in Combinatorics*, Academic Press, New York.

Fischer & Pippenger [1979]
M. Fischer and N. J. Pippenger (1979) 'Relations among complexity measures', *Jnl. of the ACM*, **26**, 361–81.

Hopcroft & Karp [1973]
J. E. Hopcroft and R. M. Karp (1973) 'An $n^{5/2}$ algorithm for maximum matching in bipartite graphs', *SIAM Jnl. of Computing*, **2**, 225–31.

Lipton & Tarjan [1979]
R. Lipton and R. E. Tarjan (1979) 'A separator theorem for planar graphs', *SIAM Jnl. of Applied Math.*, **36**, 177–89.

Lupanov [1958]
O. B. Lupanov (1958) 'On a method of circuit synthesis', *Izvestia VUZ (Radiofizika)*, **1**, 120–40 [In Russian].

Nakashima [1936]
A. Nakasima (1936) 'The theory of relay circuits', *Nippon Elec. Communication Engineering*, May 1936, 197–226.

Paterson [1975]
M. S. Paterson (1975) 'Complexity of monotone networks for Boolean matrix product', *Theoret. Comput. Sci.*, **1**, 13–20.

Paul [1977]
W. Paul (1977) 'A $2.5n$ lower bound on the complexity of Boolean functions', *SIAM Jnl. of Computing*, **6**, 427–43.

Razborov [1985a]
A. A. Razborov(1985) 'Lower bounds on the monotone complexity of some Boolean functions', *Doklady Akademii Nauk SSSR*, **281**, 798–801 [In Russian].

Razbarov [1985b]
A. A. Razborov (1985) 'A lower bound on the monotone complexity of the logical permanent', *Matem. Zametki*, **37**, 887–901 [In Russian].

Red'kin [1973]
N. P. Red'kin (1973) 'Proof of minimality of circuits consisting of functional elements', *Problemy Kibernet.*, **23**, 83–102 [In Russian]; transl. *Syst. Theo. Res.*, **23**, 85–103.

Savage [1981]
J. E. Savage (1981) 'Planar circuit complexity and the performance of VLSI algorithms', pp. 61–8 in *VLSI Systems and Computation*, Ed. H. T. Kung, B. Sproull & G. Steele, Computer Science Press.

Shannon [1938]
C. E. Shannon (1938) 'A symbolic analysis of switching and relay circuits', *Trans. AIEE*, **57**, 713–23.

Shannon [1949]
C. E. Shannon (1949) 'The synthesis of two-terminal switching circuits', *Bell Syst. Tech. Jnl.*, **28**, 59–98.

Shestakov [1938]
V. I. Shestakov (1938) 'Some mathematical methods for the construction and simplification of two-terminal electrical networks of class A', Dissertation, Lomonosov State Univ., Moscow.

Stockmeyer [1977]
L. Stockmeyer (1977) 'On the combinational complexity of certain symmetric Boolean functions', *Math. Syst. Theory*, **10**, 323–36.

Thompson [1979]
C. D. Thompson (1979) 'Area–time complexity for VLSI', pp.81–8 in *Proceedings of the 11th ACM STOC.*

Tiekenheinrich [1984]
J. Tiekenheinrich (1984) 'A $4n$ lower bound on the monotone Boolean network complexity of a one output Boolean function', *Inf. Proc. Letters*, **18**, 201–2.

Turing [1936]
A. M. Turing (1936) 'On computable numbers with an application to the *Entscheidungsproblem*', *Proc. London Mathl. Soc. Series 2*, **42**, 230–65; Corrections: ibid. **43** (1937), 544–6.

Wegener [1980]
I. Wegener (1980) 'A new lower bound on the monotone network complexity of Boolean sums', *Acta Informatica*, **13**, 109–14.

Wegener [1982]
I. Wegener (1982) 'Boolean functions whose monotone complexity is of size $n^2/\log_2 n$', *Theoret. Comput. Sci.*, **21**, 213–24.